Phytoalexins

Blackie & Son Limited
Bishopbriggs
Glasgow G64 2NZ

Furnival House
14–18 High Holborn
London WC1V 6BX

British Library Cataloguing in Publication Data

Phytoalexins.
  1. Plant diseases
  I. Bailey, John A.   II. Mansfield, John W.
  581.2   SB731

  ISBN 0-216-91162-1

Filmset by Advanced Filmsetters (Glasgow) Ltd
Printed in Great Britain by
Thomson Litho Ltd, East Kilbride, Scotland

PHYTOALEXINS

General Editors

JOHN A. BAILEY

Principal Scientific Officer,
Long Ashton Research Station,
Faculty of Agriculture and Horticulture,
University of Bristol

and

JOHN W. MANSFIELD

Lecturer in Plant Pathology,
Wye College,
University of London

Blackie
Glasgow and London

Contributors

Dr J. A. Bailey

Long Ashton Research Station, Long Ashton, Bristol BS18 9AF, United Kingdom

Dr D. T. Coxon

Agricultural Research Council, Food Research Institute, Colney Lane, Norwich NR4 7UA, United Kingdom

Professor B. J. Deverall

Department of Plant Pathology, University of Sydney, N.S.W. 2006, Australia

Dr J. L. Ingham

Department of Botany, Plant Science Laboratories, University of Reading, Whiteknights, Reading RG6 2AS, United Kingdom

Professor J. Kuć

Department of Plant Pathology, University of Kentucky, Lexington, Kentucky 40506, United States

Dr J. W. Mansfield

Department of Biological Sciences, Wye College (University of London), Ashford, Kent TN25 5AH, United Kingdom

Dr D. E. Matthews

Department of Plant Pathology, Cornell University, Ithaca, New York 14853, United States

Professor D. A. Smith

Department of Plant Pathology, University of Kentucky, Lexington, Kentucky 40506, United States

Dr A. Stoessl

Agriculture Canada, Research Institute, University Sub Post Office, London, Ontario NGA 5B7, Canada

Professor H. VanEtten

Department of Plant Pathology, Cornell University, Ithaca, New York 14853, United States

Preface

The word phytoalexin was derived from the Greek "phyton", meaning plant, and "alexein", to ward off. Phytoalexins were originally proposed to be antifungal substances produced by plants in response to fungal infection; accumulation of phytoalexins within infected tissue was believed to restrict the growth of invading fungi. It is now twenty-one years since the first phytoalexin was isolated and characterized as pisatin, a pterocarpan from garden pea, *Pisum sativum*. Subsequent work has been prolific and the accumulation of phytoalexins in infected tissues has become the most intensively studied mechanism of disease resistance in plants. In view of the rapid expansion of the literature on phytoalexins, and the contributions now being made by widely different disciplines, we considered it timely to gather together this information in a book dealing in detail with all the various aspects of these compounds. The chapters were planned to reflect the diverse interests of the research workers who have been drawn to the study of these intriguing secondary metabolites.

The early chapters emphasize the chemistry, biosynthesis and metabolism of these compounds, many of which are found only in diseased tissues. This coverage is followed by chapters on the biological significance of phytoalexins, including the mechanisms by which they accumulate, their modes of antibiotic activity and their role in regulating host-parasite interactions.

The book is written for research workers, university teachers and advanced undergraduates with an interest in plant pathology, plant biochemistry and chemistry. We hope that the ready availability of this information will encourage readers to seek answers to the numerous questions which remain and to apply this knowledge to the development of new approaches to the control of plant disease.

We wish to thank all the contributors for complying so willingly with our timetable for preparation of their chapters, and the Publishers whose continued encouragement and wise counsel has led to the prompt publication of this book. Acknowledgement is also made to Sue Briant and Pat Rowell who assisted greatly in the final preparation of many diagrams.

J.A.B.
J.W.M.

Contents

9 Mechanisms of phytoalexin accumulation   289
J. A. Bailey

10 Phytoalexins: current problems and future prospects   319
J. W. Mansfield and J. A. Bailey

1 Introduction

B. J. DEVERALL

The concept of phytoalexins

In 1940, Müller and Börger proposed that plants produced defensive substances, called phytoalexins, in response to infection. The term was derived from Greek to mean "warding-off agents in plants", and the proposal was made after deliberating two important phenomena in plant pathology. One was the active response of the cells of many plants to attempted infection; the other, the acquisition of resistance by plants after exposure to an infecting organism, and both had been the subject of previous research. Although formalized in 1940, the idea that plants might release substances inhibitory to fungi was not new.

Early investigators were well aware that many parasitic fungi were specialized for particular host plants. A fungus isolated from one type of plant was unable to parasitize closely related species. Ward, (1905), reviewing his research on the fate of specialized forms of rust fungi when placed on hosts and non-hosts, found that a common feature associated with the failure to develop on non-hosts was the rapid death of cells at the sites of attempted infection. The infection hyphae stopped growing among the dead cells, and a possible cause was suggested to be a poison released from these cells. The rapid death of plant cells at sites of attempted infection was seen in many cereal cultivars resistant to rust fungi, and was termed *hypersensitivity* by Stakman (1915). Hypersensitivity was an effective form of resistance, and was subsequently found to occur in many plants as they express resistance to fungal and bacterial parasites, but Stakman did not suggest how it might work.

Many botanists of the late nineteenth and early twentieth centuries were intrigued by the possibility that plants, like animals, might acquire resistance to infection after earlier exposure to another infection. Some perennial plants seemed to be less severely affected by disease following a previous infection— varied powdery mildew infections on leaves of oak trees in successive seasons were noted, for example. Some experiments purported to show that treating herbaceous plants (or the soil bearing them) with extracts of fungal cultures

1

rendered them more resistant to infection. Chester (1933) was however un-convinced by such reports because of inadequacies in experimental design, conceivable differences in weather conditions in successive seasons and changes in the resistance of plants with age.

The most convincing early work on the acquisition of resistance was a series of studies with orchid tissues. This started with research by Bernard on interactions between orchid seeds and fungi leading to functional seedlings with mycorrhizal roots. Not all soil-inhabiting fungi formed mycorrhizas—some did not survive the attempted infection of the seed, and some destroyed the seed. In 1909, Bernard reported that one type of fungus penetrated several layers of cells in orchid embryos before it ceased growth and disintegrated. The embryos then became resistant to infection by a normally destructive fungus. Further work tried to reveal chemical interactions between fungi and orchid cells. When small pieces of surface-sterilized tuber were placed ahead of advancing mycelium on a culture medium, the fungus failed to grow up to the pieces of tuber. By contrast, the fungus overgrew extracts of crushed tuber and pieces of tuber which had been treated at 55°C for 35 min (Bernard, 1911). It was proposed that live cells in the intact pieces of tuber responded to fungal secretions and produced substances which diffused through the culture medium and inhibited fungal growth. This suggestion was supported by Nobécourt (1923) and, although disputed by Magrou (1924), was based upon the best early experimental evidence for the production of defensive substances by plant tissues in response to encroachment by a fungus.

Before the research for the publication of 1940, Müller and his associates were involved in producing potato cultivars with resistance to *Phytophthora infestans*, the fungus which causes the late blight disease. Some cultivars were highly resistant to one race of the fungus but susceptible to another. Important differences were observed between the early stages of infection of resistant and susceptible cultivars. In susceptible potato plants, fungal hyphae grew rapidly and extensively before the host cells were killed. In resistant plants, cells died rapidly at the sites of entrance by the fungus, which was then limited to these sites. Hypersensitivity was seen as an important means of resistance to the potato blight fungus.

Müller and Börger (1940) carried out a series of experiments to establish whether the hypersensitive reaction affected the development of a race of *P. infestans* which normally parasitized the plant. For this purpose, they treated cut surfaces of potato tubers with a race which initiated hypersensitivity, and then, a day later, with a race which normally develops profusely. This second race failed to develop. When only part of the surface was treated with the first race, the second race developed profusely on the untreated part. Inhibition of the parasitic race occurred, therefore, in the hypersensitively reacting tissue, but it could have been brought about by direct physical or chemical interference by the first race rather than through changes in the physiology of the potato cells. Müller and Börger explored these possibilities. Firstly, they cut away a thin slice of tuber bearing the hypersensitively reacting cells and found that the inhibition occurred also in the exposed cells beneath. Secondly, they inoculated the two races in sequence on a cultivar susceptible to both and found no antagonism between the races in the absence of the hypersensitive reaction. For these reasons, they concluded that the inhibition was caused by a principle produced in the hypersensitive reaction. The principle was termed a *phytoalexin*.

Müller and Börger designed several experiments to investigate whether this inhibition affected unrelated fungi. Infection by a race of *Phytophthora infestans* inhibited the tuber parasite *Fusarium caeruleum*, and the growth of some saprophytic fungi was also stopped. The hypothetical phytoalexin was therefore regarded as non-specific in its effect.

Because of the historical importance of Müller and Borger's postulates about defence mechanisms in plants and the role of phytoalexins, I include a direct translation of their conclusions published in 1940.

'1. The premature death of the parasite on the tubers of the resistant W varieties is not due to any toxic "principle" already present in the tuber before the infection nor to the absence of any substance necessary for the normal development of the fungus, but to a change in the state of the host cells which come into contact with the parasite. This change of state results in a "paralysis" or the premature death of the fungus (cf. Meyer [1939]; Müller *et al.* [1939]). The principle inhibiting the development of the fungus is formed or activated only in the course of this change of state which we have termed the "defensive reaction".

2. The defensive reaction is linked with the living state of the host cell. This does not mean, however, that a tissue which is parasitized by a virulent *Phytophthora* strain, but is still alive, may not at the same time be capable of responding to the attack of an avirulent strain with the changes of state characteristic of the defensive reaction.

3. The inhibiting principle must be of a material nature. It is formed or activated in the reacting host cell and may be regarded as the end product of a "necrobiosis" released by the parasite.

4. This not yet isolated and therefore still hypothetical "defensive substance" is "non-specific". It has an inhibiting action on other parasitic fungi of the potato tuber as well as *Phytophthora infestans*. Saprophytic fungi are also inhibited in their development by this substance. However, the various parasitic species differ in their sensitivity towards this "phytoalexin".

5. The decisive factor for the fate of the parasite and hence also for the "immune" behaviour of the host is discovered only in the sensitivity of the host cells to certain material influences emanating from the *Phytophthora* fungus: the greater the sensitivity, the higher the resistance. The reaction product is accordingly not specific, but only the genotypically determined readiness of the host cell to react, which is manifested in the speed with which the hypothetical defensive substance is formed.

6. The defensive reaction is confined to the tissue colonized by the fungus and its immediate neighbourhood. There is no immunization embracing the whole individual.

7. In the resistant varieties we find an "immunization" of the portions of tissue invaded by the parasite, in the susceptible varieties the opposite is the case: the host cells invaded by a virulent *Phytophthora* strain, but still alive, also become "sensitive" to fungi which are incapable of attacking an intact potato tuber after association with the parasite for some time. Here, too, we note gradual differences in the capacity of the individual fungus species to colonize the tissue attacked by *Phytophthora*.

8. What is inherited is only the capacity to "acquire" the resistance at the place of infection and only here, but not the resistant "state" in itself. This state must first be "acquired", and this happens only after the plant has come into contact with the pathogenic agent. This serves to release the "mechanism" which transforms the portions of tissue attacked by the parasite from the "indifferent" to the "resistant" state.'

This summary considers a number of aspects of the hypersensitive reaction and induced resistance in plants. Apart from the proposals concerning phyto-alexins, statement 5 is relevant to the still unsolved problem of the role of specific fungal products in eliciting responses in plant cells and statement 6 presents conclusions which appeared to be applicable to all fungus-plant interactions until the recent demonstrations of systemic induced resistance (Kuć *et al.*, 1975; Beretta *et al.*, 1977; Sutton, 1979).

First isolations of compounds as phytoalexins

The next significant advance in research directed towards a study of phyto-alexins did not come until Müller (1958) published his work on changes which occur in infection droplets held in the seed cavities of opened bean pods. The infection droplets contained spores of the fungus *Monilinia fructicola*, a parasite of stone fruits. In the seed cavities of bean (*Phaseolus vulgaris*), the spores germinated readily and caused detectable changes in underlying cells within 12 to 14 hours. After 24 hours, the fungus had induced hypersensitivity and had advanced no further. Samples of infection droplets were collected at intervals during this process, made spore-free, and tested for their effects on new spores *in vitro*. Droplets collected after 14 hours were partly inhibitory to spore germination; those collected after 24 hours prevented germination. Droplets of water, by contrast, had become highly stimulatory to fungal growth after 24 hours' incubation in pods. The inhibitory factor could be removed from combined infection droplets by extraction with light petroleum, leaving an aqueous phase which was also highly stimulatory to fungal growth. The inhibitor behaved as a single substance in a number of fractionations and was active against unrelated fungi. Müller thus detected a chemical entity formed during a hypersensitive reaction which also had the features proposed for a phytoalexin.

The chemistry of this entity was not investigated until similar work using pea (*Pisum sativum*) had been completed. Müller (1958) had shown that a fungal inhibitor was also generated in infection droplets held in pea, as in bean. Because large volumes of combined droplets could be obtained more readily from pea than from bean, Cruickshank and Perrin (1960) decided chemical investigations might progress better using the pea system. They found that the inhibitor in infection droplets in pea could also be extracted in light petroleum and that it behaved largely as a single substance during chromatography. An antifungal compound named pisatin (1) was crystallized and subsequently characterized as a pterocarpan (Perrin and Bottomley, 1962). Soon after this, the entity in infection droplets in bean was isolated, characterized as a closely related pterocarpan and named phaseollin (2) (Cruickshank and Perrin, 1963*a*; Perrin, 1964).

1 Pisatin   2 Phascollin

The possibility existed that these antifungal compounds were products of the fungus or of the joint metabolic activities of the fungus and the host, but this was discounted when Cruickshank and Perrin (1963*b*) showed that pisatin formation was stimulated not only by exudates of the germinating spores of *Monilinia fructicola* but also by solutions of either mercuric or copper chloride. Pisatin and phaseollin were thus products of the host plant and were clearly phytoalexins in the general sense envisaged by Müller and Börger.

Concurrent work on antifungal compounds in plants

During the twenty years which elapsed between the proposal of the existence of phytoalexins and the first demonstration of a compound as a phytoalexin, other workers found that infected plants produced antifungal substances. Gäumann and Jaag (1945) returned to the problem of the defensive responses of orchid tubers and confirmed the findings of Bernard and Nobécourt mentioned earlier. Gäumann and Kern (1959*a,b*) extracted, separated and measured antifungal components in apparently healthy tubers of *Orchis militaris* and also in surface-sterilized segments inoculated with *Rhizoctonia repens*. The main antifungal activity in infected tissues was attributable to orchinol (3), later characterized by

Hardegger, Biland and Corrodi (1963). Orchinol was not detected in healthy tubers but was induced to form by several infecting fungi and chemicals. Hiura (1943) detected antifungal activity in roots of sweet potato (*Ipomoea batatas*) after infection by *Ceratocystis fimbriata*. The compound responsible for the activity was identified by Kubota and Matsuura (1953) as ipomeamarone (**4**), which was shown also to accumulate after treatment with mercuric chloride (Uritani *et al.*, 1960). Condon and Kuć (1960, 1962) isolated an active compound from infected roots of carrot (*Daucus carota*) and confirmed that it was a substituted isocoumarin, now known as methoxymellein (**5**), also found by Sondheimer (1961) in carrot roots stored at 0°C. Orchinol, ipomeamarone and methoxymellein were regarded, along with pisatin, as phytoalexins by Cruickshank (1963) in the first review of the subject.

3 Orchinol

4 Ipomeamarone

5 Methoxymellein

Although they are not all the concern of this book, it is important to point out that other types of antifungal activity occur in some plants. One type is based on active compounds present in uninfected cells. Another type depends on the creation of active substances by rapid chemical conversions from inactive precursors present in uninfected cells.

The best example of resistance based on compounds present and active in uninfected tissue is found in onion bulbs in relation to *Colletotrichum circinans*, the cause of smudge disease. The compounds are protocatechuic acid (**6**) (Link *et al.*, 1929; Angell *et al.*, 1930) and catechol (**7**) (Link and Walker, 1933) which occur in the outer scales of resistant bulbs. The compounds diffuse readily from the non-living cells in these scales into overlying droplets. Presence of the compounds only in resistant progeny of crosses between resistant and susceptible cultivars of onion supported the idea that they were the resistance factors. Pre-formed factors in living cells of other plants are known; the saponins (Defago, 1977), for example, have been implicated in resistance since the work of Turner (1960, 1961) on oat roots, and avenacins (**8** and **9**) subsequently characterized by Burkhardt *et al.* (1964) and Tschesche *et al.* (1973).

6 Protocatechuic acid   **7** Catechol   **8** Avenacin A $R_1 = CH_2OH$
$R_2 = \beta$-D-glucose 1,4β-D-glucose 1,2α-L-arabinose 1,3

9 Avenacin B $R_1 = CH_3$
$R_2 = \beta$-D-glucose 1,4β-D-glucose 1,3

There are several ways by which antifungal compounds can be formed in plant cells from inactive precursors. These include oxidations of phenols to quinones and hydrolyses of glycosides, processes probably catalysed by enzymes normally separated from substrates by compartmentalization in healthy cells. For example, wheat and maize contain a glucoside of dihydroxymethoxy-benzoxazinone and a glucosidase which interact rapidly when brought together to yield an antifungal aglucone **10** (Wahlroos and Virtanen, 1959). Non-enzymatic processes may also liberate active substances, for example in the formation of the lactone tulipalin A (**11**) in tulip bulbs (Schönbeck and Schroeder, 1972; Beijersbergen and Lemmers, 1972).

10 DIMBOA   **11** Tulipalin A

The original concept of phytoalexins includes antifungal compounds which are released from immediate precursors in infected cells but it clearly excludes compounds present *before* infection. The first compounds to be identified as phytoalexins, pisatin and phaseollin, are synthesized from remote precursors rather than released from immediate precursors—indeed, the early evidence for synthesis was the slowness with which they formed and the absence of immediate precursors. Recent evidence for synthesis is knowledge of the pathways and enzymes activated *after* infection, as discussed in later chapters. Numerous interactions between the metabolic systems of parasite and host can be conceived for the activation of, and the regulation of enzymatic steps within, these pathways. Greater interest has surrounded the many compounds now known to be synthesized by pathways activated after infection, and these alone have been regarded as phytoalexins.

Recent themes in research on phytoalexins

Research on phytoalexins in the two decades since pisatin was isolated has been prolific. Themes of chemical and biological interest can be traced in this work, and I shall review these briefly as they affect the concept of phytoalexins.

The chemical diversity of phytoalexins in the plant kingdom was apparent when pisatin and phaseollin were identified as isoflavonoids and considered to have a similar biological function to the furanosesquiterpene, ipomeamarone. The extent of this diversity was made clear when the bicyclic sesquiterpene, rishitin (**12**), was isolated from potato (Tomiyama *et al.*, 1968) and the polyacetylenic aliphatic alcohol, safynol (**13**), was isolated from safflower (Allen and Thomas, 1971). Diversity within a plant family, and also within a species, was indicated when the acetylenic aliphatic compound wyerone acid (**18**) (Letcher *et al.*, 1970) and the isoflavonoid medicarpin (**14**) (Hargreaves *et al.*, 1976) were obtained from broad bean, *Vicia faba*. Medicarpin was originally found as a phytoalexin in alfalfa, *Medicago sativa*, by Smith *et al.* (1971) but it was also found in red clover, *Trifolium pratense* (Higgins and Smith, 1972) and in *Canavalia ensiformis* (Keen, 1972). The diversity within the legume *Vicia faba* appears to be exceptional, however; most members of the Leguminosae which have been studied produce isoflavonoid phytoalexins. General uniformity of chemical types of phytoalexin within a family is also suggested by the fact that all of the studied members of the Solanaceae produce terpenoid compounds, although acetylenic phytoalexins have recently been isolated from tomato (de Wit and Kodde, 1981).

12 Rishitin

$$CH_2OHCHOHCH = CHC \equiv CC \equiv CC \equiv CCH = CHCH_3$$

13 Safynol

14 Medicarpin

Extensive research on certain species in the Leguminosae and Solanaceae has revealed that a plant may produce a number of phytoalexins and not only one, as early research indicated. Further examination of infection droplets in seed cavities of French bean pods showed the presence of a compound closely related to phaseollin; it was named phaseollidin (15) (Perrin *et al.*, 1972) and was shown to be a product of the plant and therefore a phytoalexin (Cruickshank *et al.*, 1974). When Bailey and Burden (1973) extracted bean tissue infected with either tobacco necrosis virus or the fungus *Colletotrichum lindemuthianum*, they were able to separate five phytoalexins from the extract. Two were phaseollin and phaseollidin and two others were characterized as related compounds and named phaseollinisoflavan (16) and kievitone (17) (Burden *et al.*, 1972). The

15 Phaseollidin

16 Phaseollinisoflavan

17 Kievitone

curious fact that *Vicia faba* produces the unrelated phytoalexins wyerone acid and medicarpin has already been mentioned, but recent research has revealed the predominance and variety of the acetylenic compounds related to wyerone acid in this species. Mansfield *et al.* (1980) have used high performance liquid chromatography to separate these compounds in extracts of different types of tissue after infection or treatment with mercuric chloride. Seven furanoacetylenic phytoalexins were measured; these were wyerone acid (18), wyerone (19), wyerol (20), their respective dihydro-analogues 21, 22, 23 and wyerone epoxide (24).

The proportion varied according to the tissues used, the species of infecting fungus and the time after inoculation or treatment.

$$CH_3CH_2CH=CHC\equiv CCO\underset{O}{\underset{\diagup}{\Bigl\langle}}CH=CHCOOR$$

18 Wyerone acid R = H
19 Wyerone R = CH$_3$

$$CH_3CH_2CH=CHC\equiv CCH(OH)\underset{O}{\underset{\diagup}{\Bigl\langle}}CH=CHCOOCH_3$$

20 Wyerol

$$CH_3CH_2CH_2CH_2C\equiv CCO\underset{O}{\underset{\diagup}{\Bigl\langle}}CH=CHCOOR$$

21 Dihydrowyerone acid R = H
22 Dihydrowyerone R = CH$_3$

$$CH_3CH_2CH_2CH_2C\equiv CCH(OH)\underset{O}{\underset{\diagup}{\Bigl\langle}}CH=CHCOOCH_3$$

23 Dihydrowyerol

$$CH_3CH_2CH\overset{O}{\overset{\diagup\diagdown}{-}}CHC\equiv CCO\underset{O}{\underset{\diagup}{\Bigl\langle}}CH=CHCOOCH_3$$

24 Wyerone epoxide

From potato, the terpenoid compounds lubimin (**25**) (Metlitskii *et al.*, 1971) and phytuberin (**26**) (Coxon *et al.*, 1974) have now been characterized as phytoalexins in addition to rishitin. Similar findings concerning a multicomponent response have been made in most species of plant which produce phytoalexins.

25 Lubimin   **26** Phytuberin

Different fungi are well known as inducing phytoalexin formation but it is now also clear that bacteria can initiate the process, for example, in French bean leaves (Stholasuta *et al.*, 1971) soybean leaves (Keen and Kennedy, 1974) and potato tubers (Lyon, 1972). Some viruses which cause local necrotic lesions to form in French bean and tobacco also initiate phytoalexin formation (Bailey and Burden, 1973; Bailey *et al.*, 1975; Burden *et al.*, 1975). Some nematode infections also cause phytoalexin formation (Abawi *et al.*, 1971; Kaplan *et al.*, 1980). Very commonly, cellular disorganization in the hosts at the sites of these many types of infection has been observed to precede the detection of phyto-alexins; often this disorganization was part of a hypersensitive response. From this evidence, the idea has arisen that phytoalexin formation is initiated in surrounding live cells as a consequence of cellular death at infection sites.

As mentioned earlier, application of mercuric chloride to the appropriate plant tissues causes production of the phytoalexins pisatin, orchinol, ipome-amarone and the furanoacetylenes. Pisatin formation in response to treatment with chemical reagents has been most studied and a wide range of compounds has been found to be effective including the salts of several heavy metals and some metabolic inhibitors (Perrin and Cruickshank, 1965), some antimeta-bolites and other organic molecules (Bailey, 1969; Hadwiger and Schwochau, 1970; Hadwiger, 1972). How this wide range of chemicals acts requires a unifying explanation, to be discussed in chapter 9, but it is interesting that the most effective concentrations of mercuric and copper chlorides were observed not to be phytotoxic to the tissues of pea pods (Cruickshank and Perrin, 1963b).

Several fungal products (Cruickshank and Perrin, 1968; Ayers *et al.*, 1976; Keen and Legrand, 1980) also elicit phytoalexin production and one of these, monilicolin A, has been claimed to do so without killing cells in bean pods (Paxton *et al.*, 1974). A partially purified glucan from the fungus *Phytophthora megasperma* var. *sojae* elicits phytoalexin formation in its host soybean and also in bean and potato (Cline *et al.*, 1978).

Certain types of physical injury have been shown to stimulate phytoalexin formation in some tissues. Application of dry ice to points along French bean hypocotyls caused freezing injuries and phaseollin formation whereas freezing whole hypocotyls did not initiate the process (Rahe and Arnold, 1975). Storage of carrot roots at 0°C caused methoxymellein formation (Sondheimer, 1961). Irradiation of parts of pea and soybean plants with low wavelength ultra-violet light caused phytoalexin formation (Hadwiger and Schwochau, 1971; Bridge and Klarman, 1973). The identification of so many types of biological, chemical and physical stimuli as effective phytoalexin inducers has allowed phytoalexins to be regarded as a sub-group of a class of substances termed *stress compounds* (Stoessl *et al.*, 1976).

The earliest research on the activity of phytoalexins did not encourage the idea that phytoalexins might be antibacterial. Pisatin had no effect on the growth of several species of plant parasitic bacteria (Cruickshank, 1962) and phaseollin had only a slight effect on three non-parasitic species (Cruickshank and Perrin, 1971). However, some uncharacterized compounds, which formed together with phaseollin in bean, were found to inhibit bacterial growth (Lyon and Wood, 1975), and some soybean phytoalexins (Keen and Kennedy, 1974) and rishitin from potato (Lyon and Bayliss, 1975) have also been found to be antibacterial. Although the number of investigations with bacteria are relatively few, there seems to be no reason why phytoalexins should not be regarded as antimicrobial, rather than antifungal, compounds.

The ability of some parasitic fungi to metabolize phytoalexins has been investigated. Wyerone acid was found to decrease in concentration as *Botrytis fabae* spread through leaves of broad bean from infection sites (Mansfield and Deverall, 1974). The decrease occurred first at these sites and then continued around the lesions as the fungus advanced. Wyerone acid was replaced in the lesion and then throughout the leaf by a reduced form which has less antifungal activity than the phytoalexin itself. The reductions in the acetylenic and keto-groups of the wyerone acid molecule were probably caused by the fungus in the leaf because the same process was carried out by the fungus when fed wyerone acid in culture (Mansfield and Widdowson, 1973). Phaseollin was probably metabolized in bean hypocotyls by *Fusarium solani* f. sp. *phaseoli* because a product, hydroxyphaseollone, was readily detected (VanEtten and Smith, 1975) as it was when phaseollin was fed to the fungus in culture (Heuvel and VanEtten, 1973; Heuvel *et al.*, 1974). Three other metabolic fates of phaseollin have been revealed in the presence of parasitic fungi: conversion to phaseollinisoflavan by *Stemphylium botryosum* (Heath and Higgins, 1973; Higgins *et al.*, 1974); hydroxylation to hydroxyphaseollin by *Colletotrichum lindemuthianum* (Burden *et al.*, 1974); and hydroxylation to a dihydrodiol by *Septoria nodorum* (Bailey *et al.*, 1977). One of the terpenoid phytoalexins, capsidiol, was rapidly oxidized to a less active compound by *Fusarium oxysporum* both in pepper fruit and in culture (Stoessl *et al.*, 1973). Phytoalexins have also been shown to be metabolized by higher plant cells themselves (Skipp *et al.*, 1977; Glazener and VanEtten, 1978). The significance of phytoalexin metabolism in regulating the development of parasites in their hosts will be discussed in a later chapter, but more important to the present consideration is the amount of a phytoalexin which accumulates after infection as a net result of formation and degradation. It has been suggested that one way in which some of the many apparent stimuli for phytoalexin formation act is by inhibiting degradative metabolism and thereby permitting phytoalexin accumulation (Yoshikawa, 1978).

In the course of their discussion of phytoalexins as a sub-group of stress compounds, Stoessl *et al.* (1976) drew attention to some related aspects of plant metabolism which affect the way in which phytoalexins must be viewed. The isoflavonoid phytoalexins are similar to compounds known as normal constituents of uninfected plant tissues, namely the pterocarpans in heartwoods of woody species in the Leguminosae. In a few cases, molecules known in certain species as phytoalexins, such as medicarpin and maackiain, are also known as heartwood constituents in unrelated woody species of the same family (Ingham,

1972). Many isoflavonoids are antifungal and those in heartwoods may play passive roles in the protection of these non-living tissues from fungal attack. Antifungal activity in these molecules seems to depend upon the arrangement of the aromatic rings, the active pterocarpans having a non-planar arrangement and the inactive coumestans having a planar arrangement (Perrin and Cruickshank, 1969). The related flavonoids which occur in the Leguminosae and other plant families are also generally inactive against microorganisms. Some of these flavonoids are synthesized by infected plants, alfalfa, *Medicago sativa*, for example, making molecules of this type (Olah and Sherwood, 1971) in addition to more active isoflavonoid phytoalexins. Terpenoid compounds with no known properties as defensive agents but closely related to phytoalexins are also produced in response to infection and stress in potato tubers (Katsui *et al.*, 1971; Katsui *et al.*, 1974; Coxon *et al.*, 1974). These molecules may have other functions in countering stress or they may be intermediates and by-products of biosynthetic pathways activated by stress and with no special function. Their formation indicates the general nature of the response of plants to infection and stress. The molecules which have been recognized as phyto-alexins because of their antimicrobial properties are among a number synthe-sized as part of this general response. Natural selection may favour those plants which make phytoalexins and also those plants which possess the same compounds as pre-formed factors.

In summary, the molecules recognized as phytoalexins have been revealed as the antimicrobial components of a (usually) related group of compounds synthesized by a plant in response to infection or stress. The chemistry of these compounds is usually uniform within a plant family but is diverse within the plant kingdom. Accumulation of the compounds is determined by rates of synthesis in the plant and by rates of metabolism by the infecting organism and possibly also by the plant.

In view of these findings, a group of research workers attending the NATO Advanced Study Institute on "Active Defence Mechanisms of Plants" (Sounion, Greece, April 1980) assembled informally under the leadership of Dr J. D. Paxton and proposed the following definition: "Phytoalexins are low molecular weight antimicrobial compounds that are both synthesized by and accumulate in plants after their exposure to microorganisms". In two senses, this definition is narrower than the original concept which included no reference to the chemical nature of the principle and was not restricted to synthesized com-pounds. The original concept would allow inclusion of compounds released from inactive precursors in the reacting host cell as discussed earlier. In other respects, the definition is broader, recognizing the multi-component nature of phytoalexins, including antibacterial action and being less precise about the nature of the parasite-host interaction which initiates accumulation. The definition does not require evidence that the compounds are involved in the expression of resistance, which is an unsolved problem to be discussed later in this chapter and elsewhere in this book.

Major questions remaining

One of the outstanding questions concerns the ubiquity of phytoalexin formation throughout the plant kingdom. Present knowledge of phytoalexins as
compounds synthesized from remote precursors is based on extensive studies in
two plant families, the Leguminosae and Solanaceae, and investigations of one
or a few species in each of a number of other plant families. Clearly too few
species in the other families examined and too few families altogether have been
studied to permit the assertion that phytoalexin synthesis is a universal process
in the higher plants. Information on many more plants is needed to provide
perspective on the question of the existence and nature of a universal defence
mechanism in plants. Does this involve the production of low molecular weight
antimicrobial compounds? Are these compounds synthesized from remote
precursors in some species and released from inactive precursors in others? If
the common mechanism involves these compounds, a diverse range of molecules
has been selected as defensive agents. Is it conceivable that all of these
compounds play secondary roles in defence and that a primary mechanism
exists but has not yet been detected? The way in which these questions are
viewed will be affected by knowledge of a wider selection of plants and by critical
appraisal of the role of the known compounds in defence.

The discovery of biologically active factors in an interaction between a
parasite and a host is an exciting event. It encourages the investigator to believe
that the factors are involved in regulating the interaction. Such excitement and
enthusiasm have occurred in research on phytoalexins. Legitimate doubts have
also arisen, however, about the hypothesis that phytoalexins are primary factors
in the expression of resistance to parasites. It is desirable to know whether
phytoalexins occur at those micro-sites within a plant where they might contact
a parasitic fungus or bacterium. Parasites are specialized to grow in particular
places in plants, such as intercellular spaces or between protoplasts and cell
walls. The distribution of phytoalexins in these places is not well known. Most
phytoalexins are lipophilic and it should be asked whether they occur at these
sites not only in sufficient concentration but also in an appropriate form to be
effective. It is also essential to know whether phytoalexins accumulate at the
appropriate stage during infection to cause cessation of growth.

The possibility exists that some other growth-limiting process affects the
parasite *before* phytoalexins accumulate. This possibility was emphasized by
experiments performed by Kiraly *et al.* (1972) which showed, for example, that
the growth of the potato blight fungus, *Phytophthora infestans*, can be stopped
in a susceptible cultivar by application of an antibiotic and that, as a consequence, potato cells die and phytoalexins accumulate. The implication of this
work is that fungal growth in a resistant plant is stopped by a factor, as yet
undiscovered, and that the inhibited fungus releases a toxic substance that kills
host cells. This idea was supported by the finding that killed mycelium of *P.
infestans* yielded a substance which caused hypersensitivity when added to

potato tissue. These experiments encourage the view that hypersensitivity and phytoalexin formation are consequences and not the cause of the expression of resistance.

There is a need for critical investigation of the natural sequence of events when resistance is expressed. A particularly important chapter in this book will assess the progress that has been made in solving this biological question and will discuss some of the approaches which are available to test the hypothesis that phytoalexins are involved in defence.

As discussed earlier in this Introduction, the process of phytoalexin synthesis is readily initiated by many different stimuli but little is known of the natural regulation of the process during infection. Phytoalexin accumulation has often been observed to be closely associated with the hypersensitive response of resistant host cells and to be greatly delayed in susceptible tissues. Phytoalexin formation can be initiated by substances isolated from fungal walls and from the fluids in which fungi have grown in culture. Consideration of these facts has given rise to questions about the location of phytoalexin formation at infection sites and about the way in which formation is regulated by parasites. Thus it can be asked whether phytoalexin formation occurs in dying or dead cells or in living cells. Hydrolyses and oxidations in single-step processes might occur in non-living cells. Rathmell and Bendall (1972) suggested, for example, that final interconversions of isoflavonoids in phaseollin production could be catalysed by peroxidases in dying cells. Living cells are the probable sites of more elaborate syntheses. The active live cells at infection sites might be the initially infected cells which are responding to a fungal component and producing phytoalexins before they die in the hypersensitive response. Alternatively, the active cells may be those around the infected hypersensitive cells and responding to a diffusible product of the fungus or of the dying host cells. The activating stimulus from the fungus could therefore be either a direct elicitor of phytoalexin synthesis or a toxin causing cell death.

Hypotheses about the natural regulation of phytoalexin formation and hypersensitivity must also accommodate explanations of the highly specific relationships between parasites and plants. This specialization can be very high when it is determined by complementary single genes in parasite and host. Some much-discussed interactions between rust or powdery mildew fungi and their hosts are known to be determined by dominant alleles for avirulence in the parasite and for resistance in the host (Ellingboe, 1976). In these interactions, resistance is expressed only when the complementary alleles are matched. An interpretation in physiological terms is that products of the complementary genes interact in a recognition reaction which initiates the expression of resistance. The parasite-host interactions where phytoalexins have been implicated in resistance are less well analysed in genetic terms, although that between races of *Phytophthora infestans* and cultivars of potato is thought to conform closely with the complementary gene system. The well-studied interactions between *Colletotrichum lindemuthianum* and French bean and between *Phyto-*

phthora megasperma var. *sojae* and soybean are phenotypically similar to the gene-for-gene based interactions. For each of these relationships, the problem is posed of the physiological and biochemical basis of the regulation of the expression of resistance. This regulation may involve a recognition reaction which determines whether or not hypersensitivity and phytoalexin formation will occur. The question of the involvement of specific elicitors or suppressors of these processes is an important theme for future research in physiological plant pathology.

Acknowledgements

The translation of the postulates of Müller and Börger (1940) was done by J. Hardy of the CSIRO, Australia, and sent to me by Dr J. D. Paxton. I thank S. McLeod and A. Cloud for help with the preparation of this chapter.

REFERENCES

Abawi, G. S., VanEtten, H. D. and Mai, W. F. (1971) Phaseollin production induced by *Pratylenchus penetrans* in *Phaseolus vulgaris*. *J. Nematol.*, **3**, 301.

Allen, E. H. and Thomas, C. A. (1971) Trans-trans-3,11-tridecadiene-5,7,9-triyne-1,2-diol, an antifungal polyacetylene from diseased safflower (*Carthamus tinctorius*). *Phytochemistry*, **10**, 1579–1582.

Angell, H. R., Walker, J. C. and Link, K. P. (1930) The relation of protocatechuic acid to disease resistance in the onion. *Phytopathology*, **20**, 431–438.

Ayers, A. R., Ebel, J., Finelli, F., Berger, N. and Albersheim, P. (1976) Host-pathogen interactions. IX. Quantitative assays of elicitor activity and characterization of the elicitor present in the extracellular medium of cultures of *Phytophthora megasperma* var. *sojae*. *Plant Physiol.*, **57**, 751–759.

Bailey, J. A. (1969) Effects of antimetabolites on production of the phytoalexin pisatin. *Phytochemistry*, **8**, 1393–1395.

Bailey, J. A. and Burden, R. S. (1973) Biochemical changes and phytoalexin accumulation in *Phaseolus vulgaris* following cellular browning caused by tobacco necrosis virus. *Physiol. Plant Pathol.*, **3**, 171–177.

Bailey, J. A., Burden, R. S., Mynett, A. and Brown, C. (1977) Metabolism of phaseollin by *Septoria nodorum* and other non-pathogens of *Phaseolus vulgaris*. *Phytochemistry*, **16**, 1541–1544.

Bailey, J. A., Burden, R. S. and Vincent, G. G. (1975) Capsidiol: an antifungal compound produced in *Nicotiana tabacum* and *Nicotiana clevelandii* following infection with tobacco necrosis virus. *Phytochemistry*, **14**, 597.

Beijersbergen, J. C. M. and Lemmers, C. B. G. (1972) Enzymic and non-enzymic liberation of tulipalin A (α-methylene butyrolactone) in extracts of tulip. *Physiol. Plant Pathol.*, **2**, 265–270.

Beretta, M. J. G., Martins, E. M. F. and Moraes, W. B. C. (1977) Induced protection to *Hemileia vastatrix* at a distance from the site of the inducing action in coffee plants. *Summa Phytopathol.*, **3**, 66–70.

Bernard, N. (1909) L'évolution dans la symbiose, les orchidées et leurs champignons commenseaux. *Ann. Sci. Nat. (Bot.)*, **9**, 1–196.

Bernard, N. (1911) Sur la fonction fungicide des bulbes d'Ophrydées. *Ann. Sci. Nat. (Bot.)*, **14**, 221–234.

Bridge, M. A. and Klarman, W. L. (1973) Soybean phytoalexin, hydroxyphaseollin, induced by ultraviolet irradiation. *Phytopathology*, **63**, 606–609.

Burden, R. S., Bailey, J. A. and Dawson, G. W. (1972) Structures of three new isoflavonoids from *Phaseolus vulgaris* infected with tobacco necrosis virus. *Tetrahedron Lett.*, **41**, 4175–4178.

Burden, R. S., Bailey, J. A. and Vincent, G. G. (1974) Metabolism of phaseollin by *Colletotrichum lindemuthianum*. *Phytochemistry*, **13**, 1789–1791.

Burden, R. S., Bailey, J. A. and Vincent, G. G. (1975) Glutinosone, a new antifungal sesquiterpene from *Nicotiana glutinosa* infected with tobacco mosaic virus. *Phytochemistry*, **14**, 221–223.

Burkhardt, H. J., Maizel, J. V. and Mitchell, H. K. (1964) Avenacin, an antimicrobial substance isolated from *Avena sativa*. II. Structure. *Biochemistry*, **3**, 426–431.

Chester, K. S. (1933) The problem of acquired physiological immunity in plants. *Quart. Rev. Biol.*, **8**, 119–154, 275–324.

Cline, K., Wade, M. and Albersheim, P. (1978) Host-pathogen interactions. XV. Fungal glucans which elicit phytoalexin accumulation in soybean also elicit the accumulation of phytoalexins in other plants. *Plant Physiol.*, **62**, 918–921.

Condon, P. and Kuć, J. (1960) Isolation of a fungitoxic compound from carrot tissue inoculated with *Ceratocystis fimbriata*. *Phytopathology*, **50**, 267–270.

Condon, P. and Kuć, J. (1962) Confirmation of identity of a fungitoxic compound produced by carrot root tissue. *Phytopathology*, **52**, 182–183.

Coxon, D. T., Curtis, R. F., Price, K. R. and Howard, B. (1974) Phytuberin; a novel antifungal terpenoid from potato. *Tetrahedron Lett.*, *21*, 2363–2366.

Coxon, D. T., Price, K. R., Howard, B., Osman, S. F., Kalan, E. B. and Zacharius, R. M. (1974) Two new vetispirane derivatives: stress metabolites from potato (*Solanum tuberosum*) tubers. *Tetrahedron Lett.*, **34**, 2921–2924.

Cruickshank, I. A. M. (1962) Studies on phytoalexins. IV. The antimicrobial spectrum of pisatin. *Aust. J. Biol. Sci.*, **15**, 147–159.

Cruickshank, I. A. M. (1963) Phytoalexins. *Ann. Rev. Phytopathol.*, **1**, 351–374.

Cruickshank, I. A. M., Biggs, D. R., Perrin, D. R. and Whittle, C. P. (1974) Phaseollin and phaseollidin relationships in infection-droplets on endocarp of *Phaseolus vulgaris*. *Physiol. Plant Pathol.*, **4**, 261–276.

Cruickshank, I. A. M. and Perrin, D. R. (1960) Isolation of a phytoalexin from *Pisum sativum* L. *Nature*, **187**, 799–800.

Cruickshank, I. A. M. and Perrin, D. R. (1963a) Phytoalexins of the Leguminosae. Phaseollin from *Phaseolus vulgaris* L. *Life Sci.*, **2**, 680–682.

Cruickshank, I. A. M. and Perrin, D. R. (1963b) Studies on phytoalexins. VI. Pisatin: The effect of some factors on its formation in *Pisum sativum* L., and the significance of pisatin in disease resistance. *Aust. J. Biol. Sci.*, **16**, 111–128.

Cruickshank, I. A. M. and Perrin, D. R. (1968) The isolation and partial characterization of monilicolin A, a polypeptide with phaseollin-inducing activity from *Monilinia fructicola*. *Life Sci.*, **7**, 449–458.

Cruickshank, I. A. M. and Perrin, D. R. (1971) Studies on phytoalexins. XI. The induction, antimicrobial spectrum and chemical assay of phaseollin. *Phytopath. Z.*, **70**, 209–229.

Défago, G. (1977) Rôle des saponines dans la résistance des plantes aux maladies fongiques. *Ber. Schweiz. Bot. Ges.*, **87**, 79–132.

de Wit, P. J. G. M. and Kodde, E. (1981) Induction of polyacetylenic phytoalexins in *Lycopersicon esculentum* after inoculation with *Cladosporium fulvum*. *Physiol. Plant Pathol.*, **18**, 143–148.

Ellingboe, A. H. (1976) Genetics of host-parasite interactions, in *Physiological Plant Pathology*, ed. P. H. Williams and R. Heitefuss, Springer-Verlag, Berlin, 761–778.

Gäumann, E. and Jaag, O. (1945) Über induzierte Abwehrreaktionen bei Pflanzen. *Experientia*, **1**, 21–22.

Gäumann, E. and Kern, H. (1959a) Über die isolierung und den chemischen Nachweis des Orchinols. *Phytopath. Z.*, **35**, 347–356.

Gäumann, E. and Kern, H. (1959b) Über chemische Abwehrreaktionen bei Orchideen. *Phytopath. Z.*, **36**, 1–26.

Glazener, J. A. and VanEtten, H. D. (1978) Phytotoxicity of phaseollin to, and alteration of phaseollin by, cell suspension cultures of *Phaseolus vulgaris*. *Phytopathology*, **68**, 111–117.

Hadwiger, L. A. (1972) Increased levels of pisatin and phenylalanine ammonia lyase activity in *Pisum sativum* treated with antihistaminic, antiviral, antimalarial, tranquilizing or other drugs. *Biochem. Biophys. Res. Commun.*, **46**, 71–79.

Hadwiger, L. A. and Schwochau, M. E. (1970) Induction of phenylalanine ammonia lyase and pisatin in pea pods by polylysine, spermidine or histone fractions. *Biochem. Biophys. Res. Commun.*, **38**, 683–691.

Hadwiger, L. A. and Schwochau, M. E. (1971) Ultraviolet light-induced formation of pisatin and phenylalanine ammonia lyase. *Plant Physiol.*, **47**, 588–590.

Hardegger, E., Biland, H. R. and Corrodi, H. (1963) Synthese von 2,4-Dimethoxy-6-hydroxy-phenanthren und Konstitution des Orchinols. *Helv. Chim. Acta*, **46**, 1354–1360.

Hargreaves, J. A., Mansfield, J. W. and Coxon, D. T. (1976) Identification of medicarpin as a phytoalexin in the broad bean plant (*Vicia faba* L.). *Nature*, **262**, 318–319.

Heath, M. C. and Higgins, V. J. (1973) *In vitro* and *in vivo* conversion of phaseollin and pisatin by an alfalfa pathogen *Stemphylium botryosum*. *Physiol. Plant Pathol.*, **3**, 107–120.

Heuvel, J. van den and VanEtten, H. D. (1973) Detoxification of phaseollin by *Fusarium solani* f. sp. *phaseoli*. *Physiol. Plant Pathol.*, **3**, 327–339.

Heuvel, J. van den, VanEtten, H. D., Serum, J. W., Coffen, D. L. and Williams, T. H. (1974) Identification of 1α-hydroxy-phaseollone, a phaseollin metabolite produced by *Fusarium solani*. *Phytochemistry*, **13**, 1129–1131.

Higgins, V. J. and Smith, D. G. (1972) Separation and identification of two pterocarpanoid phytoalexins produced by red clover leaves. *Phytopathology*, **62**, 235–238.

Higgins, V. J., Stoessl, A. and Heath, M. C. (1974) Conversion of phaseollin to phaseollinisoflavan by *Stemphylium botryosum*. *Phytopathology*, **64**, 105–107.

Hiura, M. (1943) Studies in storage and rot of sweet potato. *Sci. Rep. Gifu Agric. Coll. Jpn.*, **50**, 1–5.

Ingham, J. L. (1972) Phytoalexins and other natural products as factors in plant disease resistance. *Bot. Rev.*, **38**, 343–424.

Kaplan, D. T., Keen, N. T. and Thomason, I. J. (1980) Association of glyceollin with the incompatible response of soybean roots to *Meloidogyne incognita*. *Physiol. Plant Pathol.*, **16**, 309–318.

Katsui, N., Matsunaga, A. and Masamune, T. (1974) The structure of lubimin and oxylubimin, antifungal metabolites from diseased potato tubers. *Tetrahedron Lett.*, **51/52**, 4483–4486.

Katsui, N., Matsunaga, A., Imaizumi, K., Masumune, T. and Tomiyama, K. (1971) The structure and synthesis cf rishitinol, a new sesquiterpene alcohol from diseased potato tubers. *Tetrahedron Lett.*, **2**, 83–86.

Keen, N. T. (1972) Accumulation of wyerone in broad bean and demethylhomopterocarpin in jack bean following inoculation with *Phytophthora megasperma* var. *sojae*. *Phytopathology*, **62**, 1365.

Keen, N. T. and Kennedy, B. W. (1974) Hydroxyphaseollin and related isoflavanoids in the hypersensitive resistance reaction of soybeans to *Pseudomonas glycinea*. *Physiol. Plant Pathol.*, **4**, 173–185.

Keen, N. T. and Legrand, M. (1980) Surface glycoproteins: evidence that they may function as the race specific phytoalexin elicitors of *Phytophthora megasperma* f. sp. *glycinea*. *Physiol. Plant Pathol.*, **17**, 175–192.

Kiraly, Z., Barna, B. and Ersek, T. (1972) Hypersensitivity as a consequence, not the cause, of plant resistance to infection. *Nature*, **239**, 456–458.

Kubota, T. and Matsuura, T. (1953) Chemical studies on the black rot disease of sweet potato. *J. Chem. Soc. Jpn. (Pure Chem. Sect.)*, **74**, 248–251.

Kuć, J., Shockley, G. and Kearney, K. (1975) Protection of cucumber against *Colletotrichum lagenarium* by *Colletotrichum lagenarium*. *Physiol. Plant Pathol.*, **7**, 195–199.

Letcher, R. M., Widdowson, D. A., Deverall, B. J. and Mansfield, J. W. (1970) Identification and activity of wyerone acid as a phytoalexin in broad bean (*Vicia faba*) after infection by *Botrytis*. *Phytochemistry*, **9**, 249–252.

Link, K. P., Dickson, A. D. and Walker, J. C. (1929) Further observations on the occurrence of protocatechuic acid in pigmented onion scales and its relation to disease resistance in the onion. *J. Biol. Chem.*, **84**, 719–725.

Link, K. P. and Walker, J. C. (1933) The isolation of catechol from pigmented onion scales and its significance in relation to disease resistance in onions. *J. Biol. Chem.*, **100**, 379–383.

Lyon, F. M. and Wood, R. K. S. (1975) Production of phaseollin, coumestrol and related compounds in bean leaves inoculated with *Pseudomonas* spp. *Physiol. Plant Pathol.*, **6**, 117–124.

Lyon, G. D. (1972) Occurrence of rishitin and phytuberin in potato tubers inoculated with *Erwinia carotovora* var. *atroseptica*. *Physiol. Plant Pathol.*, **2**, 411–416.

Lyon, G. D. and Bayliss, C. E. (1975) The effect of rishitin on *Erwinia carotovora* var. *atroseptica* and other bacteria. *Physiol. Plant Pathol.*, **6**, 177–186.

Magrou, J. (1924) L'immunité humorale chez les plantes. *Rev. Pathol. Vég. Entomol. Agr. Fr.*, **11**, 189–192.

Mansfield, J. W. and Deverall, B. J. (1974) Changes in wyerone acid concentrations in leaves of *Vicia faba* after infection by *Botrytis cinerea* or *B. fabae*. *Ann. appl. Biol.*, **77**, 227–235.

Mansfield, J. W., Porter, A. E. A. and Smallman, R. V. (1980) Dihydrowyerone derivatives as components of the furanoacetylenic phytoalexin response of tissues of *Vicia faba. Phytochemistry*, 19, 1057–1061.

Mansfield, J. W. and Widdowson, D. A. (1973) The metabolism of wyerone acid (a phytoalexin from *Vicia faba* L.) by *Botrytis fabae* and *B. cinerea. Physiol. Plant Pathol.*, 3, 393–404.

Metlitskii, L. V., Ozeretskovskaya, O. L., Vul'fson, N. S. and Chalova, L. I. (1971) Chemical nature of lubimin, a new phytoalexin of potatoes. *Dokl. Akad. Nauk SSSR*, 200, 1470–1472.

Meyer, G. (1939) Zellphysiologische und anatomische Untersuchungen über die Reaktion der Kartoffelknolle auf den Angriff der *Phytophthora infestans* bei Sorten verschiedener Resistenz. *Arb. Biol. Reichsants.*, 23, 97–132.

Müller, K. O. (1958) Studies on phytoalexins. I. The formation and immunological significance of phytoalexin produced by *Phaseolus vulgaris* in response to infections with *Sclerotinia fructicola* and *Phytophthora infestans. Aust. J. Biol. Sci.*, 11, 275–300.

Müller, K. O. and Börger, H. (1940) Experimentelle Untersuchungen über die *Phytophthora*—Resistenz der Kartoffel. *Arb. Biol. Anst. Reichsanst. (Berl.)*, 23, 189–231.

Müller, K. O., Meyer, G. and Klinkowski, M. (1939) Physiologisch-genetische Untersuchungen über die Resistenz der Kartoffel gegenüber *Phytophthora infestans. Naturwissenschaften*, 27, 765–768.

Nobécourt, P. (1923) Sur la production d'anticorp par les tubercules des Ophrydées. *C. R. Acad. Sci.*, 177, 1055–1057.

Olah, A. F. and Sherwood, R. T. (1971) Flavones, isoflavones and coumestans in alfalfa infected by *Ascochyta imperfecta. Phytopathology*, 61, 65–69.

Paxton, J., Goodchild, D. J. and Cruickshank, I. A. M. (1974) Phaseollin production by live bean endocarp. *Physiol. Plant Pathol.*, 4, 167–171.

Perrin, D. R. (1964) The structure of phaseollin. *Tetrahedron Lett.*, 1, 29–35

Perrin, D. R. and Bottomley, W. (1962) Studies on phytoalexins. V. The structure of pisatin from *Pisum sativum* L. *J. Am. Chem. Soc.*, 84, 1919–1922.

Perrin, D. R. and Cruickshank, I. A. M. (1965) Studies on phytoalexins. VII. Chemical stimulation of pisatin formation in *Pisum sativum* L. *Aust. J. Biol. Sci.*, 18, 803–816.

Perrin, D. R. and Cruickshank, I. A. M. (1969) The antifungal activity of pterocarpans towards *Monilinia fructicola. Phytochemistry*, 8, 971–978.

Perrin, D. R., Whittle, C. P. and Batterham, T. J. (1972) The structure of phaseollidin. *Tetrahedron Lett.*, 17, 1673–1676.

Rahe, J. E. and Arnold, R. M. (1975) Injury-related phaseollin accumulation in *Phaseolus vulgaris* and its implications with regard to specificity of host-parasite interaction. *Can. J. Bot.*, 53, 921–928.

Rathmell, W. G. and Bendall, D. S. (1972) The peroxidase-catalysed oxidation of a chalcone and its possible physiological significance. *Biochem. J.*, 127, 125–132.

Schönbeck, F. and Schroeder, C. (1972) Rôle of antimicrobial substances (tuliposides) in tulips attacked by *Botrytis* spp. *Physiol. Plant Pathol.*, 2, 91–100.

Skipp, R. A., Selby, C. and Bailey, J. A. (1977) Toxic effects of phaseollin on plant cells. *Physiol. Plant Pathol.*, 10, 221–227.

Smith, D. G., McInnes, A. G., Higgins, V. J. and Millar, R. L. (1971) Nature of the phytoalexin produced by alfalfa in response to fungal infection. *Physiol. Plant Pathol.*, 1, 41–44.

Sondheimer, E. (1961) Possible identity of a fungitoxic compound from carrot roots. *Phytopathology*, 51, 71–72.

Stakman, E. C. (1915) Relation between *Puccinia graminis* and plants highly resistant to its attack. *J. Agric. Res.*, 4, 193–200.

Stholasuta, P., Bailey, J. A., Severin, V. and Deverall, B. J. (1971) Effect of bacterial inoculation of bean and pea leaves on the accumulation of phaseollin and pisatin. *Physiol. Plant Pathol.*, 1, 177–184.

Stoessl, A., Stothers, J. B. and Ward, E. W. B. (1976) Sesquiterpenoid stress compounds of the Solanaceae. *Phytochemistry*, 15, 855–872.

Stoessl, A., Unwin, C. H. and Ward, E. W. B. (1973) Postinfectional fungus inhibitors from plants: fungal oxidation of capsidiol in pepper fruit. *Phytopathology*, 63, 1225–1231.

Sutton, D. C. (1979) Systemic cross protection in bean against *Colletotrichum lindemuthianum. Australas. Plant Pathol.*, 8, 4–5.

Tomiyama, K., Sakuma, T., Ishizaka, N., Sato, N., Katsui, N., Takasugi, M. and Masamune, T. (1968) A new antifungal substance isolated from resistant potato tuber tissue infected by pathogens. *Phytopathology*, 58, 115–116.

Tschesche, R., Chandra Jha, H. and Wulff, G. (1973) Uber Triterpene—Zur Struktur des Avenacins. *Tetrahedron*, **29**, 629–633.

Turner, E. M. (1960) The nature of the resistance of oats to the take-all fungus. III. Distribution of the inhibitors in oat seedlings. *J. Exp. Bot.*, **11**, 403–412.

Turner, E. M. (1961) An enzymic basis for pathogenic specificity in *Ophiobolus graminis*. *J. Exp. Bot.*, **12**, 169–173.

Uritani, I., Uritani, M. and Yamada, H. (1960) Similar metabolic alterations induced in sweet potato by poisonous chemicals and by *Ceratostomella fimbriata*. *Phytopathology*, **50**, 30–34.

VanEtten, H. D. and Smith, D. A. (1975) Accumulation of antifungal iso-flavanoids and 1α-hydroxy-phaseollone, a phaseollin metabolite, in bean tissue infected with *Fusarium solani* f. sp. *phaseoli*. *Physiol. Plant Pathol.*, **5**, 225–237.

Wahlroos, O. and Virtanen, A. I. (1959) The precursors of 6-methoxybenzoxazolinone in maize and wheat plants, their isolation and some of their properties. *Acta Chem. Scand.*, **13**, 1906–1908.

Ward, H. M. (1905) Recent researches on the parasitism of fungi. *Ann. Bot.*, **19**, 1–54.

Yoshikawa, M. (1978) Diverse modes of action of biotic and abiotic phytoalexin elicitors. *Nature*, **275**, 546–547.

2   Phytoalexins from the Leguminosae

JOHN L. INGHAM

Introduction

The Leguminosae (Fabaceae) is one of the three largest families of flowering plants, being exceeded in the number of described species (about 17 000) only by the Orchidaceae (18 000 species) and the Compositae (25 000 species). Economically, the legumes are of enormous importance, providing food for both man and animals, as well as timber and a variety of other commodities including oils, gums and resins, and substances with medicinal or insecticidal properties, while many leguminous species are also valuable as garden ornamentals. Legumes are found throughout the tropical, subtropical and temperate regions of the world, and range in form from immense trees through shrubs and sub-shrubs to tiny prostrate herbs. Taxonomically, the family is normally divided into three subfamilies: the primitive and comparatively small Caesalpinioideae and Mimosoideae (both mainly confined to the tropics and sub-tropics), and the very large climax group Papilionoideae (Faboideae or Lotoideae) which contains virtually all the legumes of commerce and which is well represented in temperate zones.

Because of its sheer size and economic significance, the subfamily Papilionoideae has aroused considerable phytochemical interest. It is not surprising, therefore, to find that a large proportion of phytoalexin studies has involved papilionate legumes (there are currently no reports of phytoalexin production by any species belonging to the agriculturally unimportant subfamilies Caesalpinioideae and Mimosoideae) and that consequently a rich assortment of phytoalexins has been discovered in these plants. A notable feature of the Papilionoideae is the widespread ability of its constituent species to produce isoflavonoids, a group of compounds which appear to occur only sporadically elsewhere in the plant kingdom. All isoflavonoids contain the 1,2-diphenylpropane skeleton shown in Figure 2.1 (compare the biogenetically related flavonoids, which although ubiquitous in higher plants are based on the more structurally limiting 1,3-diphenylpropane system) but in every class apart from

21

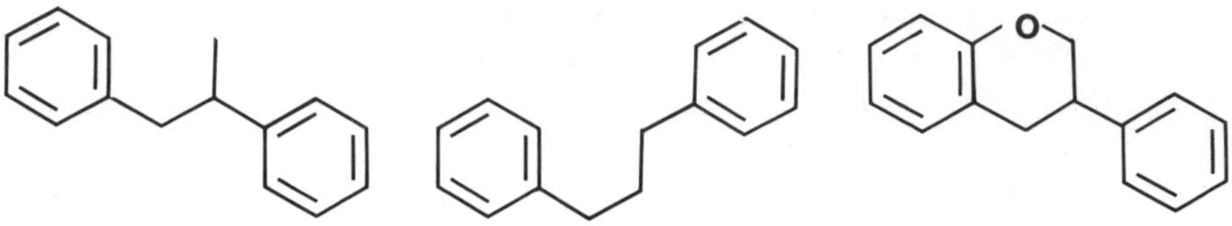

| 1,2-Diphenylpropane | 1,3-Diphenylpropane | 3-Phenylchroman |
| (isoflavonoids) | (flavonoids) | (most isoflavonoids) |

Figure 2.1   Structural skeletons of isoflavonoid and flavonoid compounds

the exceptionally rare, constitutive α-methyldeoxybenzoins (Wong, 1975), this structure is modified to give a central O-heterocyclic (pyran) ring. In essence, therefore, isoflavonoids can conveniently be regarded as derivatives of 3-phenylchroman (Figure 2.1).

At this point it is worth mentioning that substantial quantities of some isoflavonoids occur in the apparently healthy seeds, roots or heartwood of certain tropical, subtropical and temperate legumes (Wong, 1975) where they may perhaps deter feeding insects (Russell et al., 1978; Sutherland et al., 1980), or act as prohibitins (Ingham, 1973). Whilst such compounds are normally, but not invariably, absent from herbaceous tissues, surveys within the Papilionoideae have revealed that leaves, pods, hypocotyls and other non-woody plant parts have the capacity to produce isoflavonoid phytoalexins (and in rare instances a limited range of biologically active non-isoflavonoid derivatives) as a defensive response following inoculation with a wide range of fungi, bacteria and other microorganisms. A few of these compounds are chemically indistinguishable, although sometimes stereochemically distinct, from iso-flavonoids found constitutively in heartwood or other tissues, e.g. the well-known pterocarpans[1] medicarpin and maackiain (see Table 2.1, structures **28** and **31** respectively) and the isoflavan vestitol (**67**). The vast majority, however, have no known source apart from infected or diseased plants. Many Papilionoideae exhibit a multiple phytoalexin response producing three or more closely related compounds although one of these frequently predominates. From time to time, legumes differing widely in terms of their geographic origin and taxonomic relationships have been found to accumulate the same compound or compounds, e.g., **28**, **31** and **67**, but often the ability to form a given phytoalexin is restricted to a few closely related species or genera, and many phytoalexins have as yet been obtained from only one species.

Further studies involving the Papilionoideae will add more structures to the extensive number of compounds already described as well as revealing new sources of known phytoalexins. The variety of isoflavonoid types, coupled with functional group diversity and the comparatively large number of substitution sites possible on the phenylchroman nucleus (Figure 2.1), lead one to suspect that in the Papilionoideae there exists an isoflavonoid reservoir that has hardly

[1] Also referred to as pterocarpanoids and, less frequently, as pterocarpinoids.

been tapped. This view is supported by the fact that in recent years the number of reported isoflavonoid phytoalexins has increased dramatically—only two, pisatin (39) and phaseollin (46), were characterized before 1970, with the vast majority being identified since 1975. The present review gives structures for more than 100 induced isoflavonoid and non-isoflavonoid derivatives now known from legumes. It is important to appreciate that not all the compounds recorded have been shown to have a role in resistance (see chapter 8). The less fungitoxic compounds have been included, however, as they may be precursors of the more biologically active phytoalexins. Details of plant sources are catalogued (Tables 2.1 and 2.2), as are fungitoxic, spectroscopic and other data sufficient to serve as initial aids to compound identification.

Chemical diversity

Since 1960, careful examination of well over 500 papilionate legumes has revealed that compounds belonging to six isoflavonoid classes accumulate in tissues inoculated with fungi or other microorganisms. They include isoflavones, isoflavanones and coumestans,[2] but whilst these may be of considerable importance in certain species or genera (e.g. compounds 5, 8, 13 and 15 in *Laburnum anagyroides* and other members of the tribe Genisteae, 14 in *Hardenbergia violacea*, 26, 82 in *Phaseolus lunatus* and *P. vulgaris* and 11, 12 in *Centrosema*), they are overshadowed by the pterocarpan and isoflavan phytoalexins. All the relevant isoflavonoid structures are shown in Table 2.1, sections A to E (see p. 39). The compounds are presented systematically in order of increasing molecular weight with isoflavones, isoflavanones, pterocarpans and the single reported induced pterocarpene[3] preceding isoflavans and coumestans. Where two or more isoflavonoids have the same molecular weight (e.g. 4 to 6, 33 to 38, 49 to 55 and 74 to 78), they are arranged according to the nature and position of the A-ring substituents. Thus, hydroxylation precedes methoxylation, and monosubstitution is considered before disubstitution. These criteria are applied to ring B (or D) for compounds with identical A-rings.

Legume isoflavonoid phytoalexins are invariably oxygenated at C-7 and C-4′ or the equivalent C-3 and C-9 positions (see Figure 2.2 for a summary of ring nomenclature and numbering systems) with many possessing additional oxygenation elsewhere on the aromatic or heterocyclic rings. With the exception of 2-methoxyhomopterocarpin (40), all the compounds in Table 2.1 exhibit some degree of hydroxylation, this normally being associated with one or both of the aromatic (A, B/D) rings. In addition, seventeen pterocarpans (including the non-phenolic derivatives variabilin (34) and pisatin (39)) have tertiary (C-6a) hydroxyl groups which render these compounds susceptible to dehydration in the presence of mineral acids (see UV spectrophotometry, p. 30). Methoxyl substituents are also commonly encountered, these occurring frequently at C-4′

[2] Also known as coumestones and 6-oxopterocarp-6a-enes.
[3] The term pterocarpene is synonymous with dehydropterocarpan and pterocarp-6a-ene.

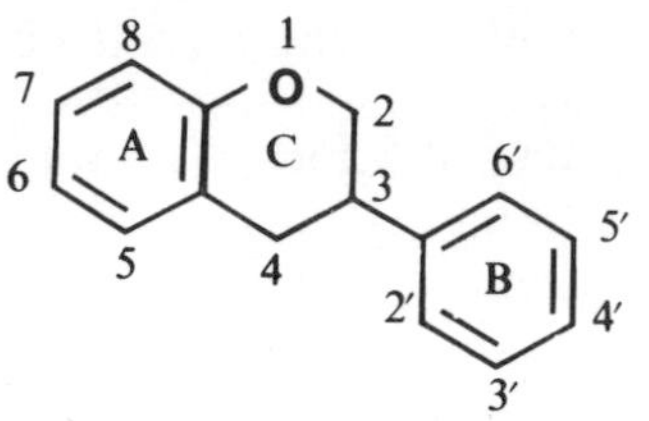

(1)

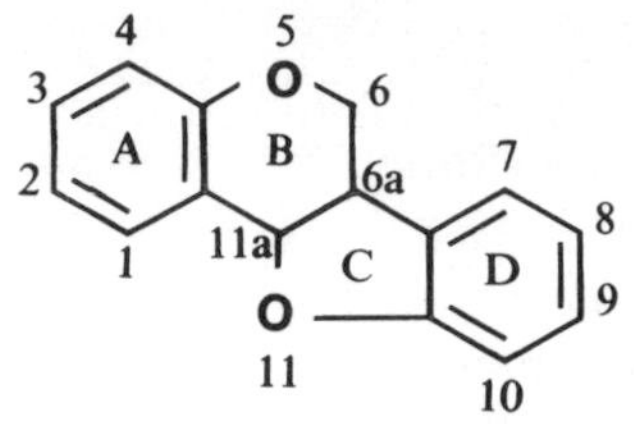

(2)

(1): Isoflavones, isoflavanones and isoflavans
(2): Pterocarpans, pterocarpenes and coumestans

Figure 2.2 Ring numbering and nomenclature systems normally applied to the known iso-flavonoid phytoalexins.

(C-9) and slightly less often at C-7 (C-3) and C-2' (isoflavones, isoflavanones and isoflavans only). Methoxylation is rare at C-6(C-2), C-3'(C-10), C-5(C-1) and C-8(C-4). None of the known induced isoflavonoids possesses hydroxyl or methoxyl attachments at either C-5'(C-8) or C-6'(C-7) although a few compounds having a C-4'/5' (C-8/9) methylenedioxy (O—CH$_2$—O) group have been recorded (**31, 39, 41** to **43, 73**). The recently discovered acanthocarpan (**42**) is the only legume phytoalexin with methylenedioxy substitution of both the A and D rings.

Apart from OH, OMe and O—CH$_2$—O groups, a growing number of iso-flavonoid phytoalexins with "complex" acyclic or cyclic substituents have been described; these attachments include isopentenyl (γ,γ-dimethylallyl) or *gem*-dimethylchromen units, and more rarely furan or modified furanoid ring systems. Prenylation may occur at C-6(C-2), C-8(C-4), C-3'(C-10) and C-5'(C-8), closure to a neighbouring oxygen atom (C-7/C-3 or C-4'/C-9) giving rise to new rings having either an angular (**25, 46, 48, 54, 55, 79, 80, 83**) or a linear (**45, 49, 51** to **53, 63, 64**) disposition. Biosynthetic studies involving the coumarin group of natural products suggest that the furan ring probably arises via loss of three carbon atoms from an intermediate hydroxyisopropyldihydrofuran moiety, and clearly a similar route may lead to the furanopterocarpan **45** which is thus included in the "complex" pterocarpan category despite its relatively low molecular weight. Modification of the prenyl group to afford an epoxide, or its attachment without cyclization to an oxygen atom (prenyloxylation), are both features not yet associated with "complex" isoflavonoid phytoalexins although such compounds do occur constitutively in a few members of the Papilionoideae, notably species of *Millettia, Psoralea* and *Tephrosia* (Gupta *et al.*, 1980; Kukla and Seshadri, 1962; Minhaj *et al.*, 1976).

Available evidence suggests that biosynthesis of "complex" pterocarpan phytoalexins may be largely restricted to species constituting the phylo-genetically-advanced pantropical tribe Phaseoleae (particularly members of the subtribes Erythrininae, Glycininae and Phaseolinae), and its near relative the Psoraleeae where glyceollin I (**54**) is of common occurrence. Prenylated isoflavones (**13** to **16**) and isoflavanones (**24** to **26**) also accumulate in some Phaseoleae, but additionally occur elsewhere as, for instance, in the essentially temperate tribes Genisteae (e.g. *Laburnum anagyroides*) and Desmodieae (e.g. *Desmodium canadense* and *Lespedeza bicolor*). Isolation of kievitone from fungus-inoculated leaves of several *Desmodium* and *Lespedeza* species (J. L. Ingham, unpublished data) is notable because it provides chemical support for the recently proposed link between the tribes Desmodieae and Phaseoleae (R. M. Polhill, pers. comm.).

Two groups of isoflavonoid phytoalexins (pterocarpans and isoflavans) contain optically active members whereas, with the possible exception of (+)-kievitone from *Vigna unguiculata* (Partridge, 1973), inducible isoflavanones appear to be optically inactive. The vast majority of pterocarpans exhibit large negative [α] values (ranging from about −150° (**46**) to −300° (**35**) at 589 nm) and can thus be assigned the 6aR;11aR absolute configuration (Ito *et al.*, 1965; Ollis, 1968; Verbit and Clark-Lewis, 1968) providing that they do not possess a C-6a hydroxyl group when, in accordance with chirality rules, the 6aS;11aS stereochemical arrangement applies (Ingham and Markham, 1980). In a similar fashion, (+)-6aH pterocarpans (none of which have been reported as phyto-alexins although several do occur constitutively in the tissues of certain Papilionoideae) are 6aS;11aS whilst the corresponding (+)-6aOH compounds (**39** and **49**) are 6aR;11aR (Figure 2.3).

For isoflavans, where very low positive or negative [α] values (often between 10° and 25° at 589 nm) are the norm, it has been found that high wavelength optical rotation data cannot be related to absolute configuration. Here, R or S assignments are normally made by comparison of ORD (optical rotatory dispersion) curves in the 200 to 300 nm region of the spectrum with those of compounds having an established stereochemistry (e.g. a 3R-isoflavan derived from a (−)-6aH pterocarpan; see above) although 2′,6′-substitution can give misleading information as a result of steric effects which influence the conformation adopted by the heterocyclic (C) ring (Pelter and Amenechi, 1969). After studying the ORD curves of several simple isoflavans, Kurosawa *et al.* (1978) concluded that 3R and 3S derivatives showed positive and negative Cotton

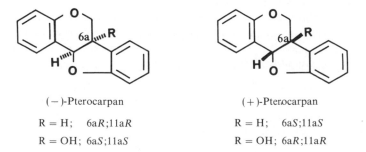

(−)-Pterocarpan   (+)-Pterocarpan

R = H;   6aR;11aR   R = H;   6aS;11aS

R = OH; 6aS;11aS   R = OH; 6aR;11aR

Figure 2.3   Stereochemistry of (+) and (−) pterocarpans having 6aH or 6aOH substituents.

effects respectively in the 260 to 300 nm spectral region. Comparable Cotton effects for optically active isoflavanones occurred between 330 and 350 nm.

Stereochemical assignments have been made for few induced isoflavans. However, when these compounds co-occur with pterocarpans of established absolute configuration, the equivalent asymmetric carbon centres (C-3 and C-6a) invariably have identical chiralities, an indication of the close biogenetic relationship between these isoflavonoid groups. Thus, 6aR;11aR-medicarpin (**28**) is found with 3R-vestitol (**67**) and 3R-sativan (**70**) in *Medicago sativa*. Demethylvestitol (**66**) from *Phaseolus vulgaris* pods is also 3R (a positive Cotton effect in the ORD curve of this isoflavan is evident at 293 nm), an observation entirely consistent with the production of 6aR;11aR pterocarpans (**27, 46, 47**) by this particular legume. Not infrequently, infected plants produce one or two major anti-microbial isoflavonoids in addition to several minor components which are apparently linked biosynthetically to the primary phytoalexin(s). The connection between these substances must, however, be regarded with some caution if [α] values have not been determined. In *P. vulgaris*, for example, it is unlikely that (+)-2′-O-methylphaseollinisoflavan[4] (**80**) arises directly from (−)-phaseollinisoflavan (**79**) since 2′-O-methylation would not be expected to affect the sign of rotation (compare (−)-vestitol (**67**) and its derived (−)-2′-O-methyl ether, sativan (**70**), in *Lotus corniculatus*). Likewise (−)-maackiain (**31**) obtained from *Pisum sativum* cannot be a precursor of (+)-pisatin (**39**) produced concurrently by the same plant; indeed, preliminary feeding experiments indicate that peas incorporate labelled (−)-maackiain into (−)-pisatin (P. M. Dewick, pers. comm.), a compound yet to be isolated from microbially-challenged *P. sativum*.

Apart from isoflavonoids, a small number of legumes also produce non-isoflavonoid phytoalexins (Table 2.2—see p. 67). They include seven fully identified ketonic furanoacetylenes (**91** to **97**) variously found in species of *Lens* and *Vicia* (both of which belong to the taxonomically advanced tribe Vicieae), and two unusual ethylated chromones (**88** and **89**) from the closely related genus *Lathyrus*. Whilst **88** and **89** could arise by degradation of an isoflavone precursor, it is probable that, like the furanoacetylenes, they originate from a branch of the acetate-polymalonate pathway and thus are quite distinct from the shikimate-polymalonate derived isoflavonoids. Formation of furano-acetylene and chromone phytoalexins is a striking, and perhaps unique, feature of the Vicieae, surveys in other papilionate tribes having failed to reveal these or similar compounds.

Three flavan and one flavanone phytoalexins are known outside the Leguminosae (see Geigert *et al.*, 1973, and chapter 4) and it is interesting, therefore, that very recently a weakly fungitoxic flavanone (**90**) was isolated from the fungus-inoculated tissues of *Lens culinaris*, *Medicago sativa* (tribe Trifolieae) and *Vicia floridana*.

---

[4] Isoflavan **80** was originally assigned the chemically incorrect common name 2′-methoxyphase-ollinisoflavan (VanEtten, 1973).

Like compounds **88** to **98**, representatives of the other non-isoflavonoid phytoalexin groups (benzofurans and stilbenes) are of similarly restricted distribution within the Papilionoideae; **85** to **87** occurring in genera such as *Coronilla* (tribe Coronilleae), *Lablab* and *Vigna* (Phaseoleae, subtribe Phaseolinae) and *Tetragonolobus* (Loteae), and **99** to **102** in *Arachis* (Aeschynomeneae) and *Trifolium* (Trifolieae). Two stilbenes (**99** and **101**) are found as *cis/trans* mixtures although the thermodynamically favoured *trans*-isomer predominates. In some microbially-infected legumes (e.g. *Lablab niger*, *Lathyrus odoratus* and *Vicia faba*), isoflavonoid and non-isoflavonoid phytoalexins are co-synthesized, but in others (notably *Arachis* and members of the genus *Vicia* excluding *V. faba*) isoflavanoids have not been found.

Isolation and purification

A wide range of techniques is available for the isolation of legume phytoalexins, the choice being governed by the plant part under examination (cotyledons as opposed to leaves for example) and whether the investigation is intended to provide qualitative or quantitative data.

The drop-diffusate method has been, and still is, a useful method of obtaining phytoalexins from leaves and pods. This procedure is described elsewhere (Harborne and Ingham, 1978; Ingham, 1981). Briefly, however, it relies on the ability of many phytoalexins to diffuse from their site of formation in leaf or pod tissues into overlying water droplets containing biotic or abiotic elicitors. After incubation for 48 to 72 h the essentially colourless fluid (diffusate) is collected and the phytoalexin(s) extracted from it with an organic solvent such as ethyl acetate. Any sugars, amino acids or similar water-soluble compounds remain in the aqueous phase. The drop-diffusate technique is also useful for the isolation of phytoalexin precursors, and for initial determination of the phytoalexins characteristic of a given legume, as it provides a preparation which requires only minimal purification, usually by silica gel thin-layer chromatography (TLC), although fractionation of extracts by column chromatography (CC) may be preferred (Biggs, 1975; Stoessl, 1972; Woodward, 1979*a,b*). Detection and quantification of several 5-deoxy isoflavonoids in diffusates from pods of *Phaseolus vulgaris* has recently been achieved by silylation, and subsequent analysis of the resulting silyl ethers using gas-liquid chromatography-mass spectrometry (GLC-MS) coupled with selective ion monitoring (Woodward, 1980*b*). A modification of the drop-diffusate procedure, the "facilitated diffusion method" (Keen, 1978) has allowed milligram quantities of anti-microbial material to be isolated from leaves of several Leguminosae following infiltration with inducing agents (Ingham *et al.*, 1981*b*; Keen and Ingham, 1980).

When phytoalexins are to be obtained from infected tissues, the problems presented by plant pigments may be overcome by the use of etiolated hypocotyls or stems (Bailey and Burden, 1973; Bailey and Ingham, 1971; Ingham, 1976*a,c*, 1977*b*, 1978*c*, 1979*f*), or alternatively by extraction of inoculated cotyledons, a technique successfully applied to the isolation of furanoacetylenes from species of *Lens* and *Vicia* (Hargreaves *et al.*, 1976*b*,*c*; Robeson, 1978*a*,*b*; Robeson and Harborne, 1980). Various isoflavonoids (e.g. **28** from *Canavalia ensiformis*, **39** from *Pisum sativum*, and **51**, **52** and **54** from *Glycine max*) and the *Arachis* stilbenes (**100** to **102**) have also been obtained in this way (Aguamah *et al.*, 1981; Keen, 1972, 1975; Keen and Ingham, 1976; Lyne *et al.*, 1976). In other cases, leaf tissue extracts may be "cleaned" by base/acid partition (a process unsuitable for the isolation of acid-labile 6a-hydroxypterocarpans) or by preparative CC prior to analytical TLC and eventual phytoalexin quantification (Fawcett *et al.*, 1971; Ingham and Dewick, 1978; Preston, 1977; Preston *et al.*, 1975; VanEtten and Bateman, 1970).

Silica gel TLC is unquestionably the purification procedure most widely employed and the literature now contains information on a bewildering array of solvent systems. A detailed appraisal of these lies beyond the scope of this review. However, a satisfactory approach adopted by the author involves the use of a general purpose TLC solvent such as chloroform:methanol (CM, 50:1, 2 or 3), to give an initial separation of the constituents in extracts followed by elution and, when necessary, further purification in one or more TLC mixtures such as benzene:methanol (BM, 9:1) or *n*-pentane:diethyl ether:glacial acetic acid (PEA, 75:25:3, 5 or 6). Solvents like CM (10:1 or 12:1) or diethyl ether:*n*-hexane (3:1) may also be useful, particularly when specific compounds are being sought. The former carries pigments to the solvent front leaving behind di, tri and tetrahydroxylated isoflavonoids as well as other highly polar material whereas in the latter system, chlorophyll remains quite close to the origin with non-polar compounds being located at comparatively high R_f. Multiple development of chromatograms in either BM or PEA may be required to effect complete resolution of geometrical isomers (e.g. *cis* and *trans*-**99**), structural isomers (e.g. **67**+**68**; **74**+**75**) or other mixtures (e.g. **8**+**26**; **11**+**12**; **28**+**31**; **39**+**42**; **67**+**77**) which co-chromatograph or separate incompletely in CM systems. PEA and variations of it (e.g. PEA+methanol, 75:25:3:2 or 4) have proved to be of value for the purification of many isoflavonoid and non-isoflavonoid compounds (apart from furanoacetylenes) and can be applied to 6a-hydroxypterocarpans without any appreciable loss of material. Comparative R_f values (TLC on silica gel) for some isoflavonoid and non-isoflavonoid phytoalexins are given in Table 2.3 (see p. 73).

Occasionally some dual or multi-component mixtures prove resistant to TLC resolution and it may be necessary to employ CC (e.g. on Sephadex LH-20 to separate phaseollin (**46**) from 2'-*O*-methylphaseollinisoflavan (**80**); VanEtten, 1973) or high pressure liquid chromatography (HPLC). The latter technique is used routinely for separation of the prenylated *Arachis* stilbenes, **100** to **102** (Aguamah *et al.*, 1981) and the pterocarpanoid glyceollins, **51**, **52**, **54** (Ingham *et al.*, 1981*b*). Recently HPLC revealed the presence, in diseased broad bean (*Vicia faba*) tissues, of dihydrowyerone acid (**92**), dihydrowyerone (**94**) and dihydro-wyerol (**96**); previous use of paper chromatography, TLC and GLC failed to separate these furanoacetylenes from their unsaturated analogues **91**, **93** and **95** (Porter *et al.*, 1979). As yet, there are no reports of the satisfactory resolution of **47** and **81** found in *Vigna unguiculata* (Preston, 1975), whilst the non-phenolic pterocarpans **34** and **39** have only been separated with great difficulty (Bilton *et al.*, 1976).

Detection

Comparatively few legume phytoalexins fluoresce strongly, the most obvious exceptions being stilbenes, benzofurans and furanoacetylenes (excluding **95** and **96**) which appear as light blue or dark blue bands on chromatograms inspected under long wavelength (360 to 370 nm) ultraviolet light (Table 2.2). A pale blue or purple/blue fluorescence is also characteristic of many 5-deoxy isoflavones[5] (except **3**) as well as coumestans **82** to **84**, and pterocarpene **65** (Table 2.1). Other isoflavonoids are either non-fluorescent (pterocarpans and isoflavans) or else appear faint brown (isoflavanones **19**, **21** to **23**, **26**) or dark purple (5-hydroxy-isoflavones). When the latter compounds occur at low concentrations in plant extracts, their fluorescence is often difficult to detect on chromatograms. Under these circumstances the investigator may resort to the use of preparative or analytical TLC plates having a fluorescent substance incorporated into the adsorbent layer. Upon inspection under short wavelength (254 nm) ultraviolet light, the separated components quench the background fluorescence and appear as dark areas. The use of fluorescent TLC plates is strongly recommended for all quantitative phytoalexin studies, and indeed is the only satisfactory method of precisely locating pterocarpans **34** and **40**.

Legume phytoalexins can also be detected with reagents that react with specific molecular substituents (e.g. hydroxyl or carbonyl groups) to afford coloured products. Of the many chromogenic substances available the most generally applicable is diazotized *p*-nitroaniline (Smith, 1958) which gives yellow or orange derivatives with many phenolic isoflavonoids, stilbenes and chromones, and purple/pink or purple/brown products with some hydroxylated benzofurans. Occasionally, the colours formed may provide a structural clue. Thus, a yellow product associated with a pterocarpan (e.g. **28**, **31**, **37**, **41**, **46**, **60**) may indicate a substituent other than OH at C-9. In contrast, an orange coloration, as for instance is given by **27**, **29**, **30**, **32**, **45**, **47**, **48**, **51** to **59** and **61** to **63** strongly suggests C-9 hydroxylation (Ingham and Markham, 1980) although two exceptions have been noted (see **38**, and VanEtten *et al.*, 1975). Comparable statements can also be made about isoflavan derivatives (contrast the diazo reactions of compounds **67**, **68** and **71**, **72**) but again the results are not wholly consistent (see **77**, **79**, **80**).

Other useful reagents include acidic dinitrophenylhydrazine for carbonyl groups (orange or yellow/orange; Krebs *et al.*, 1969), ethanolic picric acid for

---

[5] The characteristic pale blue fluorescence of 5-deoxy isoflavones can often be greatly intensified by brief (1 min) exposure to ammonia vapour.

epoxides (orange; Fioriti and Sims, 1968) and chromotropic acid for methylene-dioxy substituents (purple/pink). The latter reagent, which is specific for compounds containing the $O—CH_2—O$ grouping (Gunner and Hand, 1968), can be used to detect the non-phenolic, non-fluorescent pterocarpans pisatin (**39**) and acanthocarpan (**42**). Finally, Gibbs reagent, 2,6-dichloroquinone-4-chloroimide, followed by aqueous sodium carbonate (Ingham, 1976*a*) affords deep blue, purple/blue or purple indophenol derivatives with substances having an unsubstituted CH position *para* to a phenolic hydroxyl group. The reaction is normally immediate, but for some compounds (e.g. **5, 12, 13, 21, 23, 88, 89**) in which the only suitable OH group is hydrogen-bonded to a carbonyl unit, the colour develops gradually over a period of about 30 seconds and is frequently less intense than, for example, that characteristic of vestitol (**67**) and similar phytoalexins where the appropriate hydroxyl function is unchelated. The indophenols afforded by isoflavanones/isoflavans with 2',4'-hydroxylation tend towards purple (or purple/blue) whereas 2'-hydroxylation coupled with 4'-methoxylation (or some other non-phenolic substituent at this position) leads to predominantly blue (ultramarine or Prussian blue) products.

From time to time, the Gibbs test has been criticized because of the apparently positive results given by certain compounds (e.g. 6a,7-dihydroxy-phaseollin; Van den Heuvel and Vollaard, 1976) which, in theory, should not react to this reagent. As with other colour reactions, interpretation of the Gibbs test is subjective, although it should be stressed that the writer has never encountered any compounds giving ambiguous or false responses. As a guideline, positive results are indicated by the formation of strong, stable colours which do not fade to brown or yellow/brown over several hours at room temperature. If reasonable doubt still exists, the test can be undertaken spectrophotometrically (King *et al.*, 1957) to detect the characteristic indophenol absorption maximum between 550 and 700 nm.

Ultraviolet spectrophotometry

The ultraviolet (UV) absorption spectra of legume phytoalexins (usually determined in either ethanol or methanol) can provide an immediate clue to the nature of the compounds under study and, for isoflavonoids may give information relating to the type and location of substituents on the aromatic rings. Tables 2.1 and 2.2 list the neutral UV maxima of 102 induced compounds now known to be of leguminous origin, whilst Figures 2.4 to 2.15 (p. 35 to 38) illustrate the spectral features characteristic of each chemical group as well as the absorption changes which, particularly in the pterocarpan and isoflavan series of compounds, are attendant upon variation of both ring substituents and oxygenation patterns.

In contrast to 5-deoxy isoflavones (compounds **1** to **4, 6, 7, 10**) which absorb strongly below 260 nm, isoflavones possessing 5-hydroxylation (**5, 8, 9, 11** to **16**) exhibit a single diagnostic UV peak between 260 and 270 nm with associated shoulders or weak maxima occurring at about 290 nm and/or from 320 to 340 nm (Figure 2.4). Absorption in the latter region is most intense when isoflavones have 6,7-oxygenation of ring A (see data for **6**, **7** and **10**, and Figure 2.4), but maximum absorption never approaches that of the principle low wavelength maximum. Loss of the isoflavone C-2/C-3 double bond yields an isoflavanone (**17** to **26**). Compounds of this type absorb from near 275 nm (simple 5-deoxy derivatives) to about 290 nm (5-hydroxy derivatives) with an accompanying low peak or shoulder at somewhat higher wavelength, approximately 310 nm for many 5-deoxy isoflavanones and 325 to 340 nm for the corresponding 5-hydroxy analogues (Figure 2.5). Virtually all the known induced pterocarpans (**27** to **64**) and isoflavans (**66** to **81**) absorb strongly between 280 and 295 nm with two maxima, or a maximum and one or more shoulders (Figures 2.6 to 2.12). The coumarin chromophore of coumestans (**82** to **84**) has a marked influence on their spectral properties producing maximum absorbance at comparatively high wavelength (345 nm); pterocarpenes also absorb in this region (Figure 2.13).

As mentioned previously all pterocarpan phytoalexins are oxygenated at C-3 and C-9. The neutral UV spectrum of 3,9-dihydroxypterocarpan (**27**) is essentially unaffected, or changed only slightly, by methylation at either or both of these positions (**27**, **28**, **34**), or by hydroxylation at C-6a (**30**, **34**). Similarly, the neutral maxima of compounds having an additional oxygen function (OH/OMe) or a prenyl group at C-7 (as in the phytoalexin metabolite, 6a,7-dihydroxymedicarpin; Ingham, 1976*f*) or C-10 (**32**, **33**, **47**, **50**, **56** to **58** and **60**) are only marginally different from those of pterocarpan **27**. On the other hand, the two maxima (or shoulder + maximum) between 280 and 290 nm which characterize the spectra of **27** and similar compounds, are replaced by one UV peak at about 285 nm when a pterocarpan has the 1,3,9- (**55**, **61**) or 3,4,9- (**37**, **38**) oxygenation pattern (Figure 2.6).

One of the most dramatic substituent effects is observed when pterocarpans have an 8,9-methylenedioxy group, coupled with mono-substitution (at C-3) of ring A (**31** and **39**). Here, a pronounced maximum occurs at 310 nm or thereabouts together with expected peaks between 280 and 290 nm. These latter maxima are almost entirely abolished upon further oxygenation (at C-4) as in compounds **41** to **43** (Figure 2.7). Unlike pterocarpans **31** and **39**, the neutral UV spectrum of synthetic 3-hydroxy-8,9-dimethoxypterocarpan is essentially superimposable on that of maackiainisoflavan (Figure 2.11), the high wavelength maximum of both compounds being evident at about 300 nm. Thus, whilst **31**, **39** and the synthetic pterocarpan all have the same D-ring oxygenation pattern, possession of an 8,9-methylenedioxy substituent by the former compounds probably alters their molecular geometry in such a manner as to cause not only a significant bathochromic UV shift of the 300 nm peak but also an intensification of the maximum resulting at 310 nm.

Pterocarpans and isoflavans with "complex" angular or linear *gem*-dimethylchromen attachments also absorb, but often only weakly, above 300 nm because of conjugation between the chromene ring and the aromatic chromophore (Figure 2.8). Between 305 and 320 nm, one or more sharp maxima characterize the spectra of compounds with the chromene ring arranged in a linear fashion (**49** and **51**), whereas, an angular disposition apparently results in a shoulder or series of weak shoulders (**54** and **79**), or a shallow maximum (**46** and **80**). Finally, two prominent maxima at about 253 nm are evident in the spectra of pterocarpans **45**, **63** and **64**, and can be attributed to the 2,3-furano substituent. These maxima disappear when the furano group is located at C-3/C-4 (**48**) or converted to a dihydrofuran unit as in **52** and **53** (Figure 2.9).

Fewer general statements can be made about the UV spectra of isoflavan phytoalexins because the structures reported are less variable with respect to ring substituents and oxygenation patterns than are their pterocarpan counterparts. However, whereas the simplest known isoflavan phytoalexin (7,2',4'-trihydroxyisoflavan, **66**) absorbs strongly at 283 nm (Figure 2.10), introduction of O-methyl groups at C-7, 2' or 4' (or a prenyl substituent at C-3') has a slight, but nevertheless distinct, effect leading to the appearance of a plateau (**72**), peak and shoulder (**68**) or two peaks (**67**, **70**, **71**) between 280 and 285 nm or, in the case of isoflavans **69** and **81**, between 280 and 290 nm (Figure 2.10). The few compounds with 5,7,2',4'- (**77**, **78**) or 7,2',3',4'- (**74–76**) tetra-oxygenation exhibit a maximum at or near 280 nm with an accompanying shoulder at about three-quarters intensity between 285 and 290 nm (Figure 2.11). UV spectra of the three "complex" isoflavan phytoalexins are illustrated in Figure 2.12.

Information on the structure of an isoflavonoid phytoalexin can also be gained by determining UV spectra in the presence of various reagents. For example, addition of concentrated hydrochloric acid (2 to 3 drops per ml) to an ethanolic solution of a 6a-hydroxypterocarpan causes dehydration to a pterocarpene (e.g. **65** which can easily be derived from pterocarpan **30**) with the concurrent appearance of two diagnostic maxima in the 335–360 nm region (Figure 2.13). A distinct maximum (as opposed to a gentle shoulder) at about 250 nm in the alkali spectrum (1 drop of 0.1 N aqueous sodium hydroxide per ml phytoalexin solution) of a pterocarpan is indicative of C-3 hydroxylation. Other reagents (aluminium chloride in methanol, and anhydrous sodium acetate) can respectively reveal hydroxyl groups at C-5 and C-7 of isoflavones, isoflavanones and chromones (**88**, **89**), as well as various flavonoid compounds (Ingham, 1976c; Mabry *et al.*, 1970; Robeson *et al.*, 1980). Although neither of these reagents can be applied to pterocarpans and isoflavans, it is worth noting that C-3 hydroxylated coumestans are amenable to the use of sodium acetate in structure determination (Bickoff *et al.*, 1969). The high wavelength alcoholic UV maxima of wyerone acid (**91**), wyerone (**93**) and dihydrowyerone (**94**) shift bathochromically (about 30 nm for **91**, 46 nm for **93**, and 49 nm for **94**) in the presence of piperidine (Fawcett *et al.*, 1968; J. L. Ingham, unpublished data).

The neutral UV spectra of some non-isoflavonoid legume phytoalexins are shown in Figures 2.14 and 2.15. Confusion of such compounds with isoflavonoids is unlikely, particularly when other structural details (e.g. mass spectral data) are available. However, two furanoacetylenes (**95** and **96**) absorb strongly near 310 nm as do 8,9-methylenedioxy substituted pterocarpans, and the spectra of benzofurans (**85** to **87**) and pterocarpenes (**65**) are also superficially alike; note, however, that maxima characteristic of benzofurans occur at somewhat lower wavelengths (320–335 nm) than do those of pterocarpenes (335–360 nm). Spectroscopically similar flavanones and isoflavanones can be distinguished by the Horowitz and Shinoda colour tests (Dean, 1963; Horowitz, 1957), and by mass spectrometry. The mass spectra of simple flavanones exhibit a major $M^+ - 1$ ion. This is absent (or essentially absent) from the MS of an isoflavanone.

Synthesis

Several methods have been employed for the synthesis of legume phytoalexins and no attempt will be made here to review these exhaustively. In Tables 2.1 and 2.2, those papers which report total or partial phytoalexin syntheses are marked with a superscript 3. Additionally, some of the compounds listed have been prepared during studies unrelated to, or loosely connected with, phytoalexin research; amongst these are isoflavones **1**, **2** (Dewick and Martin, 1979b), **8** (Farkas et al., 1971) and **10** (Dewick, 1978), isoflavanones **19** and **22** (Farkas et al., 1971; Neill, 1953), pterocarpan **29** (McMurry et al., 1972), isoflavans **74** to **76** (Farkas et al., 1974), benzofuran **86** (Duffley and Stevenson, 1977) and chromones **88** and **89** (Clough and Snell, 1981). Certain weakly fungitoxic isoflavones (e.g. **1**, **2** and **5**) are now commercially available.

The synthesis of isoflavonoid derivatives has been revolutionized by the discovery that simple benzyloxy isoflavones can be obtained by oxidative rearrangement of the corresponding 2′-acetoxy (or 2′-hydroxy) chalcones[6] using thallium (III) nitrate trihydrate, and treatment of the resulting acetals with concentrated hydrochloric acid (Farkas et al., 1974). The chalcones are readily accessible via base condensation of appropriately benzylated acetophenone and aldehyde starting materials. Benzylation of all chalcone hydroxyl groups (except that at C-2′ which is required for eventual formation of the isoflavone heterocyclic ring) is an essential prerequisite if rearrangement is to proceed without undesirable side-reactions taking place. The derived isoflavone benzyl ethers (perhaps after initial debenzylation and subsequent acetylation or selective methylation) can then be converted to isoflavanones (e.g. **17**, **18**, **20**; Dewick,

[6] The ring numbering/nomenclature system applied to chalcones is as follows

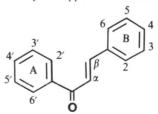

1977; Donnelly *et al.*, 1973; Woodward, 1980*a*) and isoflavans (e.g. compounds **66**, **67**, **69**, **70**, **72**, **73**, **77**, **78**; see Table 2.1 for references) by catalytic hydrogenation, a procedure which removes any remaining benzyl protecting groups. Pterocarpans result from treatment of 2'-hydroxy isoflavones with sodium borohydride (compounds **27**, **28**, **31**; Dewick, 1975, 1977; Dewick and Ward, 1978; Woodward, 1980*a*), whilst pterocarpenes (**65** and its 9-*O*-methyl ether) are readily obtained from either isoflavone or isoflavanone intermediates (Dewick, 1977; Ingham and Dewick, 1978).

Some pterocarpans (**28**, **31**, **33**, **35**, **37**) have also been prepared by mild oxidation of 2'-hydroxyisoflavans with DDQ (2,3-dichloro-5,6-dicyano-1,4-benzoquinone) (see Cornia and Merlini, 1975, and references in Table 2.1) although the method is not invariably successful. Thus pterocarpan **29** was not obtained by DDQ treatment of the corresponding 7-methoxylated isoflavan (**69**), an observation in accord with the proposed formation of a quinoid intermediate during oxidation (J. L. Ingham, unpublished data). In the presence of excess DDQ, amenable isoflavans are rapidly converted to pterocarpans, with the latter compounds then being further oxidized to yield coumestans (Dewick; 1977). Procedures for synthesis of **82** and other coumestans have been reviewed by Bickoff *et al.* (1969), and more recently by Darbarwar *et al.* (1976). Lastly, the structural similarity between 2-arylbenzofurans and isoflavonoids has been elegantly exploited in the preparation of compounds **85** to **87**. The key step in each synthesis is the formation of a deoxybenzoin via base hydrolysis of a 2'-benzyloxy isoflavone, subsequent debenzylation and acid-catalysed ring-closure affording the desired benzofuran derivative.

Certain procedures are difficult or impossible to accomplish using available laboratory techniques, introduction of a tertiary (C-6a) hydroxyl group into the heterocyclic ring system of a phenolic pterocarpan being a case in point. Indeed, only two 6a-hydroxypterocarpans, the non-phenolic compounds **34** and **39**, have actually been synthesized (Bevan *et al.*, 1964; Ingham, 1976*f*). Similarly, attempts to demethylate pterocarpans without otherwise affecting the molecule are doomed because, even under mild conditions, there is rapid fission of the bond between C-11a and the oxygen atom of ring C (see Figure 2.2). "Complex" isoflavonoids are also difficult to prepare despite methods now available for satisfactory prenylation of rings A and B of the appropriate isoflavone intermediates. However, the structure of luteone (**15**) was recently confirmed by synthesis of its 7,2',4'-trimethyl ether, and comparison of this compound with material resulting from selective methylation of the natural product (Jain *et al.*, 1979).

Acknowledgements

I should like to thank P. M. Dewick for commenting on the manuscript, and N. T. Keen, R. L. Lyne, J. W. Mansfield, L. J. Mulheirn and R. N. Strange for supplying information prior to publication. Collective thanks are also due to the many colleagues who have willingly provided me with an invaluable range of reference compounds. Financial assistance from the S.R.C. is gratefully acknowledged.

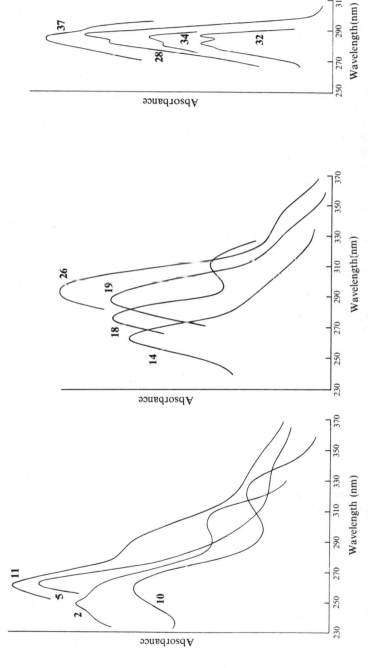

Figure 2.4 Methanolic UV spectra of the iso-flavones, formononetin (**2**), genistein (**5**), afrormosin (**10**) and cajanin (**11**).

Figure 2.5 Methanolic UV spectra of the iso-flavone, licoisoflavone A (**14**), and the isoflavanones, vestitone (**18**), dalbergioidin (**19**) and kievitone (**26**).

Figure 2.6 Ethanolic UV spectra of the ptero-carpans, medicarpin (**28**), nissolin (**32**), vari-abilin (**34**) and 4-methoxymedicarpin (**37**).

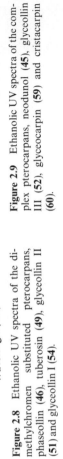

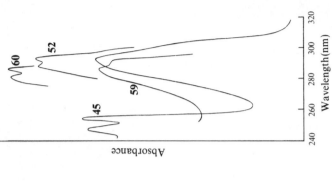

Figure 2.9 Ethanolic UV spectra of the complex pterocarpans, neodunol (**45**), glyceollin III (**52**), glyceocarpin (**59**) and cristacarpin (**60**).

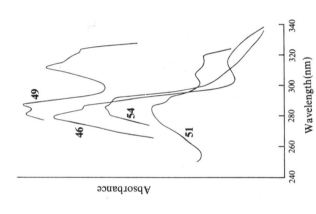

Figure 2.8 Ethanolic UV spectra of the dimethylchromen substituted pterocarpans, phaseollin (**46**), tuberosin (**49**), glyceollin II (**51**) and glyceollin I (**54**).

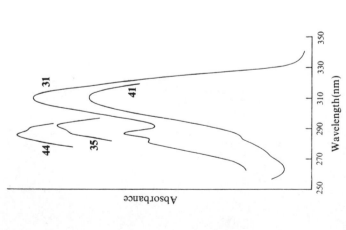

Figure 2.7 Ethanolic UV spectra of the pterocarpans, maackiain (**31**), 2-methoxymedicarpin (**35**), 4-methoxymaackiain (**41**) and 2,3,9-trimethoxy-4-hydroxypterocarpan (**44**).

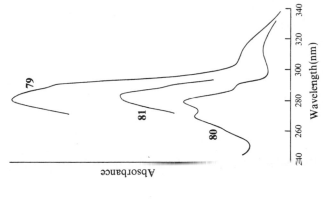

Figure 2.12 Ethanolic UV spectra of the complex isoflavans, phaseollinisoflavan (**79**), 2'-O-methylphaseollinisoflavan (**80**) and 2'-O-methylphaseollidinisoflavan (**81**).

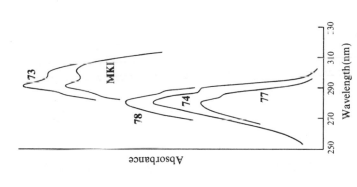

Figure 2.11 Ethanolic UV spectra of the isoflavans, astraciceran (**73**), isomucronulatol (**74**), 5-methoxyvestitol (**77**), lotisoflavan (**78**) and maackiainisoflavan (**MKI**).

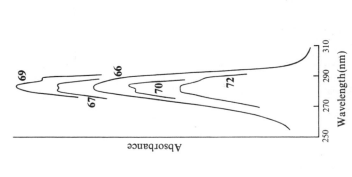

Figure 2.10 Ethanolic UV spectra of the isoflavans, demethylvestitol (**66**), vestitol (**67**), neovestitol (**69**), sativan (**70**) and arvensan (**72**).

38

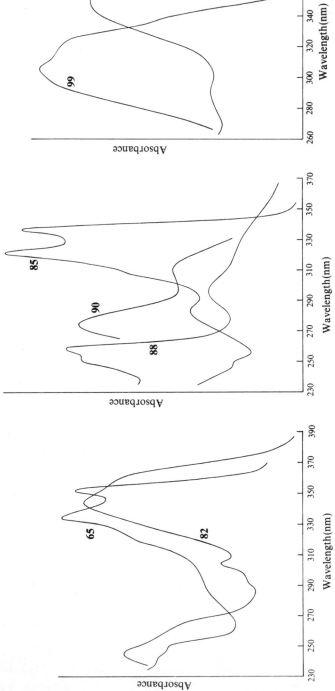

Figure 2.13 Ethanolic UV spectra of the pterocarpene, anhydroglycinol (**65**) and the coumestan, coumestrol (**82**).

Figure 2.14 UV spectra of 6-demethylvignafuran (**85**), lathodoratin (**88**) and liquiritigenin (**90**). The spectra of **85** and **88** were determined in ethanol, and that of **90** in methanol.

Figure 2.15 Ethanolic UV spectra of wyerone (**93**) and *trans*-resveratrol (**99**).

Table 2.1   Structures, sources and selected characteristics of the induced legume isoflavonoids[a-f]

Key :   Tissues ; C = cotyledons ; CC = cell (suspension) cultures ;

CT = callus tissue ; E = epicotyls ; H = hypocotyls ; L = leaves ;

P = pods ; PE = petals ; PH = phyllodes ; R = roots ; RC = root

crowns ; S = stems ; SD = seedlings ; SE = sepals ; ST = stipules.

Properties ; DPN = colour after treatment with diazotised *p*-

nitroaniline ; Gibbs = colour after treatment with Gibbs reagent ;

mp = melting point ; MW = molecular weight ; FR = fungitoxicity

rating (see footnote f).

[a] References which provide, i) mass spectral (MS)/proton magnetic resonance
(^{1}H NMR) data, ii) information on optical rotation/absolute configuration,
and/or iii) details of partial or total synthesis, are marked [1] [2] and [3]
respectively.

[b] Whenever possible the ultra-violet (UV) spectra of chromatographically
pure compounds were determined (in either ethanol [EtOH] or methanol [MeOH])
by the author using a Pye-Unicam SP 1800 recording spectrophotometer. Some
of the wavelength maxima (λ max, nm) quoted are slightly different from
those reported in the literature.

[c] Except where otherwise stated, optical rotations [α] were determined (in
EtOH, MeOH, chloroform [CHCl$_3$] or ethyl acetate [EtOAc]) at 589 nm (sodium
lamp).

[d] Data in the 'fluorescence' category refer to the appearance of compounds on
developed silica gel thin-layer chromatograms viewed under long wavelength
(360-370 nm) UV light. The effect of brief fuming with ammonia (NH$_3$) vapour is
also indicated when appropriate.

[e] For further information on fluorescence characteristics/colour reactions,
UV absorption and synthesis, the reader is referred to relevant sections in
this chapter.

[f] The fungitoxicity rating (FR) gives an indication of antifungal activity
against mycelial growth, germ-tube growth and spore germination of fungi
normally non-pathogenic, or weakly pathogenic on legumes. Ratings of 1,
2 or 3 indicate ED$_{50}$ values of below 50 μg/ml, 50 to 100 μg/ml and more
than 100 μg/ml, respectively.

Table 2.1 (*continued*)

A: ISOFLAVONES

   i) Simple Isoflavones

(1)   DAIDZEIN

Glycine max   (H)
           (L)
Phaseolus vulgaris (P)
Vigna unguiculata   (H)

MW 254 $C_{15}H_{10}O_4$

λ max MeOH, nm: 212, 238sh, 249, 262sh,
  305
DPN: Pale orange
Fluorescence: Pale blue (intensifies
  with NH_3)
mp: 323^0
FR: 3

[1]Keen *et al.*, 1972
Keen and Kennedy, 1974
Woodward, 1980a
Partridge, 1973

(2)   FORMONONETIN

Centrosema spp. (L)

MW 268 $C_{16}H_{12}O_4$

λ max MeOH, nm: 210, 238sh, 249, 262sh,
  302
DPN: Yellow/orange
Fluorescence: Pale blue (intensifies
  with NH_3)
mp: 260^0
FR: 3

Markham and Ingham, 1980

(3)   ISOFORMONONETIN

Glycine max   (L)

MW 268 $C_{16}H_{12}O_4$

λ max MeOH, nm: 212, 235sh, 249, 262,
  307sh
DPN: Pale orange
Fluorescence: Barely discernible pale
  blue
mp: 218-220^0
FR: 3

[1,3]Ingham *et al.*, 1981b

(4)   2'-HYDROXYDAIDZEIN

Phaseolus vulgaris   (P)

MW 270 $C_{15}H_{10}O_5$

λ max MeOH, nm: 214, 241, 249, 264sh,
  292, 307sh
DPN: Orange; Gibbs: Purple/blue
Fluorescence: Pale blue (intensifies
  with NH_3)
mp: 275^0
FR: 3

[1,3]Woodward, 1980a

Table 2.1 (*continued*)

(5)   GENISTEIN

MW 270 $C_{15}H_{10}O_5$

λ max MeOH, nm: 208, 262, 330sh
DPN: Orange or orange/yellow;
  Gibbs: Blue
Fluorescence: Dark purple
mp: 301-302°
FR : 2 (or 3)

Apios tuberosa (L)	Ingham and Mulheirn, unpublished
Argyrocytisus battandieri (L)	Ingham, unpublished
Cajanus cajan (S)	[1]Ingham, 1976*c*
Canavalia ensiformis (H)	Ingham, unpublished
Crotalaria juncea (L)	Ingham, unpublished
Dolichos biflorus (L,S)	Keen and Ingham, 1980
Hardenbergia violacea (L)	Ingham, unpublished
Lablab niger (H)	Ingham, 1977*b*
Laburnum anagyroides (L)	Ingham and Dewick, 1980*a*
Lupinus albus (H)	Ingham and Dewick, 1980*a*
Neonotonia wightii (S)	Ingham *et al.*, 1977
Phaseolus vulgaris (P)	[1]Biggs, 1975
	Woodward, 1979*a*
Stizolobium deeringianum (H)	Ingham, 1979*a*

(6)   DEMETHYLTEXASIN

MW 270 $C_{15}H_{10}O_5$

λ max MeOH, nm: 215, 230, 258, 329
DPN: Orange/brown
Fluorescence: Pale blue (intensifies
  with NH_3)
FR: 3

Centrosema haitiense (L)   Markham and Ingham, 1980

(7)   GLYCITEIN

MW 284 $C_{16}H_{12}O_5$

λ max MeOH, nm: 256, 319
DPN: Orange/brown
Fluorescence: Pale blue (intensifies
  with NH_3)
mp: 311-313°
FR: 3

Centrosema **spp.** (L)   Markham and Ingham, 1980

(8)   2'-HYDROXYGENISTEIN

MW 286 $C_{15}H_{10}O_6$

λ max MeOH, nm: 209, 261, 290sh, 335sh
DPN: Orange or orange/yellow; Gibbs:
  Blue
Fluorescence: Dark purple
mp: 270-272°
FR : 2 (or 3)

Table 2.1 (*continued*)

Apios tuberosa (L)	Ingham and Mulheirn, unpublished
Argyrocytisus battandieri (L)	Ingham, unpublished
Cajanus cajan (S)	[1]Ingham, 1976*c*
Crotalaria juncea (L)	Ingham, unpublished
Dolichos biflorus (L,S)	Keen and Ingham, 1980
Hardenbergia violacea (L)	Ingham, unpublished
Lablab niger (H)	Ingham, 1977*b*
Laburnum anagyroides (L)	Ingham and Dewick, 1980*a*
Lupinus albus (H)	Ingham and Dewick, 1980*a*
Neonotonia wightii (S)	Ingham *et al.*, 1977
Phaseolus vulgaris (P)	[1]Biggs, 1975
	Woodward, 1979*a*
Spartium junceum (L)	Ingham and Dewick, 1980*a*
Stizolobium deeringianum (H)	Ingham, 1979*a*

(9)   6-HYDROXYGENISTEIN

MW 286 $C_{15}H_{10}O_6$

λ max MeOH, nm: 215, 270, 340
DPN: Orange;   Gibbs: Blue
Fluorescence: Dark purple
FR : 2 (or 3)

Centrosema plumieri (L)   Markham and Ingham, 1980

(10)   AFRORMOSIN

MW 298 $C_{17}H_{14}O_5$

λ max MeOH, nm: 213, 230, 258, 322
DPN: Red/brown
Fluorescence: Light blue (intensifies
  with NH3)
mp: 236-237°
FR: 3

Centrosema spp. (L)   Markham and Ingham, 1980

(11)   CAJANIN

MW 300 $C_{16}H_{12}O_6$

λmax MeOH, nm: 216, 260, 288sh, 335sh
DPN: Orange or orange/yellow; Gibbs:
  Blue
Fluorescence: Dark purple
FR : 1 (or 2)

Cajanus cajan (S)	[1]Ingham, 1976*c*
Canavalia ensiformis (H)	Ingham, unpublished
Centrosema pascuorum (L)	Markham and Ingham, 1980
Centrosema pubescens (L)	Markham and Ingham, 1980

Table 2.1 (*continued*)

(12)   TECTORIGENIN

Centrosema spp.   (L)

MW 300 $C_{16}H_{12}O_6$

λ max MeOH, nm: 215, 268, 335
DPN: Orange; Gibbs: Blue
Fluorescence: Dark purple
mp: 230o
FR : 1 (or 2)

[1]Markham and Ingham, 1980

ii)   Complex Isoflavones

(13)   WIGHTEONE

Argyrocytisus battandieri   (L)
Laburnum anagyroides   (L)
Lupinus albus   (H)
Neonotonia wightii   (S)

MW 338 $C_{20}H_{18}O_5$

λ max MeOH, nm: 215, 268
DPN: Orange/yellow; Gibbs: Blue
Fluorescence: Dark purple
FR: 1

Ingham, unpublished
Ingham and Dewick, 1980a
Ingham et al., 1977
[1]Ingham et al., 1977

(14)   LICOISOFLAVONE A

(PHASEOLUTEONE)

Hardenbergia violacea   (L)
Phaseolus vulgaris   (P)

MW 354 $C_{20}H_{18}O_6$

λ max MeOH, nm: 210, 263, 300sh
DPN: Yellow/orange; Gibbs: Blue
Fluorescence: Dark purple
mp: 111-113o
FR: 1

Ingham, unpublished
[1]Woodward, 1979a

(15)   LUTEONE

MW 354 $C_{20}H_{18}O_6$

λ max MeOH, nm: 215, 266, 292sh,
   340sh
DPN: Orange; Gibbs: Blue
Fluorescence: Dark purple
mp: 225-227o
FR: 1

Table 2.1 (*continued*)

Argyrocytisus battandieri (L)	Ingham, unpublished
Hardenbergia violacea (L)	Ingham, unpublished
Laburnum anagyroides (L)	Ingham and Dewick, 1980*a*
Lupinus albus (H)	Ingham and Dewick, 1980*a*

(16) 2,3-DEHYDROKIEVITONE

MW 354 $C_{20}H_{18}O_6$

λ max MeOH, nm: 265, 286sh, 333sh
DPN: Orange; Gibbs: Blue or purple/blue
Fluorescence: Dark purple

Phaseolus vulgaris (P) [1]Woodward, 1979*b*

B: ISOFLAVANONES

i) Simple Isoflavanones

(17) (±)-2'-HYDROXYDIHYDRODAIDZEIN

MW 272 $C_{15}H_{12}O_5$

λ max MeOH, nm: 211, 230sh, 277, 309
DPN: Orange; Gibbs: Purple/blue

Phaseolus vulgaris (P) [1,2,3]Woodward, 1980*a*

(18) VESTITONE

MW 286 $C_{16}H_{14}O_5$

λ max MeOH, nm: 212, 230sh, 277, 311
DPN: Yellow; Gibbs: Blue
Fluorescence: Dark purple/blue
FR: 1 (or 2)

Onobrychis viciifolia (L) [1]Ingham, 1978*d*
Tipuana tipu (L) Ingham, unpublished

Table 2.1 (*continued*)

(19) (±)-DALBERGIOIDIN

MW 288 $C_{15}H_{12}O_6$

λ max MeOH, nm: 216, 225, 289, 332sh
DPN: Orange; Gibbs: Purple/blue
Fluorescence: Faint brown
mp: 227-228°
FR: 2

Dolichos biflorus (L,S)
Lablab niger (H)
Macrotyloma axillare (L)
Phaseolus vulgaris (P)
Stizolobium deeringianum (H)

Keen and Ingham, 1980
Ingham, 1977*b*
Ingham, unpublished
Woodward, 1979a
Ingham, 1979a

(20) SATIVANONE

MW 300 $C_{17}H_{16}O_5$

λ max MeOH, nm: 217, 230, 277, 312
DPN: Yellow/orange
Fluorescence: Dark purple/blue
FR: 1 (or 2)

Medicago sativa (L)

[1]Ingham, 1979*g*

(21) ISOFERREIRIN

MW 302 $C_{16}H_{14}O_6$

λ max MeOH, nm: 212, 229, 289, 330sh
DPN: Orange; Gibbs: Blue
Fluorescence: Faint brown
FR: 2

Dolichos biflorus (L,S)
Stizolobium deeringianum (H)

[1]Keen and Ingham, 1980
Keen and Ingham, 1980

(22) HOMOFERREIRIN

MW 316 $C_{17}H_{16}O_6$

λ max MeOH, nm: 210, 228, 289,
  333sh
DPN: Yellow/orange; Gibbs: Blue/
  black
Fluorescence: Faint brown
FR: 1

Argyrocytisus battandieri (L)

Ingham, unpublished

Table 2.1 (*continued*)

(23)   CAJANOL

(Revised structure)

MW 316 $C_{17}H_{16}O_6$

λ max MeOH, nm: 214, 229, 287, 336sh
DPN: Orange/yellow; Gibbs: Blue
Fluorescence: Faint brown
FR: 1

Cajanus cajan (S,R)

Stizolobium deeringianum (H)

[1]Ingham, 1976*c*
Ingham, 1979*a*
Ingham, 1979*a*

ii)   Complex Isoflavanones

(24)   (±)-5-DEOXYKIEVITONE

MW 340 $C_{20}H_{20}O_5$

λ max MeOH, nm: 286 (no other
maxima reported)
DPN: Orange; Gibbs: Purple/blue

Phaseolus vulgaris (P)

[1,2]Woodward, 1979*b*

(25)   CYCLOKIEVITONE

MW 354 $C_{20}H_{18}O_6$

λ max MeOH, nm: 271, 294sh, 307sh, 358
DPN: Orange; Gibbs: Purple/blue

Phaseolus vulgaris (P)

[1]Woodward, 1979*b*

Table 2.1 (*continued*)

(26)   (+) and (±)-KIEVITONE   MW 356 $C_{20}H_{20}O_6$

λ max MeOH, nm: 210, 294, 340sh
DPN: Orange; Gibbs: Purple/blue
Fluorescence: Faint brown
$[\alpha]$ + 6.5° (MeOH)
FR: 1 (or 2)

Dolichos biflorus (L,S)	Keen and Ingham, 1980
Lablab niger (H)	Ingham, 1977b
Macroptilium atropurpureum (L)	Ingham, unpublished
Macrotyloma axillare (L)	Ingham, unpublished
Phaseolus aureus (H)	Ingham, unpublished
Phaseolus lunatus (C)	Partridge, 1973
Phaseolus vulgaris (racemate) (C)	Gnanamanickam, 1979
(H)	[1]Burden et al., 1972
	Smith et al., 1973a
(L)	[1,2]Smith et al., 1973b
(L)	Gnanamanickam and Patil, 1977
(P)	Woodward, 1979a
(R)	Burden et al., 1974
Stizolobium deeringianum (H)	Ingham, 1979a
Vigna unguiculata (+)-isomer (C)	Keen, 1975
(H)	[1,2]Partridge, 1973
(H)	Bailey, 1973
(SD)	Martin and Dewick, 1979

C: PTEROCARPANS

i)   Simple Pterocarpans

(27)   (-)-DEMETHYLMEDICARPIN   MW 256 $C_{15}H_{12}O_4$

λ max EtOH, nm: 211, 228sh, 283, 288,
294sh
DPN: Orange
FR: 2

Erythrina crista-galli (L)	Ingham and Markham, 1980
Erythrina sandwicensis (L)	Ingham, 1980
Pachyrrhizus erosus (S)	Ingham, 1979f
Phaseolus vulgaris (P)	[1,3]Woodward, 1980a
Psophocarpus tetragonolobus (S)	Ingham and Markham, 1980

Table 2.1 (*continued*)

(28)   (−)-MEDICARPIN   MW 270 $C_{16}H_{14}O_4$
        (DEMETHYLHOMOPTEROCARPIN)

λ max EtOH, nm: 213, 228sh, 282, 287,
  293sh
DPN: Yellow
mp: 130-131°
[α] − 216° (CHCl$_3$)
FR: 1

Canavalia ensiformis (C)	Keen, 1972
	Keen, 1975
(CT)	Gustine *et al.*, 1978
(H)	Keen, 1972
	Lampard, 1974
Caragana spp. (L)	Ingham, 1979*c*
Carmichaelia flagelliformis (L)	Ingham, unpublished
Cicer arietinum (C)	Keen, 1975
(S)	Ingham, 1976*a*
Dalbergia sericea (L)	Ingham, 1979*e*
Factorovskya aschersoniana (L)	Ingham, 1979*b*
Lathyrus nissolia (PH)	Robeson and Ingham, 1979
Lathyrus spp. (L)	Robeson and Harborne, 1980
Medicago sativa (L)	[1,2]Smith *et al.*, 1971
(SD)	Dewick and Martin, 1979*a*
Medicago spp. (L)	Ingham, 1979*g*
Melilotus spp. (L)	Ingham, 1977*a*
Notospartium glabrescens (L)	Ingham, unpublished
Onobrychis spp. (L)	Ingham, 1978*d*
Ononis spp. (L)	Ingham, unpublished
Parochetus communis (L)	Ingham, 1979*b*
Robinia pseudoacacia (L)	Ingham, unpublished
Sophora japonica (L)	Ingham, unpublished
Stizolobium deeringianum (H)	Ingham, unpublished
Stylosanthes spp. (L)	Ingham, unpublished
Tipuana tipu (L)	Ingham, unpublished
Trifolium pratense (L)	Higgins and Smith, 1972
(SD)	Dewick, 1975
	[3]Dewick, 1977
Trifolium repens (L)	Cruickshank *et al.*, 1974
(R)	Sutherland *et al.*, 1980
Trifolium spp. (L)	Ingham, 1978*a*
Trigonella spp. (L)	Ingham and Harborne, 1976
Vicia faba (C)	Hargreaves *et al.*, 1977
(L)	Hargreaves *et al.*, 1976*a*
(P)	Hargreaves *et al.*, 1976*a*
	Hargreaves *et al.*, 1977
Vigna unguiculata (H)	Lampard, 1974

(29)   (−)-ISOMEDICARPIN   MW 270 $C_{16}H_{14}O_4$

λ max EtOH, nm: 212, 230sh, 283sh, 287,
  294sh
DPN: Orange
[α] −201° (CHCl$_3$)
FR: 1

Table 2.1 (*continued*)

Psophocarpus tetragonolobus (P)   [1,2] Preston, 1977

                                  (S)   Ingham and Markham, 1980

(30)   (-)-GLYCINOL

MW 272 $C_{15}H_{12}O_5$

λ max EtOH, nm: 214, 230sh, 283, 287, 293sh
DPN: Orange
FR: 2 (or 3)

Erythrina sandwicensis (L)   [1] Ingham, 1980
Glycine max (C)   Ingham *et al.*, 1981*b*
                                [1,2] Lyne and Mulheirn, 1978
                                  Weinstein *et al.*, 1980
Pueraria lobata (L)   Ingham, unpublished

(31)   (-)-MAACKIAIN

     (INERMIN : DEMETHYLPTEROCARPIN)

MW 284 $C_{16}H_{12}O_5$

λ max EtOH, nm: 210, 226sh, 282, 287, 310
DPN: Yellow
mp: 178-181°
[α] -220° (MeOH)
FR: 1

Canavalia ensiformis (H)   Ingham, unpublished
Caragana spp. (L)   Ingham, 1979*c*
Cicer arietinum (C)   Keen, 1975
                     (S)   Ingham, 1976*a*
Dalbergia sericea (L)   Ingham, 1979*e*
Lathyrus spp. (L)   Robeson and Harborne, 1980
Ononis spp. (L)   Ingham, unpublished
Pisum fulvum (L)   Robeson and Harborne, 1980
Pisum sativum (C)   Keen, 1975
                                     Robeson, 1978*a*
                     (L)   Robeson and Harborne, 1980
                     (P)   [1,2] Stoessl, 1972
                     (PE)   Ingham, 1979*d*
                     (ST)   Ingham, 1979*d*
Sophora japonica (L)   Ingham, unpublished
Stizolobium deeringianum (H)   Ingham, unpublished
Tephrosia bidwilli (L)   Ingham, unpublished
Tipuana tipu (L)   Ingham, unpublished
Trifolium pratense (L)   Higgins and Smith, 1972
                     (S,D)   Dewick, 1975
Trifolium spp. (L)   Ingham, 1978*a*
Trigonella spp. (L)   Ingham and Harborne, 1976

Table 2.1 (*continued*)

(32)   (-)-NISSOLIN

Lathyrus nissolia   (P H)

MW 286 $C_{16}H_{14}O_5$

λ max EtOH, nm: 212, 233sh, 277sh,
 281, 287
DPN: Orange
$[\alpha]$ -238° (MeOH)
FR: 1

[1]Robeson and Ingham, 1979

(33)   (-)-METHYLNISSOLIN

Lathyrus nissolia   (PH)

MW 300 $C_{17}H_{16}O_5$

λ max EtOH, nm: 211, 232sh, 277sh,
 281, 287
DPN: Yellow
$[\alpha]$ -253° (MeOH)
FR: 1

[1,3]Robeson and Ingham, 1979

(34)   (-)-VARIABILIN

  (HOMOPISATIN)

Caragana spp.   (L)
Lathyrus spp.   (L)
Lens culinaris   (-)-isomer (C,E,L)

Lens nigricans   (C)

MW 300 $C_{17}H_{16}O_5$

λ max EtOH, nm: 212, 228sh, 281, 286,
 291sh
$[\alpha]$ -230° (MeOH)
FR: 1

Ingham, 1979*c*
Robeson and Harborne, 1980
Robeson, 1978*a*
Robeson, 1978*b*
Robeson and Harborne, 1980
Robeson, 1978*b*

(35)   (-)-2-METHOXYMEDICARPIN

MW 300 $C_{17}H_{16}O_5$

λ max EtOH, nm: 217, 231sh, 288sh,
 292, 302sh
DPN: Yellow
mp: 146-148°
$[\alpha]$ -297° (EtOH)
FR: 1

Table 2.1 (*continued*)

Pisum sativum (E,RC)

[3]Ingham and Dewick, 1979
[1,2]Pueppke and VanEtten, 1975
Pueppke and VanEtten, 1976

(36)   (-)-SPARTICARPIN

MW 300 $C_{17}H_{16}O_5$

λ max EtOH,   nm: 211, 232sh, 290sh, 294, 302sh
DPN: Orange
$[\alpha]$ -170° (MeOH)
FR: 1

Spartium junceum (L)

[1,2,3]Ingham and Dewick, 1980a

(37)   (-)-4-METHOXYMEDICARPIN

MW 300 $C_{17}H_{16}O_5$

λ max EtOH,   nm: 215, 232sh, 285,292sh
DPN: Yellow
mp: 158-160°
$[\alpha]$ -190° (MeOH)
FR: 1

Trifolium cherleri (L)
Trifolium pallescens (L)

[1,3]Ingham and Dewick, 1979
Ingham, unpublished

(38)   4-HYDROXYHOMOPTEROCARPIN

MW 300 $C_{17}H_{16}O_5$

λ max EtOH,   nm: 216, 238sh, 285, 292sh
DPN: Orange or orange/yellow
Gibbs: Blue
FR: 1 (or 2)

Trifolium hybridum (L)
Trifolium pallescens (L)

Ingham, 1976e
Ingham, unpublished

Table 2.1 (*continued*)

(39)   (-) and (+)-PISATIN

MW 314 $C_{17}H_{16}O_6$

λ max EtOH, nm: 210, 230sh, 281, 286, 310
Chromotropic acid reagent: Purple/pink
mp: 72°
$[\alpha]$ +280° (EtOH) and -258° (MeOH)
FR: 1

Caragana spp. (L)	Ingham, 1979*c*
Lathyrus spp. (C,L)	Robeson, 1978*a*
	Robeson and Harborne, 1980
Pisum fulvum (L)	Robeson and Harborne, 1980
(P)	Cruickshank and Perrin, 1965
Pisum sativum (+)-isomer (C)	Keen, 1975
	Robeson, 1978*a*
(CT)	Bailey, 1970
(E,RC)	Pueppke and VanEtten, 1975
	Pueppke and VanEtten, 1976
(SD)	Sutherland *et al.,* 1980
(L)	Bailey, 1973
	Robeson and Harborne, 1980
	Shiraishi *et al.,* 1977
(P)	Cruickshank and Perrin, 1960
	[1]Perrin and Perrin, 1962
	[2]Perrin and Bottomley, 1962
	[1]Stoessl, 1972
(PE)	Ingham, 1979*d*
(R)	Burden *et al.,* 1974
(S)	Shiraishi *et al.,* 1977
(SE,ST)	Ingham, 1979*d*
Pisum spp. (P)	Cruickshank and Perrin, 1965
Tephrosia bidwilli (-)-isomer (L)	Ingham, unpublished

(40)   (-)-2-METHOXYHOMOPTEROCARPIN

MW 314 $C_{18}H_{18}O_5$

λ max EtOH, nm : 210, 230sh, 288sh 292, 302sh
mp: 122-124°
$[\alpha]$ -228° (EtOH)
FR: 1

Pisum sativum (E,RC)	[1,2]Pueppke and VanEtten, 1975
	Pueppke and VanEtten, 1976

(41)   (-)-4-METHOXYMAACKIAIN

MW 314 $C_{17}H_{14}O_6$

λ max EtOH, nm: 214, 236sh, 273sh, 283sh, 311
DPN: Yellow
mp: 159-161°
$[\alpha]$ -201° (MeOH)
FR: 1

Table 2.1 (*continued*)

Tephrosia bidwilli (L)

Ingham, unpublished

Trifolium hybridum (L)

Ingham, 1976*b*

(42)   (-)-ACANTHOCARPAN

(Revised structure)

MW 328 $C_{17}H_{12}O_7$

λ max EtOH, nm: 212, 244sh, 276sh, 286sh, 309
Chromotropic acid reagent: Purple/pink
$[\alpha]$ -259° (MeOH)
FR: 1 (or 2)

Caragana acanthophylla   (L)

Tephrosia bidwilli (L)

[1]Ingham, 1979*c*

Ingham, unpublished

(43)   (-)-TEPHROCARPIN

MW 330 $C_{17}H_{14}O_7$

λ max EtOH, nm: 212, 240sh, 273sh, 284sh, 310
DPN: Yellow
$[\alpha]$ -267° (MeOH)
FR: 1 (or 2)

Tephrosia bidwilli   (L)

Ingham, unpublished

(44)   (-)-2,3,9-TRIMETHOXY-4-HYDROXYPTEROCARPAN

MW 330 $C_{18}H_{18}O_6$

λ max EtOH, nm: 212, 235sh, 287, 292sh
DPN: Yellow; Gibbs: Blue
mp: 141-145°
$[\alpha]$ -185° (EtOH)
FR: 2

Pisum sativum (E,RC)

[1,2]Pueppke and VanEtten, 1975
Pueppke and VanEtten, 1976

Table 2.1 (*continued*)

ii) Complex Pterocarpans

(45) (-)-NEODUNOL

Pachyrrhizus erosus (S)

MW 280 $C_{17}H_{12}O_4$

λ max EtOH, nm: 213, 224sh, 241, 248, 255, 288sh, 284, 303sh
DPN: Orange
mp: 170-172°
$[\alpha]$ -207° (MeOH)
FR: 1

[1]Ingham, 1979*f*

(46) (-)-PHASEOLLIN

Phaseolus vulgaris (C)

          (CC)
          (H)

          (L)

          (P)

          (R)

          (S)
Phaseolus spp. (P)
Vigna unguiculata (H)

MW 322 $C_{20}H_{18}O_4$

λmax EtOH, nm: 210, 230, 280, 286, 315
DPN: Yellow
mp: 177-178°
$[\alpha]$ -145° (EtOH)
FR: 1

Gnanamanickam, 1979
Keen, 1975
Dixon and Bendall, 1978
Bailey and Deverall, 1971
Bailey and Ingham, 1971
Rathmell and Bendall, 1971
Bailey and Ingham, 1971
[2]Cruickshank and Perrin, 1971
Gnanamanickam and Patil, 1977
Stholasuta *et al.*, 1971
Cruickshank and Perrin, 1963
Cruickshank and Perrin, 1971
[1]Perrin, 1964
Burden *et al.*, 1974
Sutherland *et al.*, 1980
Cruickshank and Perrin, 1971
Cruickshank and Perrin, 1971
Bailey, 1973

(47) (-)-PHASEOLLIDIN

MW 324 $C_{20}H_{20}O_4$

λ max EtOH, nm: 214, 230sh, 281, 287
DPN: Orange
mp: 67-69°
$[\alpha]_{578nm}$ -178° (EtOH)
FR: 1

Table 2.1 (*continued*)

Dolichos biflorus (L,S)	Keen and Ingham, 1980
Erythrina crista-galli (L)	[2] Ingham and Markham, 1980
Erythrina sandwicensis (L)	Ingham, 1980
Lablab niger (H)	Ingham, 1977*b*
Macroptilium atropurpureum (L)	Ingham, unpublished
Phaseolus aureus (H)	Ingham, unpublished
Phaseolus vulgaris (C)	Gnanamanickam, 1979
(H)	Burden *et al.*, 1972
(L)	Gnanamanickam and Patil, 1977
(P)	[1] Perrin *et al.*, 1972
	[2] Perrin *et al.*, 1974
(R)	Burden *et al.*, 1974
Psophocarpus tetragonolobus (L,P)	Preston, 1977
(S)	Ingham, 1979
	Ingham and Markham, 1980
Vigna unguiculata (H)	Bailey, 1973
(S)	Preston, 1975

(48)　(-)-CLANDESTACARPIN

MW 336　$C_{20}H_{16}O_5$

λ max EtOH, nm : 236sh, 243sh, 381sh
287, 292sh, 310sh
DPN:　Orange
FR:　1

Glycine clandestina (L)	[1,2] Lyne *et al.*, 1981
Glycine tabacina (L)	Keen, pers. commun.
Glycine tomentella	

(49)　(+)-TUBEROSIN

MW 338　$C_{20}H_{18}O_5$

λ max EtOH,　nm: 210sh, 224, 281, 287,
312, 318sh, 324sh
DPN:　Yellow
mp:　213°
[α] +220° (MeOH)
FR:　1

Pueraria lobata (L)　　　　　　Ingham, unpublished

(50)　(-)-SANDWICENSIN

MW 338　$C_{21}H_{22}O_4$

λ max EtOH,　nm: 211, 234sh, 281, 287
DPN:　Yellow
[α] -190° (MeOH)
FR:　1

Erythrina sandwicensis (L)　　　　[1,2] Ingham, 1980

Table 2.1 (*continued*)

(51)   (−)-GLYCEOLLIN II

MW 338 $C_{20}H_{18}O_5$

λ max EtOH, nm: 210sh, 227, 275sh,
  280sh, 285, 292sh, 307, 318
DPN: Orange
mp: 89-93°
FR: 1

Glycine canescens (L)
Glycine falcata
Glycine gracilis
Glycine latrobeana
Glycine max (C)

      (L)
Glycine soja (L)

Keen, pers. comm.

Ingham *et al.*, 1981*b*
Lyne *et al.*, 1976
Ingham *et al.*, 1981*b*
Keen, pers. comm.

(52)   (−)-GLYCEOLLIN III

MW 338 $C_{20}H_{18}O_5$

λ max EtOH, nm: 218, 230sh, 281sh, 287,
  292, 298sh
DPN: Orange
mp: 149-153°
FR: 1

Glycine gracilis (L)
Glycine latrobeana
Glycine max (C)

      (L)
Glycine soja (L)

Keen, pers. comm.

Ingham *et al.*, 1981*b*
Lyne *et al,* 1976
Ingham *et al.*, 1981*b*
Keen, pers. comm.

(53)   (−)-CANESCACARPIN

MW 338 $C_{20}H_{18}O_5$

λ max EtOH, nm: 230sh, 281sh, 287,
  292, 298sh
DPN: Orange
mp: 164-167°
FR: 1

Glycine canescens (L)   [1,2] Lyne *et al.*, 1981

Table 2.1 (*continued*)

(54)   (−)-GLYCEOLLIN I

    (6a-HYDROXYPHASEOLLIN ;

     revised structure)

MW 338 $C_{20}H_{18}O_5$

λ max EtOH, nm: 210sh, 228, 281sh,
   286, 291sh, 306sh, 320sh
DPN: Orange
[α] −207° (EtOAc)
FR: 1

Glycine canescens   (L)   Keen, pers. comm.
Glycine falcata
Glycine latrobeana
Glycine max (C)   [1]Burden and Bailey, 1975
                        Ingham *et al.*, 1981*b*
                        Keen, 1975
                        Keen and Horsch, 1972
              [1,2]Lyne *et al.*, 1976
      (CT,P,R)   Keen and Horsch, 1972
      (H)   Keen and Horsch, 1972
              [1,2]Sims *et al.*, 1972
      (L)   Ingham *et al.*, 1981*b*
                        Keen and Kennedy, 1974
Glycine soja (L)   Keen, pers. comm.
Psoralea spp. (L)   Ingham, unpublished

(55)   (−)-APIOCARPIN

MW 338 $C_{20}H_{18}O_5$

λ max EtOH, nm: 214, 238sh, 286, 293sh
DPN: Orange
[α]−180° (MeOH)
FR: 1

Apios tuberosa   (L)   Ingham and Mulheirn, unpublished

(56)   (−)-SANDWICARPIN

MW 340 $C_{20}H_{20}O_5$

λ max EtOH, nm: 212, 234sh, 281, 287
DPN: Orange
[α] −278° (MeOH)
FR: 1 (or 2)

Erythrina sandwicensis (L)   [1,2]Ingham, 1980

Table 2.1 (*continued*)

(57)   (−)−DOLICHIN A

Dolichos biflorus   (L)

MW 340 $C_{20}H_{20}O_5$

λ max EtOH,   nm: 212, 232sh, 282, 288
DPN: Orange
[α] −265° (MeOH)
FR: 3

[1,2] Ingham *et al.*, 1981*a*

(58)   (−)−DOLICHIN B

Dolichos biflorus   (L)

MW 340 $C_{20}H_{20}O_5$

λ max EtOH,   nm: 213, 232sh, 282, 288
DPN: Orange
[α] −235° (MeOH)
FR: 3

[1,2] Ingham *et al.*, 1981*a*

(59)   (−)−GLYCEOCARPIN

Glycine max   (C,L)

MW 340 $C_{20}H_{20}O_5$

λ max EtOH,   nm: 214, 230sh, 282sh,
   287, 292sh
DPN: Orange
[α] −236° (MeOH)
FR: 2

[1,2] Ingham *et al.*, 1981*b*

(60)   (−)−CRISTACARPIN

MW 354 $C_{21}H_{22}O_5$

λ max EtOH,   nm: 210, 234sh, 281, 287
DPN: Yellow
[α] −220° (MeOH)
FR: 1

Table 2.1 (*continued*)

Erythrina crista-galli (L) [1,2] Ingham and Markham, 1980
Erythrina sandwicensis (L) Ingham, 1980
Psophocarpus tetragonolobus (S) Ingham and Markham, 1980

(61) (-)-1-METHOXYPHASEOLLIDIN

MW 354 $C_{21}H_{22}O_5$

λ max EtOH, nm: 214, 232sh, 284, 289sh
DPN: Orange
$[\alpha]$ -225° (CHCl$_3$)
FR: 1

Psophocarpus tetragonolobus (L,P) [1,2] Preston, 1977

(62) (-)-GLYCEOLLIN IV

MW 354 $C_{21}H_{22}O_5$

λ max EtOH, nm: 213, 230sh, 281sh, 286, 291sh
DPN: Orange
FR: 1

Glycine max (C) [1,2] Lyne and Mulheirn, 1978
Glycine soja (L) Keen, pers. comm.

(63) (-)-GLYCEOFURAN

MW 354 $C_{20}H_{18}O_6$

λ max EtOH, nm: 211, 226sh, 250, 257, 287sh, 293, 306sh
DPN: Orange
mp: 181°
$[\alpha]$ -242° (MeOH)
FR: 3

Glycine max (C,L) [1,2] Ingham *et al.*, 1981b

Table 2.1 (*continued*)

(64)   (-)-9-O-METHYLGLYCEOFURAN

MW 368 $C_{21}H_{20}O_6$

λ max EtOH, nm: 211, 225sh, 250,
257, 287sh, 293, 306sh
$[\alpha]$ -247° (MeOH)
FR: 3

Glycine max   (L)   [1,2] Ingham *et al.*, 1981b

D: PTEROCARPENE

(65)   ANHYDROGLYCINOL

MW 254 $C_{15}H_{10}O_4$

λ max EtOH, nm: 212, 231sh, 242sh,
250sh, 290sh, 335, 353
DPN: Brown
Fluorescence: Dark purple/blue
FR: 3

Tetragonolobus maritimus   (L)   [1,3] Ingham and Dewick, 1978

E: ISOFLAVANS

   i)   Simple Isoflavans

(66)   (-)-DEMETHYLVESTITOL

MW 258 $C_{15}H_{14}O_4$

λ max EtOH, nm: 209, 225sh, 283
DPN: Orange/yellow: Gibbs:
Purple/blue
$[\alpha]$ -11° (MeOH)
FR: 2

Anthyllis vulneraria   (L)   [1] Ingham, 1977c

Table 2.1 (*continued*)

Erythrina sandwicensis (L)	Ingham, 1980
Hosackia americana (L)	Ingham, unpublished
Lablab niger (H)	Ingham, 1977b
Lotus spp. (L)	Ingham, 1977c
	Ingham and Dewick, 1979
	Ingham and Dewick, 1980c
Phaseolus vulgaris (P)	[1,2,3]Woodward, 1980a
Tetragonolobus spp. (L)	Ingham, 1977c

(67) (-)-VESTITOL

HO⟶⟨ring⟩⟶O

HO⟶OCH₃

MW 272 $C_{16}H_{16}O_4$

λ max EtOH, nm: 210, 226sh, 281, 285
DPN: Yellow; Gibbs: Blue
[α] -5° and -8° (both MeOH)
FR: 1

Canavalia ensiformis (H)	Ingham, unpublished
Carmichaelia flagelliformis (L)	Ingham, unpublished
Dalbergia sericea (L)	Ingham, 1979d
Factorovskya aschersoniana (L)	Ingham, 1979b
Hosackia americana (L)	Ingham, unpublished
Lotus spp. (L)	[1,2]Bonde et al., 1973
	Ingham 1977c
	Ingham and Dewick, 1979
	Ingham and Dewick, 1980c
Lotus uliginosus (R)	[1,2]Russell et al., 1978
	Sutherland et al., 1980
Medicago sativa (SD)	[3]Dewick and Martin, 1979a
Medicago spp. (L)	Ingham, 1979y
Onobrychis spp. (L)	Ingham, 1978d
Robinia pseudoacacia (L)	Ingham, unpublished
Tetragonolobus requienii (L)	Ingham, 1977c
Tipuana tipu (L)	Ingham, unpublished
Trifolium spp. (L)	Ingham, 1978a
Trigonella spp. (L)	Ingham and Harborne, 1976

(68) ISOVESTITOL

HO⟶⟨ring⟩⟶O

CH₃O⟶OH

MW 272 $C_{16}H_{16}O_4$

λ max EtOH, nm: 209, 225sh, 282, 285sh
DPN: Orange
FR: 1

Anthyllis vulneraria (L)	[1]Ingham, 1977c
Erythrina sandwicensis (L)	Ingham, 1980
Hosackia americana (L)	Ingham, unpublished
Lablab niger (H)	Ingham, 1977b
Tetragonolobus spp. (L)	Ingham, 1977c
Trifolium spp. (L)	Ingham, 1978a

Table 2.1 (*continued*)

(69)   NEOVESTITOL

MW 272 $C_{16}H_{16}O_4$

λ max EtOH, nm: 212, 227sh, 281sh,
284, 289
DPN: Orange/yellow; Gibbs: Purple/
blue
FR: 1

Dalbergia sericea (L)

[1,3]Ingham, 1979*e*

(70)   (-)-SATIVAN
       (SATIVIN)

MW 286 $C_{17}H_{18}O_4$

λ max EtOH, nm: 208, 227, 280, 284,
290sh
DPN: Yellow
mp: 125–127°
[α] -15° and -22° (both MeOH)
FR: 1

Lotus spp. (L)

[2]Bonde *et al.*, 1973
[1]Ingham, 1977*c*
Ingham and Dewick, 1979

Lotus uliginosus (R)   Russell *et al.*, 1978
Medicago sativa (SD)   [3]Dewick and Martin, 1979*a*
                  (L)   [1,2] Ingham and Millar, 1973
Medicago spp. (L)   Ingham, 1979*g*
Trifolium spp. (L)   Ingham, 1978*a*
Trigonella spp. (L)   Ingham and Harborne, 1976

(71)   ISOSATIVAN

MW 286 $C_{17}H_{18}O_4$

λ max EtOH, nm: 214, 227sh, 281, 284,
289sh
DPN: Yellow; Gibbs: Blue
FR: 1

Medicago scutellata (L)   Ingham, 1979*g*
Trifolium spp. (L)   [1,3]Ingham, 1976*b*
                                 Ingham, 1978*a*

(72)   ARVENSAN

MW 286 $C_{17}H_{18}O_4$

λ max EtOH, nm: 210, 225, 280-284,
288sh
DPN: Orange
FR: 1

Table 2.1 (*continued*)

Trifolium arvense (L)	[1,3]Ingham and Dewick, 1977
Trifolium stellatum (L)	Ingham, 1978*a*

(73)   ASTRACICERAN

MW 300 $C_{17}H_{16}O_5$

λ max EtOH, nm: 210, 230sh, 286sh, 292, 300
DPN: Yellow
FR: 1

Astragalus cicer (L)	[1,3]Ingham and Dewick, 1980*b*
Astragalus pyrenaicus (L)	[1,3]Ingham and Dewick, 1980*h*

(74)   ISOMUCRONULATOL

MW 302 $C_{17}H_{18}O_5$

λ max EtOH, nm: 217, 228sh, 276sh, 281, 290sh
DPN: Yellow or yellow/brown; Gibbs: Blue
FR: 1

Astragalus glycyphyllos (L)	Ingham, unpublished
Astragalus penduliflorus (L)	Ingham, unpublished
Carmichaelia flagelliformis (L)	Ingham, unpublished
Colutea arborescens (L)	Ingham, unpublished
Glycyrrhiza glabra (L)	[1]Ingham, 1977*d*

(75)   MUCRONULATOL

MW 302 $C_{17}H_{18}O_5$

λ max EtOH, nm: 213, 228sh, 276sh, 281, 290sh
DPN: Yellow or yellow/brown; Gibbs: Blue
FR: 1

Astragalus cicer (L)	Ingham and Dewick, 1980*b*,
Astragalus gummifer (L)	Ingham, unpublished
Astragalus pyrenaicus (L)	Ingham, unpublished

Table 2.1 (*continued*)

(76)   LAXIFLORAN

MW 302 $C_{17}H_{18}O_5$

λ max EtOH, nm: 211, 288sh, 282, 290sh
DPN: Orange
FR: 1

HO   O
CH$_3$O   OH
OCH$_3$

Lablab niger   (H)

[1]Ingham, 1977b

(77)   5-METHOXYVESTITOL

MW 302 $C_{17}H_{18}O_5$

λ max EtOH, nm: 212, 230sh, 276sh, 280, 286sh
DPN: Orange/yellow; Gibbs: Blue
FR: 1

HO   O
OCH$_3$ HO   OCH$_3$

Lotus edulis   (L)
Lotus hispidus   (L)

Ingham and Dewick, 1980c
[1,3]Ingham and Dewick, 1979

(78)   LOTISOFLAVAN

MW 302 $C_{17}H_{18}O_5$

λ max EtOH, nm: 212, 230sh, 281, 287sh
DPN: Orange; Gibbs: Purple/blue
FR: 1

CH$_3$O   O
OCH$_3$ HO   OH

Lotus angustissimus   (L)
Lotus edulis   (L)

[1,3]Ingham and Dewick, 1980c
[1,3]Ingham and Dewick, 1980c

ii)   Complex Isoflavans

(79)   (-)-PHASEOLLINISOFLAVAN

MW 324 $C_{20}H_{20}O_4$

λ max EtOH, nm: 210, 229, 280, 290sh, 310sh
DPN: Orange; Gibbs: Blue
FR: 1

HO   O
HO   O

Table 2.1 (*continued*)

Phaseolus vulgaris   (C)   Gnanamanickam, 1979
                       (CC)   Dixon and Bendall, 1978
                       (H)   Bailey and Burden, 1973
                                              [1]Burden *et al.*, 1972
                       (L)   Gnanamanickam and Patil, 1977
                       (R)   Burden *et al.*, 1974
                                              [2]Sutherland *et al.*, 1980

(80)   (+)-2'-O-METHYLPHASEOLLINISOFLAVAN

       (2'-METHOXYPHASEOLLINISOFLAVAN)   MW 338 $C_{21}H_{22}O_4$

λ max EtOH, nm: 210sh, 228, 260sh, 270, 280, 290sh, 314
DPN: Orange
$[\alpha]$ +19.5° (EtOH)
FR: 1

Phaseolus vulgaris   (H)   [1,2,3]VanEtten, 1973
                                   VanEtten and Smith, 1975
                       (R)   Sutherland *et al.*, 1980

(81)   2'-O-METHYLPHASEOLLIDINISOFLAVAN

MW 340 $C_{21}H_{24}O_4$

λ max EtOH, nm: 212, 230sh, 275sh, 281sh, 284, 290sh
DPN: Orange
FR: 1

Vigna unguiculata (S)   [1]Preston, 1975

F: COUMESTANS

    i) Simple Coumestans

(82)   COUMESTROL   MW 268 $C_{15}H_{18}O_5$

λ max EtOH, nm: 210, 246, 267sh, 293sh, 305, 345, 360sh
DPN: Faint orange
Fluorescence: Violet blue
mp: 385°
FR: 3

Table 2.1 (*continued*)

Glycine max (H)	Keen *et al.*, 1972
(L)	Keen and Kennedy, 1974
Phaseolus lunatus (R)	[1]Rich *et al.*, 1977
Phaseolus vulgaris (C)	Ghanamanickam, 1979
(CC)	Dixon and Bendall, 1978
(H)	Rathmell and Bendall, 1971
(L)	Gnanamanickam and Patil, 1977
	Lyon and Wood, 1975
(P)	Woodward, 1980*a*
Vigna unguiculata (H)	Partridge, 1973

ii) Complex Coumestans

(83)   SOJAGOL

MW 336 $C_{20}H_{16}O_5$

λ max MeOH, nm: 255, 305, 347
Fluorescence: Violet blue
mp: 284-286°
FP.: 3

Glycine max (H)	[1]Keen *et al.*, 1972
(L)	Keen and Kennedy, 1974

(84)   PSORALIDIN

MW 336 $C_{20}H_{16}O_5$

λ max EtOH, nm: 211, 244, 268sh, 293sh,
     306, 349, 364sh
DPN: Faint orange
Fluorescence: Violet blue
mp: 290-292°

Phaseolus lunatus (R)   [1]Rich *et al.*, 1977

Table 2.2 Structures, sources and selected characteristics of the non-isoflavonoid legume phyto-alexins.[a,b]

Key : C = cotyledons ; E = epicotyls ; H = hypocotyls ; L = leaves ;
P = pods ; R = roots ; S = stems ; SD = seedlings.

[a]UV maxima of compounds 100 and 102 were determined in a
mixture of acetonitrile and water (MeCN/H$_2$O).

[b]For additional information see footnotes to Table 1.

A · BENZOFURANS

(85) 6-DEMETHYLVIGNAFURAN

MW 256 C$_{15}$H$_{12}$O$_4$

λ max EtOH, nm: 212, 228sh, 249sh,
282, 307sh, 321, 337
DPN: Purple/pink
Fluorescence: Dark blue
FR: 1

Anthyllis vulneraria (L)
Coronilla emerus (L)
Tetragonolobus maritimus (L)

Ingham and Dewick, 1978
Dewick and Ingham, 1980
[1,3]Ingham and Dewick, 1978

(86) VIGNAFURAN

MW 270 C$_{16}$H$_{14}$O$_4$

λ max EtOH, nm: 210, 228sh, 248sh,
283, 306sh, 320, 335
DPN: Purple/brown
Fluorescence: Dark blue
FR: 1

Lablab niger (H)
Vigna unguiculata (L)
(SD)

Ingham, 1977*b*
[1,3]Preston *et al.*, 1975
[3]Martin and Dewick, 1979

(87) ISOPTEROFURAN

MW 286 C$_{16}$H$_{14}$O$_5$

λ max EtOH, nm: 213, 245sh, 285,
294sh, 304sh, 318, 332
DPN: Purple/brown
Fluorescence : Dark blue
FR: 1

Coronilla emerus (L)

[1,3]Dewick and Ingham, 1980

Table 2.2 (*continued*)

B: CHROMONES

(88)   LATHODORATIN

MW 206 $C_{11}H_{10}O_4$

λ max EtOH,   nm: 212, 231, 252,
     259, 296, 328sh
DPN: Orange/yellow; Gibbs: Blue
Fluorescence: Dark purple/blue
mp: 201-203°
FR: 1

Lathyrus hirsutus   (L)
Lathyrus odoratus   (C,E,L,P,R)

Robeson, 1978a
[1]Robeson *et al.*, 1980

(89)   METHYL-LATHODORATIN

MW 220 $C_{12}H_{12}O_4$

λ max EtOH,   nm: 210, 232, 251, 258,
     292, 326sh
DPN: Orange/yellow; Gibbs: Blue
Fluorescence: Faint yellow/brown
mp: 68-70°
FR: 1 (or 2)

Lathyrus odoratus   (C,L,P)

Robeson, 1978a
Robeson *et al.*, 1980

C: FLAVANONES

(90)   LIQUIRITIGENIN

MW 256 $C_{15}H_{12}O_4$

λ max MeOH,   nm: 217, 231, 275,
     312
DPN: Orange/yellow
Fluorescence: Barely discernible
  pale blue
FR: 3

Lens culinaris   (C,L)
Medicago sativa   (L)
Vicia floridana   (L)

Robeson, 1978a
[1]Ingham, 1979g
Robeson, 1978a

Table 2.2 (*continued*)

D: FURANOACETYLENES

R = (furan ring)—CH=HC–COOR' or R"

R' = H

R" = CH$_3$

(91) WYERONE ACID

$H_3C-CH_2-CH=HC-C\equiv C-\overset{\underset{\|}{O}}{C}-R(R')$

MW 244 C$_{14}$H$_{12}$O$_4$

λ max MeOH, nm: 222, 284, 356
Fluorescence: Deep blue
FR: 1

Vicia faba (C)	Hargreaves *et al.*, 1977
	Mansfield *et al.*, 1980
(L)	Hargreaves *et al.*, 1977
	Letcher *et al.*, 1970
	Mansfield *et al.*, 1980
(P)	Hargreaves *et al.*, 1976d
	Hargreaves *et al.*, 1977
	Mansfield *et al.*, 1980
Vicia galilea (P)	Hargreaves *et al.*, 1976a
Vicia narbonensis (P)	Hargreaves *et al.*, 1976a

(92) DIHYDROWYERONE ACID

$H_3C-(CH_2)_3-C\equiv C-\overset{\underset{\|}{O}}{C}-R(R')$

MW 246 C$_{14}$H$_{14}$O$_4$

λ max MeOH, nm: 346
Fluorescence: Deep blue
FR: 1

Vicia faba (C,L,P)	Mansfield *et al.*, 1980

(93) WYERONE

$H_3C-CH_2-CH=HC-C\equiv C-\overset{\underset{\|}{O}}{C}-R(R'')$

MW 258 C$_{15}$H$_{14}$O$_4$

λ max EtOH, nm: 224, 292, 351
Fluorescence: Deep blue
mp: 65-66°
FR: 1

Lens spp. (C)	Robeson, 1978*b*
Vicia faba (C)	[1]Hargreaves *et al.*, 1976*b*
(L)	Fawcett *et al.*, 1971
	Other references as given for (91)
(P)	References as given for (91)
(S)	Keen, 1972
(SD)	[1,3]Fawcett *et al.*, 1968
Vicia galilea (P)	Hargreaves *et al.*, 1976a
Vicia narbonensis (P)	Hargreaves *et al.*, 1976a
Vicia spp. (C)	Robeson and Harborne, 1980

Table 2.2 (*continued*)

(94)   DIHYDROWYERONE

$$H_3C-\left(CH_2\right)_3-C\equiv C-\underset{O}{\overset{\|}{C}}-R\left(R''\right)$$

MW 260 $C_{15}H_{16}O_4$

λ max EtOH, nm: 235, 340
Fluorescence: Deep blue
mp: 80-81°
FR: 1

Lens culinaris (C)
Vicia faba (C,L,P)
        (SD)
Vicia sativa (C)

Robeson, 1978*b*
Mansfield *et al.*, 1980
[1,3]Fawcett *et al.*, 1968
Robeson, 1978*a*

(95)   WYEROL

$$H_3C-CH_2-CH=HC-C\equiv C-\underset{OH}{\overset{|}{C}}H-R\left(R''\right)$$

MW 260 $C_{15}H_{16}O_4$

λ max MeOH, nm: 312
mp: 55-56°
FR: 2 (or 3)

Vicia faba (C,L,P)

        (SD)

[1]Hargreaves *et al.*, 1976*b*
Hargreaves *et al.*, 1977
Mansfield *et al.*, 1980
[1,3]Fawcett *et al.*, 1968

(96)   DIHYDROWYEROL

$$H_3C-\left(CH_2\right)_3-C\equiv C-\underset{OH}{\overset{|}{C}}H-R\left(R''\right)$$

MW 262 $C_{15}H_{18}O_4$

λ max MeOH, nm: 312
FR: 2 (or 3)

Vicia faba (C,L,P)

Mansfield *et al.*, 1980

(97)   WYERONE EPOXIDE

$$H_3C-CH_2-\overset{\overset{\displaystyle O}{\diagup\diagdown}}{CH-CH}-C\equiv C-\underset{O}{\overset{\|}{C}}-R\left(R''\right)$$

MW 274 $C_{15}H_{14}O_5$

λ max EtOH, nm: 238, 347
Picric acid reagent: Orange
Fluorescence: Deep blue
mp: 74-76°
FR : 1

Lens spp. (C)
Vicia faba (C,L,P)

Vicia spp. (C)

Robeson, 1978*b*
[1,3]Hargreaves *et al.*, 1976*c*
Mansfield *et al.*, 1980
Robeson and Harborne, 1980

(98)   PA-4

Structure undetermined
(Possible wyerone derivative)

Vicia faba (P)

MW 290

λ max MeOH, nm: 347
Fluorescence: Deep blue
FR: 1

[1]Hargreaves *et al.*, 1977

Table 2.2 (*continued*)

E : STILBENES

R = - CH = HC ⟨ring⟩ — OH, with R'or R"

R' = H

R" = OH

(99) *CIS*- & *TRANS*-RESVERATROL

HO —⟨ring⟩— R (R')
HO

Arachis hypogaea (H)
Trifolium campestre (L)
Trifolium dubium (L)

MW 228 $C_{14}H_{12}O_3$

λ max EtOH, nm: (*trans*) 210, 220sh, 238sh, 296sh, 307, 322sh, 340sh
λ max EtOH, nm: (*cis*) 210, 220sh, 285
DPN: Orange (both isomers)
Fluorescence: (*trans*) Pale blue (intensifies with NH_3)
FR: 2 (or 3), isomer mixture
[1]Ingham, 1976d
Ingham, 1978 a
Ingham, 1978 a

(100)

HO —⟨ring⟩— R (R')
HO

Arachis hypogaea (C)

MW 296 $C_{19}H_{20}O_3$

λ max $MeCN/H_2O$, nm: 219, 241sh, 327sh, 331, 346sh, 364sh
DPN: Orange
Fluorescence: Pale blue
FR: 1

[1]Aguamah *et al.*, 1981

(101) *CIS*- & *TRANS*-4-ISOPENTENYLRESVERATROL

HO —⟨ring⟩— R (R')
HO

Arachis hypogaea (C)

MW 296 $C_{19}H_{20}O_3$

λ max $MeCN/H_2O$, nm: (*trans*) 220, 295sh, 307, 324, 340sh
λ max $MeCN/H_2O$, nm: (*cis*) about 290
DPN: Orange (both isomers)
Fluorescence: (*trans*) Pale blue
FR: 1

[1]Aguamah *et al.*, 1981
[1]Keen and Ingham, 1976

Table 2.2 (*continued*)

(102)

MW 312 $C_{19}H_{20}O_4$

λ max MeCN/H$_2$O, nm: 220, 245sh,
    310sh, 340, 346sh
DPN: Orange
Fluorescence: Pale blue
FR: 1

Arachis hypogaea (C)

[1]Aguamah *et al.*, 1981

Table 2.3 TLC data for some isoflavonoid and non-isoflavonoid phytoalexins

Compound	R_f value (x 100)[a] in solvent system				
	CM[b]	C[c]	PEA[b]	BM[b]	EH[b]
Isoflavones					
Daidzein (1)	15	–	3	16	–
Formononetin (2)	40	–	14	31	–
Genistein (5)	22	–	14	18	–
2'-Hydroxygenistein (8)	11		0	13	–
Cajanin (11)	33	–	12	29	–
Tectorigenin (12)	31	–	16	21	39
Isoflavanones					
Dalbergioidin (19)	9	–	5	12	–
Kievitone (26)	11	–	7	15	–
Pterocarpans					
Demethylmedicarpin (27)	25	–	22	21	–
Medicarpin (28)	68	19	54	57	71
Isomedicarpin (29)	68	–	54	57	–
Glycinol (30)	8	–	5	10	–
Maackiain (31)	68	–	52	57	–
Variabilin (34)	81	–	37	63	–
2-Methoxymedicarpin (35)	87	–	49	71	–
4-Methoxymedicarpin (37)	85	–	49	69	–
Pisatin (39)	81	–	37	63	63
2-Methoxyhomopterocarpin (40)	93	–	51	87	–
Acanthocarpan (42)	81	–	32	–	–
2,3,9-Trimethoxy-4-hydroxy- pterocarpan (44)	85	–	20	61	–
Phaseollin (46)	70	–	65	61	77
Phaseollidin (47)	51	–	43	35	63
Tuberosin (49)	32	–	22	28	–
Glyceollin I (54)	30	–	22	27	47

Table 2.3 (*continued*)

	CM	C	PEA	BM	EH
Isoflavans					
Demethylvestitol (66)	7	–	7	12	–
Vestitol (67)	30	–	25	26	47
Isovestitol (68)	30	–	22	26	–
Sativan (70)	71	–	55	61	68
Isosativan (71)	77	–	61	63	68
Arvensan (72)	70	–	53	57	–
Isomucronulatol (74)	51	–	30	35	–
Mucronulatol (75)	49	–	20	31	–
5-Methoxyvestitol (77)	30	–	18	28	–
Phaseollinisoflavan (79)	38	–	41	31	–
Coumestans					
Coumestrol (82)	20	–	11	19	–
Psoralidin (84)	30	–	18	24	–
Non-Isoflavonoid Compounds					
Vignafuran (86)	66	–	51	57	–
Lathodoratin (88)	58	–	59	48	–
Wyerone acid (91)	–	O	–	–	–
Wyerone (93)	–	60	–	–	–
Wyerol (95)	–	32	–	–	–
Wyerone epoxide (97)	–	53	–	–	–
PA-4 (98)	–	15	–	–	–
trans-Resveratrol (99)	5	–	5	12	–

Key : BM = benzene : methanol (9:1) ; C = chloroform ; CM = chloroform : methanol (50:2) ; EH = diethyl ether : *n*-hexane (3:1) ; PEA = *n*-pentane : diethyl ether : glacial acetic acid (75:25:3).

[a]Rf values (x 100) on glass-backed, pre-coated silica gel TLC plates (Merck 5715) ; layer thickness, 0.25 mm ; developing distance, 15 cm.

[b]Partially saturated tank atmosphere ; chromatograms were developed 20 – 30 mins after preparation of the solvent mixtures.

[c]Saturated tank atmosphere (data from Hargreaves *et al.*, 1977).

REFERENCES

Aguamah, G. E., Langcake, P., Leworthy, D. P., Page, J. A., Pryce, R. J. and Strange, R. N. (1981). Isolation and characterization of novel stilbene phytoalexins from *Arachis hypogaea*. *Phytochemistry*, **20**, 1381–1393.

Bailey, J. A. (1970) Pisatin production by tissue cultures of *Pisum sativum* L. *J. Gen. Microbiol.*, **61**, 409–415.

Bailey, J. A. (1973) Production of antifungal compounds in cowpea (*Vigna sinensis*) and pea (*Pisum sativum*) after virus infection. *J. Gen. Microbiol.*, **75**, 119–123.

Bailey, J. A. and Burden, R. S. (1973) Biochemical changes and phytoalexin accumulation in *Phaseolus vulgaris* following cellular browning caused by tobacco necrosis virus. *Physiol. Plant Pathol.*, **3**, 171–177.

Bailey, J. A. and Deverall, B. J. (1971) Formation and activity of phaseollin in the interaction between bean hypocotyls (*Phaseolus vulgaris*) and physiological races of *Colletotrichum lindemuthianum*. *Physiol. Plant Pathol.*, **1**, 435–449.

Bailey, J. A. and Ingham, J. L. (1971) Phaseollin accumulation in bean (*Phaseolus vulgaris*) in response to infection by tobacco necrosis virus and the rust *Uromyces appendiculatus*. *Physiol. Plant Pathol.*, **1**, 451–456.

Bevan, C. W. L., Birch, A. J., Moore, B. and Mukerjee, S. K. (1964) A partial synthesis of ($\pm$)-pisatin: some remarks on the structure and reactions of pterocarpin. *J. Chem. Soc.*, 5991–5995.

Bickoff, E. M., Spencer, R. R., Witt, S. C. and Knuckles, B. E. (1969) *Studies on the chemical and biological properties of coumestrol and related compounds*. U.S.D.A. Technical Bulletin, No. 1408.

Biggs, D. R. (1975) Post-infectional compounds from the French bean, *Phaseolus vulgaris*; isolation and identification of genistein and 2′,4′,5,7-tetrahydroxyisoflavone. *Aust. J. Chem.*, **28**, 1389–1392.

Bilton, J. N., Debnam, J. R. and Smith, I. M. (1976) 6a-hydroxypterocarpans from red clover. *Phytochemistry*, **15**, 1411–1412.

Bonde, M. R., Millar, R. L. and Ingham, J. L. (1973) Induction and identification of sativan and vestitol as two phytoalexins from *Lotus corniculatus*. *Phytochemistry*, **12**, 2957–2959.

Burden, R. S. and Bailey, J. A. (1975) Structure of the phytoalexin from soybean. *Phytochemistry*, **14**, 1389–1390.

Burden, R. S., Bailey, J. A. and Dawson, G. W. (1972) Structures of three new isoflavanoids from *Phaseolus vulgaris* infected with tobacco necrosis virus. *Tetrahedron Lett.*, 4175–4178.

Burden, R. S., Rogers, P. M. and Wain, R. L. (1974) Investigations on fungicides. XVI. Natural resistance of plant roots to fungal pathogens. *Ann. appl. Biol.*, **78**, 59–63.

Clough, J. M. and Snell, B. K. (1981) Confirmation of the structures of the phytoalexins lathodoratin and methyl-lathodoratin by synthesis. *Phytochemistry*, **20**, 1752.

Cornia, M. and Merlini, L. (1975) A possible chemical analogy for pterocarpan biosynthesis. *J. Chem. Soc. Chem. Commun.*, 428–429.

Cruickshank, I. A. M. and Perrin, D. R. (1960) Isolation of a phytoalexin from *Pisum sativum* L. *Nature*, **187**, 799–800.

Cruickshank, I. A. M. and Perrin, D. R. (1963) Phytoalexins of the Leguminosae. Phaseollin from *Phaseolus vulgaris*. L. *Life Sci.*, 680–682.

Cruickshank, I. A. M. and Perrin, D. R. (1965) Studies on phytoalexins. IX. Pisatin formation by cultivars of *Pisum sativum* L. and several other *Pisum* species. *Aust. J. Biol. Sci.*, **18**, 829–835,

Cruickshank, I. A. M. and Perrin, D. R. (1971) Studies on phytoalexins. XI. The induction, antimicrobial spectrum and chemical assay of phaseollin. *Phytopath. Z.*, **70**, 209–229.

Cruickshank, I. A. M., Veeraraghavan, J. and Perrin, D. R. (1974) Some physical factors affecting the formation and/or net accumulation of medicarpin in infection droplets on white clover leaflets. *Aust. J. Plant Physiol.*, **1**, 149–156.

Darbarwar, M., Sundaramurthy, V. and Subba Rao, N. V. (1976) Coumestans. *J. Sci. Indust. Res. India*, **35**, 297–312.

Dean, F. M. (1963) *Naturally Occurring Oxygen Ring Compounds*. Butterworths, London.

Dewick, P. M. (1975) Pterocarpan biosynthesis: chalcone and isoflavone precursors of demethylhomopterocarpin and maackiain in *Trifolium pratense*. *Phytochemistry*, **14**, 979–982.

Dewick, P. M. (1977) Biosynthesis of pterocarpan phytoalexins in *Trifolium pratense*. *Phytochemistry*, **16**, 93–97.

Dewick, P. M. (1978) Biosynthesis of the 6-oxygenated isoflavone afrormosin in *Onobrychis viciifolia*. *Phytochemistry*, **17**, 249–250.

Dewick, P. M. and Ingham, J. L. (1980) Isopterofuran, a new 2-arylbenzofuran phytoalexin from *Coronilla emerus*. *Phytochemistry*, **19**, 289–291.

Dewick, P. M. and Martin, M. (1979*a*) Biosynthesis of pterocarpan and isoflavan phytoalexins in *Medicago sativa*: the biochemical interconversion of pterocarpans and 2'-hydroxyisoflavans. *Phytochemistry*, **18**, 591–596.

Dewick, P. M. and Martin, M. (1979*b*) Biosynthesis of pterocarpan, isoflavan and coumestan metabolites of *Medicago sativa*: chalcone, isoflavone and isoflavanone precursors. *Phytochemistry*, **18**, 597–602.

Dewick, P. M. and Ward, D. (1978) Isoflavone precursors of the pterocarpan phytoalexin maackiain in *Trifolium pratense*. *Phytochemistry*, **17**, 1751–1754.

Dixon, R. A. and Bendall, D. S. (1978) Changes in phenolic compounds associated with phaseollin production in cell suspension cultures of *Phaseolus vulgaris*. *Physiol. Plant Pathol.*, **13**, 283–294.

Donnelly, D. M. X., Thompson, J. C., Whalley, W. B. and Ahmad, S. (1973) *Dalbergia* species. Part IX. Phytochemical examination of *Dalbergia stevensonii* Standl. *J. Chem. Soc. Perkin. Trans.*, I, 1737–1744.

Duffley, R. P. and Stevenson, R. (1977) Synthesis of pterofuran and vignafuran. *J. Chem. Soc. Perkin Trans.*, I, 802–804.

Farkas, L., Gottsegen, A., Nógrádi, M. and Antus, S.(1971) Synthesis of the natural isoflavanones ferreirin, dalbergioidin, and ougenin. *J. Chem. Soc. C*, 1994–2000.

Farkas, L., Gottsegen, A., Nógrádi, M. and Antus, S. (1974) Synthesis of sophorol, violanone, lonchocarpan, claussequinone, philenopteran, leiocalycin, and some other natural isoflavonoids by the oxidative rearrangement of chalcones with thallium(III) nitrate. *J. Chem. Soc. Perkin Trans.*, I, 305–312.

Fawcett, C. H., Spencer, D. M., Wain, R. L., Fallis, A. G., Jones, E. R. H., Le Quan, M., Page, C. B., Thaller, V., Shubrook, D. C. and Whitham, P. M. (1968) Natural acetylenes. Part XXVII. An antifungal acetylenic furanoid keto-ester (wyerone) from shoots of the broad bean (*Vicia faba* L.; Fam. Papilionaceae). *J. Chem. Soc.*, C, 2455–2462.

Fawcett, C. H., Firn, R. D. and Spencer, D. M. (1971) Wyerone increase in leaves of broad bean (*Vicia faba* L.) after infection by *Botrytis fabae*. *Physiol. Plant Pathol.*, **1**, 163–166.

Fioriti, J. A. and Sims, R. J. (1968) A spray reagent for the identification of epoxides on thin layer plates. *J. Chromatog.*, **32**, 761–763.

Geigert, J., Stermitz, F. R., Johnson, G., Maag, D. D. and Johnson, D. K. (1973) Two phytoalexins from sugarbeet (*Beta vulgaris*) leaves. *Tetrahedron*, **29**, 2703–2706.

Gnanamanickam S. S. (1979) Isolation of isoflavonoid phytoalexins from seeds of *Phaseolus vulgaris*. *Experientia*, **35**, 323.

Gnanamanickam, S. S. and Patil, S. S. (1977) Accumulation of antibacterial isoflavonoids in hypersensitively responding bean leaf tissues inoculated with *Pseudomonas phaseolicola*. *Physiol. Plant Pathol.*, **10**, 159–168.

Gunner, S. W. and Hand, T. B. (1968) The detection of the methylenedioxy group on thin-layer chromatographic plates. *J. Chromatog.*, **37**, 357–358.

Gupta, B. K., Gupta, G. K., Dhar, K. L. and Atal, C. K. (1980) Psoralidin oxide, a coumestan from the seeds of *Psoralea corylifolia*. *Phytochemistry*, **19**, 2232–2233.

Gustine, D. L., Sherwood, R. T. and Vance, C. P. (1978) Regulation of phytoalexin synthesis in jackbean callus cultures. Stimulation of phenylalanine ammonia-lyase and *o*-methyltransferase. *Plant Physiol.*, **61**, 226–230.

Harborne, J. B. and Ingham, J. L. (1978) "Biochemical aspects of the coevolution of higher plants with their fungal parasites," in *Biochemical Aspects of Plant and Animal Coevolution*, ed. Harborne, J. B., Academic Press, London, 343–405.

Hargreaves, J. A., Mansfield, J. W. and Coxon, D. T. (1976*a*) Identification of medicarpin as a phytoalexin in the broad bean plant (*Vicia faba* L.). *Nature*, **262**, 318–319.

Hargreaves, J. A., Mansfield, J. W. and Coxon, D. T. (1976*b*) Conversion of wyerone to wyerol by *Botrytis cinerea* and *B. fabae in vitro*. *Phytochemistry*, **15**, 651–653.

Hargreaves, J. A., Mansfield, J. W., Coxon, D. T. and Price, K. R. (1976*c*) Wyerone epoxide as a phytoalexin in *Vicia faba* and its metabolism by *Botrytis cinerea* and *B. fabae in vitro*. *Phytochemistry*, **15**, 1119–1121.

Hargreaves, J. A., Mansfield, J. W. and Rossall, S. (1977) Changes in phytoalexin concentrations in tissues of the broad bean plant (*Vicia faba* L.), following inoculation with species of *Botrytis*. *Physiol. Plant Pathol.*, **11**, 227–242.

Higgins, V. J. and Smith, D. G. (1972) Separation and identification of two pterocarpanoid phytoalexins produced by red clover leaves. *Phytopathology*, **62**, 235–238.

Horowitz, R. M. (1957) Detection of flavanones by reduction with sodium borohydride. *J. Org. Chem.*, **22**, 1733–1735.

Ingham, J. L. (1973) Disease resistance in higher plants. The concept of pre-infectional and post-infectional resistance. *Phytopath. Z.*, **78**, 314–335.

Ingham, J. L. (1976*a*) Induced and constitutive isoflavonoids from stems of chickpeas (*Cicer arietinum* L.) inoculated with spores of *Helminthosporium carbonum* Ullstrup. *Phytopath. Z.*, **87**, 353–367.

Ingham, J. L. (1976*b*) Isosativan: an isoflavan phytoalexin from *Trifolium hybridum* and other *Trifolium* species. *Z. Naturforsch.*, **31c**, 331–332.

Ingham, J. L. (1976*c*) Induced isoflavonoids from fungus-infected stems of pigeon pea (*Cajanus cajan*). *Z. Naturforsch.*, **31c**, 504–508.

Ingham, J. L. (1976*d*) 3,5,4′-trihydroxystilbene as a phytoalexin from groundnuts (*Arachis hypogaea*). *Phytochemistry*, **15**, 1791–1793.

Ingham, J. L. (1976*e*) *A Comparative Study of Phytoalexins from the Leguminosae*. Ph.D. Thesis, University of Reading, England.

Ingham, J. L. (1976*f*) Fungal modification of pterocarpan phytoalexins from *Melilotus alba* and *Trifolium pratense*. *Phytochemistry*, **15**, 1489–1495.

Ingham, J. L. (1977*a*) Medicarpin as a phytoalexin of the genus *Melilotus*. *Z. Naturforsch.*, **32c**, 449–452.

Ingham, J. L. (1977*b*) Phytoalexins of hyacinth bean (*Lablab niger*). *Z. Naturforsch.*, **32c**, 1018–1020.

Ingham, J. L. (1977*c*) Isoflavan phytoalexins from *Anthyllis*, *Lotus* and *Tetragonolobus*. *Phytochemistry*, **16**, 1279–1282.

Ingham, J. L. (1977*d*) An isoflavan phytoalexin from leaves of *Glycyrrhiza glabra*. *Phytochemistry*, **16**, 1457–1458.

Ingham, J. L. (1978*a*) Isoflavonoid and stilbene phytoalexins of the genus *Trifolium*. *Biochem. Syst. Ecol.*, **6**, 217–223.

Ingham, J. L. (1978*b*) Phytoalexin production by high- and low-coumarin cultivars of *Melilotus alba* and *Melilotus officinalis*. *Can. J. Bot.*, **56**, 2230–2233.

Ingham, J. L. (1978*c*) Phaseollidin, a phytoalexin of *Psophocarpus tetragonolobus*. *Phytochemistry*, **17**, 165.

Ingham, J. L. (1978*d*) Flavonoid and isoflavonoid compounds from leaves of sainfoin (*Onobrychis viciifolia*). *Z. Naturforsch.*, **33c**, 146–148.

Ingham, J. L. (1979*a*) A revised structure for the phytoalexin cajanol. *Z. Naturforsch.*, **34c**, 159–161.

Ingham, J. L. (1979*b*) Isoflavonoid phytoalexins of *Parochetus communis* and *Factorovskya ascher-soniana*. *Z. Naturforsch.*, **34c**, 290–292.

Ingham, J. L. (1979*c*) Phytoalexin production by species of the genus *Caragana*. *Z. Naturforsch.*, **34c**, 293–295.

Ingham, J. L. (1979*d*) Phytoalexin production by flowers of garden pea (*Pisum sativum*). *Z. Naturforsch.*, **34c**, 296–298.

Ingham, J. L. (1979*e*) Isoflavonoid phytoalexins from leaflets of *Dalbergia sericea*. *Z. Naturforsch.* **34c**, 630–631.

Ingham, J. L. (1979*f*) Isoflavonoid phytoalexins of yam bean (*Pachyrrhizus erosus*). *Z. Naturforsch.*, **34c**, 683–688.

Ingham, J. L. (1979*g*) Isoflavonoid phytoalexins of the genus *Medicago*. *Biochem. Syst. Evol.*, **7**, 29–34.

Ingham, J. L. (1980) Induced isoflavonoids of *Erythrina sandwicensis*. *Z. Naturforsch.*, **35c**, 384–386.

Ingham, J. L. (1981) Phytoalexin induction and its taxonomic significance in the Leguminosae (subfamily Papilionoideae). *Proc. Internat. Legume Conf. Kew, 1978*, (Adv. Legume Systematics), 599–626.

Ingham, J. L. and Dewick, P. M. (1977) Isoflavonoid phytoalexins from leaves of *Trifolium arvense*. *Z. Naturforsch.*, **32c**, 446–448.

Ingham, J. L. and Dewick, P. M. (1978) 6-demethylvignafuran as a phytoalexin of *Tetragonolobus maritimus*. *Phytochemistry*, **17**, 535–538.

Ingham, J. L. and Dewick, P. M. (1979) A new isoflavan phytoalexin from leaflets of *Lotus hispidus*. *Phytochemistry*, **18**, 1711–1714.

Ingham, J. L. and Dewick, P. M. (1980*a*) Sparticarpin: a pterocarpan phytoalexin from *Spartium junceum*. *Z. Naturforsch.*, **35c**, 197–200.

Ingham, J. L. and Dewick, P. M. (1980*b*) Astraciceran: a new isoflavan phytoalexin from *Astragalus cicer*. *Phytochemistry*, **19**, 1767–1770.

Ingham, J. L. and Dewick, P. M. (1980*c*) Isolation of a new isoflavan phytoalexin from two *Lotus* species. *Phytochemistry*, **19**, 2799–2800.

Ingham, J. L. and Harborne, J. B. (1976) Phytoalexin induction as a new dynamic approach to the study of systematic relationships among higher plants. *Nature*, **260**, 241–243.

Ingham, J. L. and Markham, K. R. (1980) Identification of the *Erythrina* phytoalexin cristacarpin and a note on the chirality of other 6a-hydroxypterocarpans. *Phytochemistry*, **19**, 1203–1207.

Ingham, J. L. and Millar, R. L. (1973) Sativin: an induced isoflavan from the leaves of *Medicago sativa* L. *Nature*, **242**, 125–126.

Ingham, J. L., Keen, N. T. and Hymowitz, T. (1977) A new isoflavone phytoalexin from fungus-inoculated stems of *Glycine wightii*. *Phytochemistry*, **16**, 1943–1946.

Ingham, J. L., Keen, N. T., Markham, K. R. and Mulheirn, L. J. (1981*a*) Dolichins A and B, two new pterocarpans from bacteria-treated leaves of *Dolichos biflorus*. *Phytochemistry*, **20**, 807–809.

Ingham, J. L., Keen, N. T., Mulheirn, L. J. and Lyne, R. L. (1981*b*) Inducibly-formed isoflavonoids from leaves of soybean (*Glycine max*). *Phytochemistry*, **20**, 795–798.

Itô, S., Fujise, Y. and Mori, A. (1965) Absolute configuration of pterocarpinoids. *J. Chem. Soc. Chem. Commun.*, 595–596.

Jain, A. C., Kumar, A. and Gupta, R. C. (1979) Constitution of luteone and parvisoflavones-A and -B and synthesis of their methyl ethers and related isoflavones. *J. Chem. Soc. Perkin Trans.*, **I**, 279–282.

Keen, N. T. (1972) Accumulation of wyerone in broad bean and demethylhomopterocarpin in jack bean after inoculation with *Phytophthora megasperma* var. *sojae*. *Phytopathology*, **62**, 1365–1366.

Keen, N. T. (1975) The isolation of phytoalexins from germinating seeds of *Cicer arietinum*, *Vigna sinensis*, *Arachis hypogaea*, and other plants. *Phytopathology*, **65**, 91–92.

Keen, N. T. (1978) Phytoalexins: efficient extraction from leaves by a facilitated diffusion technique. *Phytopathology*, **68**, 1237–1239.

Keen, N. T. and Horsch, R. (1972) Hydroxyphaseollin production by various soybean tissues: a warning against use of "unnatural" host-parasite systems. *Phytopathology*, **62**, 439–442.

Keen, N. T. and Ingham, J. L. (1976) New stilbene phytoalexins from American cultivars of *Arachis hypogaea*. *Phytochemistry*, **15**, 1794–1795.

Keen, N. T. and Ingham, J. L. (1980) Phytoalexins from *Dolichos biflorus*. *Z. Naturforsch.*, **35c**, 923–926.

Keen, N. T. and Kennedy, B. W. (1974) Hydroxyphaseollin and related isoflavanoids in the hypersensitive resistance reaction of soybean to *Pseudomonas glycinea*. *Physiol. Plant Pathol.*, **4**, 173–185.

Keen, N. T., Zaki, A. I. and Sims, J. J. (1972) Biosynthesis of hydroxyphaseollin and related isoflavanoids in disease-resistant soybean hypocotyls. *Phytochemistry*, **11**, 1031–1039.

King, F. E., King, T. J. and Manning, L. C. (1957) An investigation of the Gibbs reaction and its bearing on the constitution of jacareubin. *J. Chem. Soc.*, 563–566.

Krebs, K. G., Heusser, D. and Wimmer, H. (1969) "Spray reagents" in *Thin-Layer Chromatography*, ed. E. Stahl, Allen and Unwin Ltd., London, 854–909.

Kukla, A. S. and Seshadri, T. R. (1962) Constitution and synthesis of maxima isoflavones-A and -B. *Tetrahedron*, **18**, 1443–1448.

Kurosawa, K., Ollis, W. D., Redman, B. T., Sutherland, I. O., Alves, H. M. and Gottlieb, O. R. (1978) Absolute configurations of isoflavans. *Phytochemistry*, **17**, 1423–1426.

Lampard, J. F. (1974) Demethylhomopterocarpin: an antifungal compound in *Canavalia ensiformis* and *Vigna unguiculata* following infection. *Phytochemistry*, **13**, 291–292.

Letcher, R. M., Widdowson, D. A., Deverall, B. J. and Mansfield, J. W. (1970) Identification and activity of wyerone acid as a phytoalexin in broad bean (*Vicia faba*) after infection by *Botrytis*. *Phytochemistry*, **9**, 249–252.

Lyne, R. L. and Mulheirn, L. J. (1978) Minor pterocarpinoids of soybean. *Tetrahedron Lett.*, 3127–3128.

Lyne, R. L., Mulheirn, L. J. and Leworthy, D. P. (1976) New pterocarpinoid phytoalexins of soybean. *J. Chem. Soc. Chem. Commun.*, 497–498.

Lyne, R. L., Mulheirn, L. J. and Keen, N. T. (1981) Novel pterocarpinoids from *Glycine* species. *Tetrahedron Lett.*, 2483–2484.

Lyon, F. M. and Wood, R. K. S. (1975) Production of phaseollin, coumestrol and related compounds in bean leaves inoculated with *Pseudomonas* spp. *Physiol. Plant Pathol.*, **6**, 117–124.

Mabry, T. J., Markham, K. R. and Thomas, M. B. (1970) *The Systematic Identification of Flavonoids*, Springer, Berlin.

McMurry, T. B. H., Martin, E., Donnelly, D. M. X. and Thompson, J. C. (1972) 3-Hydroxy-9-methoxy and 3-methoxy-9-hydroxypterocarpans. *Phytochemistry*, **11**, 3283–3286.

Mansfield, J. W., Porter, A. E. A. and Smallman, R. V. (1980) Dihydrowyerone derivatives as components of the furanoacetylenic phytoalexin response of tissues of *Vicia faba*. *Phytochemistry*, **19**, 1057–1061.

Markham, K. R. and Ingham, J. L. (1980) Tectorigenin, a phytoalexin of *Centrosema haitiense* and other *Centrosema* species. *Z. Naturforsch.*, **35c**, 919–922.

Martin, M. and Dewick, P. M. (1979) Biosynthesis of the 2-arylbenzofuran phytoalexin vignafuran in *Vigna unguiculata*. *Phytochemistry*, **18**, 1309–1317.

Minhaj, N., Khan, H., Kapoor, S. K. and Zaman, A. (1976) Extractives of *Millettia auriculata*—III. *Tetrahedron*, **32**, 749–751.

Neill, K. G. (1953) The synthesis of homoferreirin. *J. Chem. Soc.*, 3454–3455.

Ollis, W. D. (1968) "New structural variants among the isoflavonoid and neoflavonoid classes", in *Recent Advances in Phytochemistry*, Vol. 1, eds. T. J. Mabry, R. E. Alston and V. C. Runeckles, Appleton-Century-Crofts, New York, 329–378.

Partridge, J. (1973) *Fungus Disease Defense Responses in Legumes*, Ph.D. Thesis, University of California (Riverside), U.S.A.

Pelter, A. and Amenechi, P. I. (1969) Isoflavonoid and pterocarpinoid extractives of *Lonchocarpus laxiflorus*. *J. Chem. Soc.*, C, 887–896.

Perrin, D. D. and Perrin, D. R. (1962) The N.M.R. spectrum of pisatin. *J. Amer. Chem. Soc.*, **84**, 1922–1925.

Perrin, D. R. (1964) The structure of phaseolin. *Tetrahedron Lett.*, 29–35.

Perrin, D. R. and Bottomley, W. (1962) Studies on phytoalexins. V. The structure of pisatin from *Pisum sativum* L. *J. Amer. Chem. Soc.*, **84**, 1919–1922.

Perrin, D. R., Whittle, C. P. and Batterham, T. J. (1972) The structure of phaseollidin. *Tetrahedron Lett.*, 1673–1676.

Perrin, D. R., Biggs, D. R. and Cruickshank, I. A. M. (1974) Phaseollidin, a phytoalexin from *Phaseolus vulgaris*: isolation, physicochemical properties and antifungal activity. *Aust. J. Chem.*, **27**, 1607–1611.

Porter, A. E. A., Smallman, R. V. and Mansfield, J. W. (1979) Analysis of furanoacetylenic phytoalexins from the broad bean plant by high-performance liquid chromatography. *J. Chromatog.*, **172**, 498–504.

Preston, N. W. (1975) 2'-*O*-Methylphaseollidinisoflavan from infected tissue of *Vigna unguiculata*. *Phytochemistry*, **14**, 1131–1132.

Preston, N. W. (1977) Induced pterocarpans of *Psophocarpus tetragonolobus*. *Phytochemistry*, **16**, 2044–2045.

Preston, N. W., Chamberlain, K. and Skipp, R. A. (1975) A 2-arylbenzofuran phytoalexin from cowpea (*Vigna unguiculata*). *Phytochemistry*, **14**, 1843–1844.

Pueppke, S. G. and VanEtten, H. D. (1975) Identification of three new pterocarpans from *Pisum sativum* infected with *Fusarium solani* f. sp. *pisi*. *J. Chem. Soc. Perkin Trans.*, I, 946–948.

Pueppke, S. G. and VanEtten, H. D. (1976) Accumulation of pisatin and three additional antifungal pterocarpans in *Fusarium solani*-infected tissues of *Pisum sativum*. *Physiol. Plant Pathol.*, **8**, 51–61.

Rathmell, W. G. and Bendall, D. S. (1971) Phenolic compounds in relation to phytoalexin biosynthesis in hypocotyls of *Phaseolus vulgaris*. *Physiol. Plant Pathol.*, **1**, 351–362.

Rich, J. R., Keen, N. T. and Thomason, I. J. (1977) Association of coumestans with the hypersensitivity of Lima bean roots to *Pratylenchus scribneri*. *Physiol. Plant Pathol.*, **10**, 105–116.

Robeson, D. J. (1978a) *A Comparative Study of Phytoalexin Induction in the Tribe Vicieae*, Ph.D. Thesis, University of Reading, England.

Robeson, D. J. (1978b) Furanoacetylene and isoflavonoid phytoalexins in *Lens culinaris*. *Phytochemistry*, **17**, 807–808.

Robeson, D. J. and Harborne, J. B. (1980) A chemical dichotomy in phytoalexin induction within the tribe Vicieae of the Leguminosae. *Phytochemistry*, **19**, 2359–2365.

Robeson, D. J. and Ingham, J. L. (1979) New pterocarpan phytoalexins from *Lathyrus nissolia*. *Phytochemistry*, **18**, 1715–1717.

Robeson, D. J., Ingham, J. L. and Harborne, J. B. (1980) Identification of two chromone phytoalexins in the sweet pea, *Lathyrus odoratus*. *Phytochemistry*, **19**, 2171–2173.

Russell, G. B., Sutherland, O. R. W., Hutchins, R. F. N. and Christmas, P. E. (1978) Vestitol: a phytoalexin with insect feeding-deterrent activity. *J. Chem. Ecol.*, **4**, 571–579.

Shiraishi, T., Oku, H., Ouchi, S. and Tsuji, Y. (1977) Local accumulation of pisatin in tissues of pea seedlings infected by powdery mildew fungi. *Phytopath. Z.*, **88**, 131–135.

Sims, J. J., Keen, N. T. and Honwad, V. K. (1972) Hydroxyphaseollin, an induced antifungal compound from soybeans. *Phytochemistry*, **11**, 827–828.

Smith, D. A., VanEtten, H. D. and Bateman, D. F. (1973a) Kievitone: the principal antifungal component of "Substance II" isolated from *Rhizoctonia*-infected bean tissues. *Physiol. Plant Pathol.*, **3**, 179–186.

Smith, D. A., VanEtten, H. D., Serum, J. W., Jones, T. M., Bateman, D. F., Williams, T. H. and Coffen, D. L. (1973b) Confirmation of the structure of kievitone, an antifungal isoflavanone isolated from *Rhizoctonia*-infected bean tissues. *Physiol. Plant Pathol.*, **3**, 293–297.

Smith, D. G., McInnes, A. G., Higgins, V. J. and Millar, R. L. (1971) Nature of the phytoalexin produced by alfalfa in response to fungal infection. *Physiol. Plant Pathol.*, **1**, 41–44.

Smith, I. (1958) "Phenolic acids", in *Chromatographic Techniques*, ed. I. Smith, Heinemann Ltd., London, 189–205.

Stholasuta, P., Bailey, J. A., Severin, V. and Deverall, B. J. (1971) Effect of bacterial inoculation of bean and pea leaves on the accumulation of phaseollin and pisatin. *Physiol. Plant Pathol.*, **1**, 177–183.

Stoessl, A. (1972) Inermin associated with pisatin in peas inoculated with the fungus *Monilinia fructicola*. *Can. J. Biochem.*, **50**, 107–108.

Sutherland, O. R. W., Russell, G. B., Biggs, D. R. and Lane, G. A. (1980) Insect feeding deterrent activity of phytoalexin isoflavonoids. *Biochem. Syst. Ecol.*, **8**, 73–75.

Van den Heuvel, J. and Vollaard, P. J. (1976) Metabolism of phaseollin by different races of *Colletotrichum lindemuthianum*. *Neth. J. Plant Pathol.*, **82**, 103–108.

VanEtten, H. D. (1973) Identification of a second antifungal isoflavan from diseased *Phaseolus vulgaris* tissue. *Phytochemistry*, **12**, 1791–1792.

VanEtten, H. D. and Bateman, D. F. (1970) Isolation of phaseollin from *Rhizoctonia*-infected bean tissue. *Phytopathology*, **60**, 385–386.

VanEtten, H. D. and Smith, D. A. (1975) Accumulation of antifungal isoflavonoids and 1a-hydroxyphaseollone, a phaseollin metabolite in bean tissue infected with *Fusarium solani* f. sp. *phaseoli*. *Physiol. Plant Pathol.*, **5**, 225–237.

VanEtten, H. D., Pueppke, S. G. and Kelsey, T. C. (1975) 3,6a-Dihydroxy-8,9-methylenedioxy-pterocarpan as a metabolite of pisatin produced by *Fusarium solani* f. sp. *pisi*. *Phytochemistry*, **14**, 1103–1105.

Verbit, L. and Clark-Lewis, J. W. (1968) Optically active aromatic chromophores—VIII. Studies in the isoflavonoid and rotenoid series. *Tetrahedron*, **24**, 5519–5527.

Weinstein, L. I., Hahn, M. G. and Albersheim, P. (1981) Isolation and biological activity of glycinol, a pterocarpan phytoalexin synthesized by soybeans. *Plant Physiol.*, in press.

Wong, E. (1975) "The isoflavonoids" in *The Flavonoids*, eds. J. B. Harborne, T. J. Mabry and H. Mabry, Chapman and Hall, London 743–800.

Woodward, M. D. (1979a) Phaseoluteone and other 5-hydroxyisoflavonoids from *Phaseolus vulgaris*. *Phytochemistry*, **18**, 363–365.

Woodward, M. D. (1979b) New isoflavonoids related to kievitone from *Phaseolus vulgaris*. *Phytochemistry*, **18**, 2007–2010.

Woodward, M. D. (1980a) Phaseollin formation and metabolism in *Phaseolus vulgaris*. *Phytochemistry*, **19**, 921–927.

Woodward, M. D. (1980b) Detection and estimation of phaseollin and other 5-deoxyisoflavonoids in inoculation droplets by selected ion monitoring. *Physiol. Plant Pathol.*, **17**, 17–31.

3 Phytoalexins from the Solanaceae

JOSEPH KUĆ

Introduction

The Solanaceae contains a myriad of chemical compounds, many with exceedingly interesting chemical structures and many which inhibit the growth of microorganisms. We have a high level of certainty about the structure of some of the compounds in the Solanaceae. The same level of certainty is lacking when considering the role of compounds in disease resistance. The evidence for a relationship between the accumulation of compounds after infection and their role in determining disease resistance is circumstantial. Nevertheless, because it is circumstantial does not necessarily mean it does not have some validity. I have chosen to use the term phytoalexin to include not only low molecular weight antimicrobial compounds synthesized from remote precursors (see chapter 1), but also those that may be released from closely related products. By using such a broad definition, I hope to provide a more overall assessment of the role of individual compounds in resistance of solanaceous species to disease. Along with their chemical structures, I will consider the compounds with respect to their biosynthesis, metabolism, accumulation, and distribution within the Solanaceae, but this will be done from an integrative perspective, leaving detailed considerations of individual aspects of the phytoalexins to my colleagues in other chapters.

The Leguminosae, Solanaceae and Graminae include the major food crops throughout the world. Most phytoalexin research has been conducted with plants in the Leguminosae and Solanaceae, and relatively few reports on the accumulation of phytoalexins in plants of the Gramineae (e.g. momilactones of rice) are available. This has been due in large part to the ease of extraction and chemical characterization of phytoalexins from the Leguminosae and Solanaceae. However, it also raises questions as to whether phytoalexins are universally important in disease resistance or whether new phytoalexins, not amenable to the extraction procedures employed previously, remain to be discovered.

Solanum tuberosum, the potato, ranks third in world production behind wheat and rice. The *per capita* consumption in the United States is estimated at 55 kg per year and the figures for countries in Northern and Eastern Europe are even higher. In addition, the potato makes an important contribution to diet in areas of Central and South America, and it provides a source of high quality protein in addition to carbohydrate. The family Solanaceae contains other important food crops, e.g. tomato, pepper and eggplant, but also includes plants that are highly poisonous, e.g. tobacco, bitter nightshade and henbane. Even potato and tomato plants, however, may be classified as poisonous since the foliage of both, and parts other than the tuber of the potato, can be toxic to some mammals including man.

Three main classes of phytoalexins have been reported in the Solanaceae: the phenylpropanoid phytoalexins derived from the shikimic acid pathway, the terpenoid phytoalexins derived from the acetate-mevalonate pathway, and more recently, the acetylenes and polyacetylenes derived from the acetate-malonate pathway. The isoflavonoid and related pterocarpan phytoalexins common in the Leguminosae have not been reported in the Solanaceae. Similarly the norsesqui- and sesquiterpenoid phytoalexins of the Solanaceae (Kuć *et al.*, 1979; Kuć *et al.*, 1976; Kuć and Lisker, 1978; Stoessl *et al.*, 1976, 1977*b*). Convolvulaceae and Malvaceae (Kuć and Lisker, 1978) are not apparent in the Leguminosae. Phenylpropanoid phenolics appear ubiquitous in plants though surprisingly a case for them has not been made as phytoalexins in the Leguminosae. The polyacetylene phytoalexins, however, have been reported in the Compositae (Thomas and Allen, 1970), the Leguminosae (Letcher *et al.*, 1976), Umbelliferae (Harding and Heale, 1980) and Solanaceae (de Wit and Kodde, 1981). Of course, it is difficult to determine the extent of family-related specificity for phytoalexins, since it is not possible to ascertain from reading the literature whether a phytoalexin is not reported in a plant family because it does not exist in the family or because it has not been looked for. For many years the only phytoalexins listed for the Umbelliferae for example were 6-methoxy mellein and the chlorogenic acids (Kuć, 1976*a*).

Assuming that sesquiterpenoid phytoalexins do not occur in the Leguminosae and pterocarpan phytoalexins do not occur in the Solanaceae, the intriguing question becomes, why is this the case? Clearly, the Leguminosae have a functional acetate-mevalonate pathway and the Solanaceae a functional shiki-mate acetate-malonate pathway. The specificities undoubtedly occur in fringe pathways and are involved in unique enzymes and/or the organization of such enzymes.

Corollary questions which arise are: does a pathogen of potatoes degrade, or is it for some other reason insensitive to, phytoalexins in potato but not carrots or beans, and does a pathogen unique for potato elicit phytoalexin accumulation in all other plants it does not parasitize? To my knowledge these questions remain unanswered but they are addressed by my colleagues in this book.

Potato (*Solanum tuberosum*)

Phytoalexin research in the Solanaceae started with the classical studies of the *Phytophthora infestans*–potato interaction by K. O. Müller and his colleagues (Müller and Börger, 1940; see chapter 1). Their experimentation offered evidence that low molecular weight fungitoxic chemicals were produced by potato tubers after infection with *Phytophthora infestans* and that the accumulation of these chemicals was related to the resistance of the tuber to the fungus. They further demonstrated that the inoculation of tubers with strains of *P. infestans* to which the tubers were resistant protected tubers from disease caused by subsequent inoculation with strains to which the potatoes were susceptible. An excellent review of the fascinating papers by Müller and his colleagues is found in the thesis by Henfling (1979).

Phenolics

Reference to chemically defined phytoalexins in potato occurred many years later (Johnson and Schaal, 1952, 1955; Kuć, 1957). These reports were concerned with two phenylpropanoid-related phytoalexins, caffeic acid (**1a**) and chlorogenic acid (**2**).

Johnson and Schaal suggested that the endogenous phenols in potato peel and those produced around wounds were important in the resistance of tubers to scab, and they reported a positive correlation between the content of chlorogenic acid and resistance to potato scab. They further reported that

1a Caffeic acid (R = OH)
1c Ferulic acid (R = OCH₃)
1d *p*-Coumaric acid (R = H)

2 Chlorogenic acid

3 Scopolin (R = CH₃; G = glucose)

1b Quinic acid

6 α-Solamarine (R = β-solatriosyl)
7 β-Solamarine (R = β-chacotriosyl)

17 Tomatine (R = lycotetraosyl)

16 Glutinosone

4 α-Solanine (R = B-solatriosyl)
5 α-Chaconine (R = β-chacotriosyl)

oxidation products of chlorogenic and caffeic acids were produced after infection and that these products were even more fungitoxic than the parent
phenols. Kuć (1957) verified the presence of high levels of chlorogenic and caffeic
acids in potato peel but also reported the accumulation of the acids and their
oxidation products in slices inoculated with several non-pathogens of potato. It
was apparent, however, that phenols and their oxidation products were not the
only fungitoxic compounds present in the peel or produced after infection or
injury.

Steroid glycoalkaloids

At approximately the same time that the papers concerning chlorogenic and
caffeic acid appeared, reports also appeared on the possible role of steroid glycoalkaloids, found in peel or produced around sites of injury in tubers, in disease
resistance (McKee, 1955, 1959). α-Solanine (**4**) and α-chaconine (**5**) are the
main steroid glycoalkaloids in potato. They are found in healthy tubers as well
as foliage (Shih, 1972) and their concentration increases under stress conditions
in tubers (Locci and Kuć, 1967), α-Solamarine (**6**) and β-solamarine (**7**) have
been reported as additional steroid glycoalkaloids in the cultivar Kennebec.
Tubers of the wild potato species, *S. acaule*, contain appreciable quantities of
demissine and tomatine (**17**) (Shih and Kuć, 1974). The steroid glycoalkaloids
are largely localized in the peel below the primary periderm of healthy tubers
(outer 2 mm of tissue), but their concentration increases markedly around sites
of injury to levels which may reach or surpass that in the peel. They are the
major extractable fungitoxic compounds in the peel and in tuber tissue around
sites of mechanical injury (Allen and Kuć, 1968; McKee, 1955). Analyses of four
cultivars indicated an average of 0.55 and 0.03 mg/g fresh weight in the peel and
peeled tubers respectively. Flowers, fruit, sprouts and foliage contain a considerably higher content of steroid glycoalkaloids than do whole tubers (Shih, 1972).
The accumulation of the glycoalkaloids in tubers is suppressed by inoculation
with various pathogens and nonpathogens (Ishizaka and Tomiyama, 1972; Shih
and Kuć, 1973; Shih *et al.*, 1973). After inoculation with nonpathogens,
incompatible races of *P. infestans*, or treatment with elicitor from *P. infestans*,
the suppression is concomitant with an increase in the accumulation of simpler
norsesqui- and sesquiterpenoid phytoalexins, e.g., rishitin, lubimin, solavetivone
and phytuberin, which will be discussed later in the chapter. With all fungi
tested, α-chaconine was more fungitoxic than α-solanine (Allen and Kuć, 1968;
McKee, 1959). An interesting aspect of the fungitoxicity of the steroid glycoalkaloids is that the protonated forms (approximately pH 5.5 or lower) are less
active than the free bases (McKee, 1959; Allen and Kuć, 1968). The difference is
greater than 100-fold between pH 5.5 and 7.4, and emphasizes the importance
of clearly defining the conditions for measuring the activity of phytoalexins and
the difficulty of relating assays *in vitro* to activity *in vivo*.

The contribution of steroid glycoalkaloids in disease resistance of the potato is uncertain. Locci and Kuć (1967) and Allen and Kuć (1968) suggested that the compounds are part of a general mechanism for the disease resistance of potato, but that the mechanism for resistance is multicomponent. The evidence is strong that the steroid glycoalkaloids and norsesqui- and sesquiterpenoids are synthesized via the acetate-mevalonate pathways (Shih and Kuć, 1973; Stoessl et al., 1976; Stoessl et al., 1977b; Stoessl et al., 1978). A number of pertinent questions arise. What is the mechanism for the enhancement of norsesqui- and sesquiterpenoid accumulation and the suppression of steroid glycoalkaloid accumulation following infection? It is difficult to believe that the accumulation of the two classes of compounds is under separate and completely independent control. Does the phenomenon of suppression enhance disease resistance of the tuber? Does a similar phenomenon occur in leaves? Is there a significance in disease resistance to the production of steroid glycoalkaloids by P. infestans (Mass et al., 1977)?

Suberin and lignin

If decisions as to the resistance or susceptibility of potato cultivars to late blight were based on the resistance of whole tubers or slices aged for more than 48 hours, it would appear that all potato cultivars are immune (Shih et al., 1973). This may be due in part to the high concentration of steroid glycoalkaloids in the peel and top 1 mm of aged sliced tubers. Unlike the norsesqui- and sesquiterpenoids which accumulate to maximum levels 72 to 96 hours after infection and then decline to levels found in uninfected tissue, the concentration of steroid glycoalkaloids remains constant after reaching a maximum (Kuć, 1975). Two other processes associated with wound healing, suberization and lignification, may however also be important in disease resistance.

Suberin is a polymer containing phenolics and long-chain decarboxylic, hydroxy and perhaps epoxy fatty acids (Kolattukudy, 1975). It forms a highly lipophilic coating within and on the surface of tissues and may function as a physical barrier to microbial penetration, as well as excluding water from or to the infection site and excluding microbial toxins from the host. The fatty acids released by action of acyl hydrolases may also function as phytoalexins and the peroxides, hydroperoxides, and epoxides generated by the action of lipoxygenases after injury or infection may add further to the toxicity of the environment to infectious agents (Galliard, 1975).

The covalent attachment of antimicrobial materials to suberin polymers would provide localized protection for the plants when needed—i.e. in the presence of a pathogen's degradative enzymes and those released by the host during infection or injury, but the insoluble form of the polymers would prevent general toxicity to the plant.

Lignification, either directly on to cell wall components or on to suberin would also form a potential barrier to infection (Kolattukudy, 1975; Henderson and Friend, 1979). The phenolic oxidation products and free radicals generated during the lignification process would add still further to the toxicity of the environment to infectious agents. Is it surprising, therefore, that farmers centuries ago recognized the importance of allowing potato seed pieces to "wound heal" before planting in the field? Clarke (1973) presented evidence that scopolin (3), a phenolic derived from intermediates of lignin precursors, accumulates more in susceptible than resistant interactions of potato and *P. infestans*. He suggested that channelling of lignin precursors to scopolin may be critical in preventing lignin formation and thereby enhance development of the pathogen. The accumulation of scopolin and its aglycone, scopoletin, was also suggested to inhibit two important enzymes for lignin formation, glucose 6-phosphate dehydrogenase and peroxidase (Clarke, 1973). The author considers the relation of scopolin to lignification and disease resistance to be unclear. Evidence is accumulating, however, which indicates the importance of lignification not only in the *P. infestans*–potato interaction (Friend, 1976; Henderson and Friend, 1979) but also in disease resistance and immunization of non-solanaceous plants (Hammerschmidt and Kuć, 1980; Ride, 1980; Vance *et al.*, 1980).

Norsesqui and sesquiterpenoids

The group of compounds in potato that have received the greatest recent attention are the norsesqui- and sesquiterpenoid stress metabolites (SSM). This is due in large part to the fact that they accumulate after infection and often accumulate more rapidly and in greater magnitude in resistant interactions (Kuć *et al.*, 1979; Kuć *et al.*, 1976; Kuć and Lisker, 1978). Much of the research with SSM has been with R-gene resistance in the potato–*P. infestans* interaction. The importance of SSM in disease resistance should only be considered, however, in the context of the other disease resistance mechanisms associated with wound healing. Thus, the rapid accumulation of SSM may temporarily restrict the development of an infectious agent. However, unless the infectious agent is killed its ultimate restriction may also depend upon the normal wound-healing process because the concentrations of SSM drop markedly in infected tissues ca. 96 hours after infection, though *P. infestans* remains contained (Kuć, 1975). It would appear, therefore, that SSM in potato are an important adjunct to disease resistance; an adjunct that may be critical or at times a limiting factor until the tissue can return to comparative normality.

On a quantitative basis the main SSMs in potato are rishitin (8) (Tomiyama *et al.*, 1968; Katsui *et al.*, 1968), lubimin (9) (Katsui *et al.*, 1974; Stoessl *et al.*, 1974), solavetivone (10) (Coxon *et al.*, 1974), phytuberin (11) (Varns, 1970; Coxon *et al.*, 1977), phytuberol (12) Currier, 1975; Price *et al.*, 1976; Coxon *et al.*, 1977), and anhydro-β-rotunol (13) (Coxon *et al.*, 1974). Investigations of the role of SSM in disease resistance and the mechanism for their elicitation and suppression were materially aided by the development of a rapid semi-micro method for their separation and quantitation (Henfling and Kuć, 1979). Rishitin, lubimin and solavetivone generally comprise 85% or more of the total

12 Phytuberol

13 Anhydro-β-rotunol

11 Phytuberin

10 Solavetivone

8 Rishitin

9 Lubimin

SSM accumulating. Though rishitin is often the major SSM which accumulates after infection, lubimin or solavetivone can, in some interactions and under some experimental conditions, accumulate to higher concentrations. Using the same potato cultivar and race of *P. infestans* in different experiments, the author has observed either rishitin, lubimin or solavetivone as the major SSM. It appears that slight changes in the physiological state of the tubers and environment can profoundly influence which of the terpenoids predominates. This is illustrated by the enhanced accumulation of phytuberin and phytuberol in potato tubers treated with ethylene or ethrel prior to inoculation with incompatible races of *P. infestans* or treatment with elicitor prepared from the fungus (Henfling *et al.*, 1978). The tissue tested, temperature (Lisker and Kuć, 1978; Currier and Kuć, 1975) and microbial or host degradation (Ward and Stoessl, 1977; Ward *et al.*, 1977; Stoessl *et al.*, 1977b; Stoessl and Stothers, 1980; Ishiguri *et al.*, 1978) all qualitatively and quantitatively influence SSM accumulation. In a report by Price *et al.* (1976) examples of fungus–potato interactions are cited in which phytuberin and phytuberol are the main SSMs. Rishitin and phytuberol also accumulate in potato tissue infected with bacteria (Lyon *et al.*, 1975).

The accumulation of 14 to 16 additional norsesqui- and sesquiterpenoids has also been reported in infected potato tubers. They include rishitinol (Katsui *et al.*, 1971), 3-hydroxylubimin (Katsui *et al.*, 1974), isolubimin (Stoessl *et al.*, 1978), 10-epilubimin, 15-dihydrolubimin, 15-dihydro-10-epilubimin (Stoessl *et al.*, 1977b), cyclodehydroisolubimin (Coxon and Price, 1979), and 2-epilubimin, 15-dihydro-2-epilubimin (Stoessl and Stothers, 1980). The structures of these compounds are presented in chapter 5. These compounds generally accumulate at low concentrations and their accumulation may be due either to synthesis by the host or degradation by host or pathogen of terpenoids synthesized by the host. In some cases these constituents, or a precursor, may be synthesized by a pathogen. The fungitoxic properties of the compounds and hence their contribution to resistance have not been fully investigated. It is possible that the degradation of phytoalexins, however, may be a mechanism for their detoxification and hence important in determining disease reaction as well as minimizing phytotoxicity. The isolation and characterization of these compounds has been vital in studies of the mechanisms for biosynthesis of SSM and their degradation by host or pathogen (see chapters 5 and 6). Though the metabolism of phytoalexins and the role of phytoalexins in resistance will be considered in detail in chapters 6 and 8, a short consideration of the subject appears pertinent at this time.

The elicitation of SSM in potato is under tighter metabolic control than is the elicitation of pterocarpan phytoalexins in legumes. Mercuric chloride and cupric chloride elicit little or no accumulation of SSM (Varns, 1970; Varns *et al.*, 1971a; unpublished data, and personal communication from Professor N. F. Haard). Autoclaved sonicates and cell wall preparations of oomycetes, in which glucan-cellulose is the main cell wall component, elicited the accumulation of

SSM and browning of slices, whereas autoclaved sonicates and cell wall preparations of other fungi and heat-treated bacteria failed to elicit browning or accumulation of more than a trace of the compounds (Lisker and Kuć, 1977; and author's unpublished data). Living fungi with glucan-cellulose or glucan-chitin as main cell wall components elicited SSM accumulation and browning, whereas other fungi and bacteria did not. Tissue of tomato, potato, cucumber and French bean, and 15 carbohydrates, 9 lipids, 2 saponins, poly-L-lysine and three purified glucans prepared from *P. infestans, P. cinnamomi* and *P. megasperma* had little or no activity though some caused considerable necrosis. Though the accumulation of phytoalexins has not been established as a mechanism unique for disease resistance in plants and the term "stress metabolites" more clearly depicts the compounds (Kuć, 1972b), it is evident that more than necrosis and browning is required for appreciable accumulation of SSM in potato.

Cell walls of *P. infestans* contain an elicitor of SSM accumulation in potato (Henfling *et al.*, 1980). This was demonstrated by differential centrifugation of sonicates of *P. infestans* and the use of zoospores, cystospores and sporangial ghosts of the fungus. Evidence had accumulated that this elicitor has lipoidal properties though it is often associated with polysaccharide and protein (Currier, 1975; Bostock and Kuć, 1980). Recently, Bostock *et al.* (1981b) demonstrated that eicosapentaenoic and arachidonic acids extracted from *P. infestans* elicit the accumulation of high levels of SSM in potato tubers. The acids were present in all active preparations including cell wall preparations from the fungus. Their work indicates that the acids are the main elicitors in the fungus, and their finding is a marked departure from the generally accepted reports of fungal cell wall carbohydrates as elicitors. It is apparent that the two compounds *per se* cannot serve as "specific" elicitors and their presence in the fungus cannot in itself explain specificity in R-gene resistance of potato to *P. infestans*. This is consistent with the observation that crude elicitor preparations from six races of *P. infestans* elicited the accumulation of high levels of terpenoids in 15 potato cultivars tested, including those reported to lack R-genes for resistance to *P. infestans* (Sato *et al.*, 1968a;Varns, 1970; Varns and Kuć, 1971; Varns *et al.*, 1971a,b). Less than 3% of the fatty acids exist free, and the liberation of the acids and/or their conversion to other metabolites may be critical in the *P. infestans*–potato interaction.

The key to R-gene specificity in the interaction may be the ability of compatible races of the fungus to suppress hypersensitive cell death, necrosis and SSM accumulation (Tomiyama, 1966; Varns and Kuć, 1971, 1972; Doke, 1975). Inoculation of tubers with compatible races of *P. infestans* markedly suppressed SSM accumulation resulting from subsequent inoculation of tissue with incompatible races or treatment with elicitor preparations (Varns and Kuć, 1971, 1972). Contact with elicitor preparations for 15 to 30 minutes was sufficient to commit tissue to a hypersensitive response (Doke, Lisker and Kuć, unpublished data) and inoculation with the compatible race must precede inoculation with the incompatible race for suppression (Varns and Kuć, 1971, 1972).

Factors which inhibit the hypersensitive reaction of potato tuber tissue (Kennebec, R_1) to *Phytophthora infestans* were isolated from mycelia and zoospores of race 1234 (compatible) and race 4 (incompatible) of the fungus (Garas *et al.*, 1979; Doke *et al.*, 1979). They were partially characterized as glucans containing β-1,3 and β-1,6 linkages and 17 to 23 glucose units. The glucans from both mycelia and zoospores included a non-anionic glucan and an anionic glucan; one or two residues of the latter were esterified with a phosphoryl monoester.

Death of host cells, browning, the leakage of electrolytes, and the accumulation of rishitin and lubimin (hypersensitive reaction) in tuber slices inoculated with race 4 or treated with an elicitor from the fungus were suppressed by pretreatment of slices with the glucans. The glucans from the compatible race were more active in suppressing the hypersensitive reaction than those from the incompatible race, and the anionic glucan was more active than the non-anionic glucan.

Crude elicitors from races 4 and 1234 lost terpenoid-eliciting activity when mixed with a microsomal fraction prepared from potato tuber tissue. The glucans from the compatible race, but not the incompatible race, markedly reduced the loss resulting from the reaction between the crude elicitor and the microsomal fraction.

The data suggest that the compatible interaction between potato tissue and *P. infestans* may be caused by a suppression of the hypersensitive response of the host tissue by water-soluble glucans from the fungus. The concept of resistance suppression as a mechanism for susceptibility was strengthened by the report that glucans, with the chemical properties and suppressor activity described for the glucans isolated from mycelia and zoospores, were produced extracellularly by germinating cystospores (Doke *et al.*, 1980). Further strong evidence for the importance of suppressors was recently provided by Doke and Tomiyama (1980*a*). They reported the suppression of the hypersensitive response of potato tuber protoplasts to hyphal wall components isolated from the fungus by water-soluble glucans isolated from *P. infestans*. Marked suppression to crude elicitor preparations occurred in 19 interactions with glucans from races of the fungus which are compatible, and considerably less suppression occurred in 33 interactions with glucans from races which are incompatible. The reaction of Pentland Ace (R_3) to compatible races 1234 and 3 were the only exceptions. In total, 7 races of the fungus and 9 cultivars were tested, and the glucans from the 7 races of the fungus markedly suppressed the hypersensitive reaction of Irish Cobbler (r) which is susceptible to all known races of the fungus.

The possibility that protoplasmic membrane sites may serve as recognition sites for elicitor (Doke *et al.*, 1979) was supported by the recent paper by Doke and Tomiyama (1980*b*) which reported that crude elicitor from fungal walls of *P. infestans* was toxic to protoplasts of potato tuber tissues. There was no significant difference in the physiological activities of the crude elicitor preparations from 7 races of the fungus though differences were apparent with different cultivars regardless of their R-genes for resistance.

The implication of these data is that SSM probably have a role in containing *P. infestans* in incompatible R-gene reactions of the host and pathogen. Suppression of SSM accumulation, browning, host cell death, and leakage of electrolytes from host cells is consistent with susceptibility. Compounds from *P. infestans* can rapidly elicit all the symptoms of a hypersensitive reaction, but, limited by current extraction procedures and methods for bioassay, they do not have R-gene specificity. It is likely that arachidonic and eicosapentaenoic acids, either liberated or bound to other fungal components, are the active elicitors of the hypersensitive response. The free acids are active and all active fractions contain the acids. The story, however, is clouded by several considerations which do not disprove its underlying theme but do complicate the plot and leave room for subsequent chapters or even books.

The suppressor appears to be a low molecular weight polysaccharide, whereas the active non-specific elicitors appear to be highly unsaturated fatty acids. Evidence from our laboratory indicates that the fatty acid elicitors are bound to carbohydrate and this could explain the apparent anomaly. The fatty acids might cause host cell damage and SSM accumulation and the carbohydrate might enhance elicitor activity and contribute to specificity. The binding of elicitor to carbohydrate is supported by recent work of Garas and Kuć (1981) which demonstrated that elicitor activity from crude preparations was quantitatively precipitated from solution by potato lectins. Though SSM were reported in leaf tissue (Metlitskii *et al.*, 1970), and Tomiyama and colleagues have used leaf petiole sections in some of their studies, this author is unaware of critical studies relating the accumulation of the phytoalexins in whole leaves to resistance. Critical studies with leaves are essential since late blight is a disease of leaves as well as tubers.

A recent report by Henfling *et al.* (1981) demonstrated that treatment of tubers held in cold storage with abscisic acid made them susceptible to *P. infestans* regardless of their R-gene designation, and this induced susceptibility was associated with a marked suppression of SSM accumulation. Young tubers harvested early in the growing season, however, showed the typical compatible and incompatible reactions to appropriate races of *P. infestans* even though the incompatible reactions did not accumulate SSM earlier or in higher concentrations than the compatible interactions (Henfling, 1979; Bostock *et al.*, 1981*a*). Young potato sprouts rapidly accumulate SSM after inoculation by compatible or incompatible races of the fungus (Lisker and Kuć, 1978). These studies serve to emphasize the dangers of considering phytoalexin accumulation without taking into account the concentrations occurring at micro-sites within infected tissues and also without an awareness of other mechanisms or conditions which may influence or contribute to disease resistance. The concept of limiting factors may be critical. Thus, a tuber which suberizes very rapidly may not have to accumulate high concentrations of SSM as rapidly and maintain them as long as one that suberizes more slowly. A young tuber harvested in the field may have suberization or lignification as the limiting factor for resistance rather than SSM accumulation. Certainly the last chapter of the potato–*P. infestans* story has not been written. It might make interesting reading to refer to earlier reviews prepared by this author (Kuć, 1972a,b).

Pepper (*Capsicum frutescens*)

Phenolics accumulate in peppers after infection by fungi, bacteria or chilling injury but the presence of steroid glycoalkaloids in the outer tissues of pepper fruit or their accumulation as a result of wounding or infection has not been reported.

The phytoalexin in pepper that has generated most interest is the bicyclic sesquiterpene capsidiol (14) (Birnbaum *et al.*, 1974). It accounts for about 33% of the total ether extractives obtained from diffusates of pepper fruit injected with *Monilinia fructicola* (Stoessl *et al.*, 1972). Other fungi and bacteria also elicit accumulation of capsidiol (Stoessl *et al.*, 1972; Ward *et al.*, 1973).

A series of extensive studies of the rate and magnitude of capsidiol accumulation coupled with ultrastructural studies support a role for the compound in the restriction of fungal growth and development in infected pepper fruit (Jones *et al.*, 1975). Twenty-four hours after inoculation with a nonpathogen, capsidiol accumulated in pepper fruits to levels that prevented growth of fungi *in vitro* (Jones *et al.*, 1975). It also accumulated around sites of localized infections in leaves (Ward, 1976).

Capsidiol is fungitoxic, but can be degraded by fungi to the generally less toxic derivative capsenone (Stoessl *et al.*, 1973). This and the capacity to tolerate high concentrations of capsidiol for other reasons may allow some pathogenic fungi to develop in fruit. Capsidiol is also rapidly metabolized by healthy pepper fruits (Ward *et al.*, 1977; Stoessl *et al.*, 1977a) and the amount which accumulates after infection is, therefore, a function of the rate of synthesis and of degradation by both host and pathogen. The compound is synthesized via the acetate-mevalonate pathway (Baker and Brooks, 1976; Stoessl *et al.*, 1977b). Data concerning mechanisms leading to accumulation of capsidiol are not available. Clearly it can accumulate in pepper fruit and foliage to levels which restrict the growth of many fungi *in vitro*. It is also clear that its accumulation or the sensitivity of fungi to inhibition by capsidiol cannot explain the resistance or susceptibility of pepper to all fungi, bacteria and viruses. It also appears likely to this author that mechanisms other than the accumulation of SSM contribute to the resistance of pepper or may be the most important mechanisms for disease resistance.

This point is strongly supported by the data of Molot *et al.* (1981), who report that *Phytophthora capsici* invaded the decapitated stems of susceptible cultivars of pepper (Yolo Wonder and Clairon) at a constant rate for 9 days following inoculation, while it became progressively inhibited in the respective isogenic resistant lines Phyo 636 and Fidelio. Capsidiol concentration at the advancing edge of the mycelium, although somewhat higher in the resistant lines, never exceeded the ED_{80} for activity *in vitro*; its kinetics were similar in the four cultivars with maximum accumulation at the fourth day and a marked decrease afterwards. However, only in the resistant lines did infection progressively induce a durable state of resistance inside the tissues as demonstrated by a subsequent inoculation. Upon infection, induced tissues accumulated less capsidiol than non-preinfected ones. The inhibition of fungal growth and level of induced resistance were closely correlated, but bore no overall relation to capsidiol concentration at the front of the mycelium.

Tobacco (*Nicotiana tabacum*)

The literature contains many reports of the accumulation of phenolic compounds and their oxidation products in tobacco as a result of infection by fungi, bacteria or viruses, and also (in the more recent literature) their accumulation in response to inoculation with tobacco mosaic virus (TMV) or *Peronospora hyoscyami* f. sp. *tabacina*. The hypersensitive local lesion reaction of *N. tabacum* cv. Xanthi to TMV leads to the production and accumulation of numerous phenolics including 3-caffeoyl quinic acid (chlorogenic acid) (**2**), 4- and 5-caffeoyl quinic acids (see **1a,b**), 3-, 4-, and 5-feruloylquinic acids (see **1b,c**), 3-p-coumaryl quinic acid (see **1b,d**), 1-caffeoyl, feruloyl and *o*-coumaroyl esters of glucose, 1-*O*-coumaryl gentiobiose, scopolin (**3**) and rutin (Tanguy and Martin, 1972). The increase in phenolics became evident after the appearance of local lesions and when the synthesis of virus was practically completed. An increase in temperature inhibited the hypersensitive reaction and resulted in a marked decrease of the phenolics. Increases of chlorogenic acid, p-coumaryl quinic acid, scopolin and rutin also occurred in cv. Samson infected systemically with TMV. In general, however, fewer compounds and considerably less accumulation occurred in systemically infected foliage. The authors concluded that phenolic compounds do not appear to be responsible for the necrotic lesions and their accumulation is a secondary effect of viral infection. There does not appear to be a direct relationship between virus multiplication and levels of phenolic compounds. A more recent report indicates that the activities of phenylalanine ammonia lyase (PAL), cinnamic-4-hydrolase (CAH), caffeic acid-*o*-methyltransferase (OMT) and peroxidase increase in tobacco foliage infected with local lesion TMV (Legrand *et al.*, 1976). The data indicate that enzymes of the phenylpropanoid pathway are good biochemical markers of the necrotic reaction. Necrosis and increased enzyme activity occur at approximately the same time, the increased enzyme activities are proportional to the number of developing necrotic lesions, and the greater the increase in enzyme activities the larger the final size of the lesion. Almost all the cells which show a stimulation of PAL, CAH and OMT activities are destroyed by necrobiosis.

Ravise and Tanguy (1973) reported that resistance of young tobacco plantlets to *Phytophthora* spp. is associated with a marked increase of phenolic cinnamic acid derivatives, including many of the compounds already discussed. They reported further that extracts containing the phenols were toxic for the strains of *Phytophthora* tested and that the activities of pectinolytic enzymes produced by the fungi were inhibited by about 1 μg of extract per ml.

Tobacco plants infected with *P. hyoscyami* f. sp. *tabacina* (the incitant of blue mould) were severely stunted and accumulated high levels of scopoletin, the aglycone of scopolin (**3**), in the upper part of the stem (Reuveni and Cohen, 1978). The scopoletin concentration increased during the first 10 days of pathogenesis and declined thereafter. Plants infected with *P. tabacina* also contained higher amounts of *p* coumaric acid (**1d**), *o*-coumaric acid, and other unidentified phenolics, than did uninfected plants. Tobacco plants sprayed with ethrel were also stunted and accumulated phenolics, and the authors suggested that the changes in phenolics and the stunting were caused by the increase of ethylene in infected plants. The report by Edreva (1977) also emphasized the similarity of the physiological response, including the accumulation of phenolics, to infection with that to a physiological disease and injury. He concluded that the accumulation of phenolics results from damage to cell integrity regardless of the causative stimulus. The author made the interesting suggestion that quinone formation is an alarm signal to plant cells, a common means of translating information about the beginning of membrane damage caused by a variety of stimuli. The quinone-protein interaction and brown pigment formation limit the damage by the toxic quinones which are generated, and the metabolism of the affected tissue returns to normal, coincident with the containment of an infectious agent and tissue repair.

Five bicyclic sesquiterpenes are established as stress metabolites of *Nicotiana* species: capsidiol (**14**), in *N. tabacum* infected with TNV (Bailey *et al.*, 1975), *Peronospora tabacina* (Cruickshank *et al.*, 1976), or *Phytophthora parasitica* var. *nicotianae* (Helgeson *et al.*, 1978); phytuberin (**11**), in *N. tabacum* infected with *Pseudomonas lachrymans* (Hammerschmidt and Kuć, 1979); glutinosone (**16**) in *N. glutinosa* infected with TMV (Burden *et al.*, 1975); rishitin (**8**) in *N. tabacum* infected with *P. parasitica* var. *nicotianae* (Budde and Helgeson, 1980); and solavetivone (**10**) in *N. tabacum* infected with TMV (Uegaki *et al.*, 1981). Recent data suggest that *P. hyoscyami* directs the accumulation of a stereo-isomer of phytuberin in tobacco which is highly inhibitory to conidial germination of the fungus (Cohen *et al.*, 1981) whereas phytuberin from potato or tobacco infected with *P. lachrymans* was inactive.

In addition, phytuberol (**12**) was isolated from *N. tabacum* but it is unclear whether the tobacco was diseased or healthy (Takagi *et al.*, 1979). Coxon *et al.* (1977) and Takagi *et al.* (1979) reported the formation of phytuberol derivatives by the silicic acid-catalysed addition of ethanol or methanol across the 2,3 double bond. This may account in part for the many derivatives of phytuberin and phytuberol which have plagued this author and undoubtedly other investigators during purification of the compounds from solutions containing ethanol or methanol. Acetone also easily forms an addition product with the two hydroxyl groups of rishitin. This raises the possibility that some of the phytoalexins reported to date may, in fact, be artifacts produced during extraction and purification.

Finally, Guedes *et al.* (1981) recently reported the accumulation of rishitin, capsidiol, lubimin, solavetivone, phytuberin and phytuberol in tobacco foliage inoculated with the nonpathogen of tobacco, *P. lachrymans*. Accumulation of the terpenoids coincided with the appearance of necrosis, was detected in and immediately around necrotic tissue, and all of the compounds decreased markedly in concentration after necrotic lesions reached maximum development.

Tomato (*Lycopersicon esculentum*)

The tomato has not been investigated as thoroughly as the potato or pepper with respect to phytoalexin accumulation. Infection of foliage with various fungi and bacteria often leads to the accumulation of cinnamic acid derivatives, including chlorogenic acid, but the relationship between the accumulation of phenolics and disease resistance in tomato remains unclear. The relationship between the content of the major steroid glycoalkaloid, tomatine (**17**), and disease resistance is equally unclear. Tomatine may be a factor in the resistance of tomato to disease caused by *Septoria lycopersici* (Arneson and Durbin, 1968)

$$CH_3-[CH_2]_5-CH_2-CH=CH-\overset{\overset{\displaystyle H}{\overset{\displaystyle O}{|}}}{CH}-C\equiv C-C\equiv C-\overset{\overset{\displaystyle H}{\overset{\displaystyle O}{|}}}{CH}-CH=CH_2$$

18 Falcarindiol

$$CH_3-[CH_2]_5-CH_2-CH=CH-CH_2-C\equiv C-C\equiv C-\overset{\overset{\displaystyle H}{\overset{\displaystyle O}{|}}}{CH}-CH=CH_2$$

19 Falcarinol

$$CH_3-[CH_2]_5-\overset{\overset{\displaystyle H}{\overset{\displaystyle O}{|}}}{CH}-CH=CH-\overset{\overset{\displaystyle H}{\overset{\displaystyle O}{|}}}{CH}-C\equiv C-C\equiv C-H$$

20 *cis*-Tetradeca-6-ene-1,3-dyne-5,8 diol

23 9-Hydroxynerolidol

22 9-Oxonerolidol

21 Aubergenone

14 Capsidiol

25 2,3-Dihydro-germacrene

24 11-Hydroxy-9,10 dehydronerolidol

and *Pseudomonas solanacearum* (Mohanakumaran *et al.*, 1969). It may also be associated with the resistance of tomato plants either to *Verticillium* wilt (Tjamos and Smith, 1974) or *Fusarium* wilt (McCance and Drysdale, 1975; Langcake *et al.*, 1972). In the latter report, tomatine was found to increase in stems and roots after infection.

The main sesquiterpenoid isolated from tomato is rishitin (8) (Sato *et al.*, 1968*b*). More recently this was isolated from fruit, foliage and roots of tomato (de Wit and Flach, 1979; Hutson and Smith, 1980; Grzelinska and Sierakowska, 1978). Although SSM accumulation generally appeared higher in resistant than susceptible tomato plants with leaf mould, *Verticillium* or *Fusarium* wilt, the total amount detected was low and further research is necessary to determine if rishitin contributes to resistance.

The polyacetylenic phytoalexins falcarindiol (18) and probably falcarinol (19) have recently been reported to accumulate in tomato fruits and leaves inoculated with *C. fulvum*, whereas cis-tetradeca-6-ene-1,3-dyne-5,8 diol (20) accumulated only in the fruit (de Wit and Kodde, 1981).

Egg plant (*Solanum melongena*)

Five main SSM, lubimin (9), aubergenone (21), 9-oxonerolidol (22), 9-hydroxy-nerolidol (23) and 11-hydroxy-9,10 dehydronerolidol (24), have been isolated from eggplant fruits inoculated with *M. fructicola* and other fungi (Ward *et al.*, 1975; Stoessl *et al.*, 1975; Murai *et al.*, 1978). A function in resistance has not yet been established; however, it would be consistent with the concentrations which accumulate in infected tissues and with their fungitoxicity.

Jimsonweed (*Datura stramonium*)

Four main terpenoid stress metabolites, lubimin (9), 3-hydroxylubimin, capsidiol (14) and 2,3 dihydroxygermacrene (25), have been isolated from immature capsules of jimsonweed (Ward *et al.*, 1976; Birnbaum *et al.*, 1976).

Summary

The Solanaceae produce phenolic, terpenoid and acetylenic stress metabolites (SSM) which, especially those from potato and pepper, accumulate to concentrations which are toxic to a broad spectrum of fungi and bacteria. The rate of accumulation, localization and toxicity of the SSM, as related to the development of fungi, provide evidence that they have a role in disease resistance.

The evidence for a role in resistance to bacterial diseases is not as clear. Evidence is not available that SSM, phenolics or polyacetylenes have a role in the resistance of plants to viral diseases and yet resistance to diseases caused by viruses can be clearcut and is as much under genetic control as is resistance to fungal and bacterial diseases. It is surprising and disappointing that mechanisms for resistance to viral diseases are not receiving more attention. The work with the potato–*P. infestans* interaction provides evidence that SSM have a role in containing a pathogen in a highly specific interaction (R-gene) for which information concerning the genetic relationships of fungal race and cultivar is available. The degradation of SSM by host and pathogens broadens the scope of their possible implication in disease resistance.

In my view, however, it is unrealistic to assume that the SSM phytoalexins are the only ones contributing to resistance. The accumulation of acetylenic phyto-alexins in tomato infected with *C. fulvum* may be critical in restricting develop-ment of the fungus in the host's race-specific interaction. In plants, as in animals, disease resistance may depend upon the close coordination of numerous mechanisms, and susceptibility is usually evident within narrow physiological and environmental parameters. In potato, the accumulation and oxidation of phenols, lignification, and suberization may also be involved in disease resist-ance. A great deal of importance and emphasis in research has been placed on race-specific interactions as related to phytoalexin accumulation. However, the phytoalexins are notable for their lack of specificity with respect to anti-microbial activity (Kuć, 1976*a,b*). Current thinking explains race specificity on the basis of either specific elicitors or suppressors. In the potato–*P. infestans* interaction, evidence supports suppression as a critical factor. The processes of wounding and wound healing have received relatively little attention and appear out of "scientific fashion", even though they are intimately associated with the accumulation of stress metabolites.

Phytoalexins often accumulate to high physiological concentrations, and their function as factors in resistance depends more upon the concentration which accumulates than their activity as anti-microbial compounds. Accumula-tion is only made possible by the diversion of considerable quantities of normal metabolites and energy to their biosynthesis. In nonphotosynthesizing plant tissues such metabolites arise primarily from carbohydrate, probably starch. There appears to be a deregulation of metabolism and at the same time a diversion of metabolites into regulated pathways, segments of which are largely unused by the healthy plant. The accumulation of phytoalexins is regulated by the rate of their synthesis in the host and metabolism by host or pathogen. Some phytoalexins are synthesized *de novo* and their accumulation is not due to hydrolysis or other degradative changes but rather to organized and meta-bolically regulated pathways. Data are not available for the SSM, as they are with the shikimate-acetate-malonate derived phytoalexins, to suggest that the diversion is accompanied by the appearance of specific enzymes. Clearly, a central precursor is acetyl CoA, and factors influencing its synthesis would be vital for the accumulation of SSM, polyacetylenes, pterocarpans, and iso-flavonoids. On the other hand, the phenylpropanoid-derived phenolics, and in part the pterocarpans and isoflavonoids, arise from transformations originating with fructose-6-phosphate and 3-phospho-glyceraldehyde. What factors start the flow of these raw materials? What causes the deregulation of metabolism?

Why the specific diversion of metabolites? Why the organized synthesis in the apparent midst of disorganization? In most instances studied in the Solanaceae, injury itself is not sufficient to catalyse the accumulation of SSM. Why? What is the specific metabolic role of elicitors and suppressors? How does metabolism return to "normal"? Clearly the study of stress metabolites is sufficiently justified from the chemical and biochemical perspective; their possible role as resistance factors merely enhances their interest.

As a result of our knowledge of the specificity of one aspect of disease resistance in animals, the antibody–antigen interaction, have we over-emphasized specificity when considering disease resistance in plants? Have we applied with sufficient vigour all the available biochemical and physiological information to the control of plant disease, which is, after all, the ultimate concern of plant pathology? Perhaps, with John Maynard Keynes, we may conclude that "the difficulty lies, not in the new ideas, but in escaping from the old ones...."

Acknowledgements

Journal Paper no. 81-11-101 of the Kentucky Agricultural Experiment Station, Lexington, Kentucky 40546.

The author's work reported in this chapter was supported in part by grants from the Rockefeller Foundation, Ciba-Geigy Corporation, the Science and Education Administration of the United States Department of Agriculture Competitive Research Grants Office (7800505), the Kentucky Tobacco and Health Research Institute (KTRB 21138), and the Alexander von Humboldt Foundation.

REFERENCES

Allen, E. and Kuć, J. (1968) α-Solanine and α-chaconine as fungitoxic compounds in extracts of irish potato tubers. *Phytopathology*, **58**, 776–781.

Arneson, P. and Durbin, R. (1968) The sensitivity of fungi to α-tomatine. *Phytopathology*, **58**, 536–537.

Bailey, J. A., Burden, R. S. and Vincent, G. G. (1975) Capsidiol: an antifungal compound produced in *Nicotiana tabacum* and *N. clevelandii* following infection with tobacco necrosis virus. *Phytochemistry*, **14**, 597.

Baker, F. C. and Brooks, C. J. W. (1976) Biosynthesis of the sesquiterpenoid capsidiol in sweet pepper fruits inoculated with fungal spores. *Phytochemistry*, **15**, 689–694.

Birnbaum, G. I., Huber, C. P., Post, M. L. and Stothers, J. B. (1976) Sesquiterpenoid stress compounds of *Datura stramonium*: biosynthesis of the three major metabolites from [1,2¹³C] acetate and the X-ray structure of 3-hydroxylubimin. *J. Chem. Soc. Chem. Comm.*, 330–331.

Birnbaum, G. I., Stoessl, A., Grover, S. H. and Stothers, J. B. (1974) The complete stereostructure of capsidiol, X-ray analysis and ¹³C nuclear magnetic resonance of eremophilane derivatives having trans-vicinal methyl groups. *Can. J. Chem.*, **52**, 993–1005.

Bostock, R. M., Henfling, J. W. D. M. and Kuć, J. (1981a) Lack of correlation between resistance and the accumulation of sesquiterpene stress metabolites in potatoes inoculated with *Phytophthora infestans* during the growing season. *Phytopathology*, submitted.

Bostock, R. M. and Kuć, J. (1980) A lipophilic fraction from *Phytophthora infestans* elicits the accumulation of sesquiterpenoid stress metabolites in potato tuber. *Phytopathology*, **70**, 688.

Bostock, R. M., Kuć, J. and Laine, R. A. (1981b) Eicosapentaenoic and arachidonic acids from *Phytophthora infestans* elicit fungitoxic sesquiterpenes in potato. *Science*, **212**, 67–69.

Budde, A. D. and Helgeson, J. P. (1981) Phytoalexins in tobacco callus tissues challenged by zoospores of *Phytophthora parasitica* var *nicotiana*. *Phytopathology*, **71**, 206.

Burden, R. S., Bailey, J. A. and Vincent, G. G. (1975) Glutinosone, a new antifungal sesquiterpene from *Nicotiana glutinosa* infected with tobacco mosaic virus. *Phytochemistry*, **14**, 221–223.

Clarke, D. D. (1973) The accumulation of scopolin in potato tissue in response to infection. *Physiol. Plant Pathol.*, **3**, 347–358.

Cohen, Y., Bialer, M. and Kuć, J. (1981) Phytuberin, inhibitory to *Peronspora hyoscyami* f. sp. *tabacina* in blue mold infected tobacco leaves. *Physiol. Plant Pathol.*, submitted.

Coxon, D. T. and Price, K. R. (1979) Cyclodehydroisolubimin: a new tricyclic sesquiterpene from potato tubers inoculated with *Phytophthora infestans*. *J. Chem. Soc. Chem. Comm.*, 2348–349.

Coxon, D. T., Price, K. R., Howard, B., Osmon, S. F., Kalan, E. B. and Zacharius, R. M. (1974) Two new vetispirane derivatives: stress metabolites from potato (*Solanum tuberosum*) tubers. *Tetrahedron Lett.*, 2921–2924.

Coxon, D. T., Price, K. R., Howard, B. and Curtis, R. F. (1977) Metabolites from microbially infected potato. Part 1, Structure of phytuberin. *J. Chem. Soc. Perkin*, 53–59.

Cruickshank, I. A. M., Perrin, D. R. and Mandryk, M. (1976) Capsidiol produced by tobacco infected with *Peronospora hyoscyami* f. sp. *tabacina*. *Ann. Rep. Div. Plant Ind. CSIRO Australia 1975*, 65.

Currier, W. W. (1975) Characterization of the induction and suppression of terpenoid accumulation in the potato–*Phytophthora infestans* interaction. Ph.D. Thesis, Purdue Univ., Lafayette, Indiana (U.S.A.).

Currier, W. and Kuć, J. (1975) Effect of temperature on rishitin and steroid glycoalkaloid accumulation in potato tuber. *Phytopathology*, **65**, 1194–1197.

deWit, P. J. G. M. and Flach, W. (1979) Differential accumulation of phytoalexins in tomato leaves but not in fruits after inoculation with *Cladosporium fulvum*. *Physiol. Plant Pathol.*, **15**, 257–267.

deWit, P. J. G. M. and Kodde, E. (1981) Induction of polyacetylenic phytoalexins in *Lycopersicon esculentum* after inoculation with *Cladosporium fulvum*. *Physiol. Plant Pathol.*, **18**, 143–148.

Doke, N. (1975) Prevention of the hypersensitive reaction of potato cells to infection with an incompatible race of *Phytophthora infestans* by constituents of the zoospores. *Physiol. Plant Pathol.*, **7**, 1–7.

Doke, N., Garas, N. A. and Kuć, J. (1979) Partial characterization and aspects of the mode of action of a hypersensitive-inhibiting factor (HIF) isolated from *Phytophthora infestans*. *Physiol. Plant Pathol.*, **15**, 127–140.

Doke, N., Garas, N. and Kuć, J. (1980) Effect on host hypersensitivity of suppressors released during the germination of cystospores of *Phytophthora infestans*. *Phytopathology*, **70**, 35–39.

Doke, N. and Tomiyama, K. (1980a) Suppression of the hypersensitive response of potato tuber protoplasts to hyphal wall components by water soluble glucans isolated from *Phytophthora infestans*. *Physiol. Plant Pathol.*, **16**, 177–186.

Doke, N. and Tomiyama, K. (1980b) Effect of hyphal wall components from *Phytophthora infestans* on protoplasts of potato tuber tissues. *Physiol. Plant Pathol.*, **16**, 169–176.

Edreva, A. (1977) Comparative biochemical studies of an infectious disease (blue mould) and a physiological disorder of tobacco. *Physiol. Plant Pathol.*, **11**, 149–161.

Friend, J. (1976) "Lignification in infected tissue", in *Biochemical Aspects of Plant-Parasite Interaction*, eds. J. Friend and D. Threlfall, Academic Press, 291–303.

Galliard, T. (1975) "Degradation of plant lipids by hydrolytic and oxidative enzymes", in *Recent Advances in Chemistry and Biochemistry of Plant Lipids*, eds. T. Galliard and E. Mercer, Academic Press, 319–357.

Garas, N. A., Doke, N. and Kuć, J. (1979) Suppression of the hypersensitive reaction in potato tubers by mycelial components from *Phytophthora infestans*. *Physiol. Plant Pathol.*, **15**, 117–126.

Garas, N. A. and Kuć, J. (1981) Lectin from potato lyses zoospores of *Phytophthora infestans* and precipitates the elicitor of terpenoid accumulation in potato. *Physiol. Plant Pathol.*, **18**, 227–237.

Grzelinska, A. and Sierakowska, J. (1978) Isolation of rishitin from tomato plants. *Phytopathol. Z.*, **91**, 320–321.

Guedes, M. E. M., Bostock, R., Hammerschmidt, R. and Kuć, J. (1981) The accumulation of six sesquiterpenoid phytoalexins in tobacco infected with *Pseudomonas lachrymans*. *Phytochemistry*, submitted.

Hammerschmidt, R. and Kuć, J. (1979) Isolation and identification of phytuberin from *Nicotiana tabacum* previously infiltrated with an incompatible bacterium. *Phytochemistry*, **18**, 874–875.

Hammerschmidt, R. and Kuć, J. (1980) Enhanced peroxidase activity and lignification in the induced systemic protection of cucumber. *Phytopathology*, **70**, 689.

Harding, V. K. and Heale, J. B. (1980) Isolation and identification of the antifungal compounds accumulating in the induced resistance response of carrot root slices to *Botrytis cinerea*. *Physiol. Plant Pathol.*, **17**, 277–289.

Helgeson, J. P., Budde, A. D. and Haberlach, G. T. (1978) Capsidiol: a phytoalexin produced by tobacco callus tissues. *Plant Physiol.*, **61**, suppl. 53.

Henderson, S. J. and Friend, J. (1979) Increase in PAL and lignin-like compounds as race-specific responses of potato tubers to *Phytophthora infestans*. *Phytopathol. Z.*, **94**, 323–334.

Henfling, J. W. D. M. (1979) Aspects of the elicitation and accumulation of terpene phytoalexins in the potato–*Phytophthora infestans* interaction. Ph.D. Thesis, Univ. of Kentucky, Lexington, Ky (U.S.A.).

Henfling, J. W. D. M., Bostock, R. M. and Kuć, J. (1980) Cell walls of *Phytophthora infestans* contain an elicitor of terpene accumulation in potato tubers. *Phytopathology*, **70**, 772–776.

Henfling, J. W. D. M., Bostock, R. M. and Kuć, J. (1981) Effect of abscisic acid on rishitin and lubimin accumulation and resistance to *Phytophthora infestans* and *Cladosporium cucumerinum* in potato tuber tissue slices. *Phytopathology*, **70**, 1074–1078.

Henfling, J. W. D. M. and Kuć, J. (1979) A semi-micro method for the quantitation of sesquiterpenoid stress metabolites in potato tuber tissue. *Phytopathology*, **69**, 609–612.

Henfling, J. W. D. M., Lisker, N. and Kuć, J. (1978) Effect of ethylene on phytuberin and phytuberol accumulation in potato tuber slices. *Phytopathology*, **68**, 857–862.

Hughes, D. L. and Coxon, D. T. (1974) Phytuberin: revised structure from the X-ray crystal analyses of dihydrophytuberin. *J. Chem. Soc.*, 822–823.

Hutson, R. A. and Smith, I. M. (1980) Phytoalexins and tyloses in tomato cultivars infected with *Fusarium oxysporum* f. sp. *lycopersici* or *Verticillium albo-atrum*. *Physiol. Plant Pathol.*, **17**, 245–257.

Ishiguri, Y., Tomiyama, K., Doke, N., Murai, A., Katsui, N., Yagihashi, F. and Masamune, T. (1978) Induction of rishitin-metabolizing activity in potato tuber disks by wounding and identification of rishitin metabolites. *Phytopathology*, **68**, 720–725.

Ishizaka, N. and Tomiyama, K. (1972) Effect of wounding or infection by *Phytophthora infestans* on the contents of terpenoids in potato tubers. *Plant and Cell Physiol.*, **13**, 1053–1063.

Johnson, G. and Schaal, L. (1952) Relation of chlorogenic acid to scab resistance in potatoes. *Science*, **115**, 627–629.

Johnson, G. and Schaal, L. (1955) The inhibitory effect of phenolic compounds on growth of strepto-myces scabies as related to the mechanism of scab resistance. *Phytopathology*, **45**, 626–628.

Jones, D., Graham, W. and Ward, E. W. B. (1975) Ultrastructural changes in pepper cells in an incompatible interaction with *Phytophthora infestans*. *Phytopathology*, **65**, 1274–1285 (see also 1286–1287, 1409–1417, 1417–1418).

Katsui, N., Matsunaga, A., Imaizumi, K., Masamune, T. and Tomiyama, K. (1971) The structure and synthesis of rishitinol, a new sesquiterpene alcohol from diseased potato tubers. *Tetrahedron Lett.*, 83–86.

Katsui, N., Matsunaga, A. and Masamune, T. (1974) The structure of lubimin and oxylubimin, antifungal metabolites from diseased potato tubers. *Tetrahedron Lett.*, 4483–4486.

Katsui, N., Murai, A., Takasugi, M., Imaizumi, K. and Masamune, T. (1968) The structure of rishitin, a new antifungal compound from diseased potato tubers. *J. Chem. Soc. Chem. Comm.*, 43–44.

Kolattukudy, P. (1975) "Biochemistry of cutin, suberin, and waxes, the lipid barriers on plants", in *Recent Advances in the Chemistry and Biochemistry of Plant Lipids*, eds. T. Galliard and E. Mercer, Academic Press, 203–246.

Kuć, J. (1957) A biochemical study of the resistance of potato tuber to attack by various fungi. *Phytopathology*, **47**, 676–680.

Kuć, J. (1972a) Phytoalexins. *Ann. Rev. Phytopathology*, **10**, 207–232.

Kuć, J. (1972b) "Compounds accumulating in plants after infection", in *Microbial Toxins*, eds. S. Ajl, G. Weinbaum, S. Kadis, Vol. 8, Academic Press, 211–247.

Kuć, J. (1975) Teratogenic constituents of potatoes. *Rec. Adv. Phytochem.*, **9**, 139–150.

Kuć, J. (1976a) "Phytoalexins", in *Encylopedia of Plant Physiology*, eds. R. Heitefuss and P. Williams, Vol. 4, Springer-Verlag, 632–652.

Kuć, J. (1976b) "Phytoalexins in the specificity of plant-parasite interaction", in *Specificity in Plant Diseases*, eds. R. K. S. Wood and A. Graniti, Plenum Press, 253–268.

Kuć, J., Currier, W. W. and Shih, M. J. (1976) "Terpenoid phytoalexins", in *Biochemical Aspects of Plant-Parasite Relationship*, eds. J. Friend and D. R. Threlfall, Academic Press, 225–237.

Kuć, J., Henfling, J., Garas, N. and Doke, N. (1979) Control of terpenoid metabolism in the potato-*Phytophthora infestans* interaction. *J. Food Protection*, **42**, 508–511.

Kuć, J. and Lisker, N. (1978) "Terpenoids and their role in wounded and infected plant storage tissue", in *Biochemistry of Wounded Plant Tissues*, ed. G. Kahl, Walter de Gruyter & Co., 203–242.

Langcake, P., Drysdale, R. and Smith, H. (1972) Post-infectional production of an inhibitor of *Fusarium oxysporum* f. sp. *lycopersici* by tomato plants. *Physiol. Plant Pathol.*, **2**, 17–25.

Legrand, M., Fritiz, B. and Hirth, L. (1976) Enzymes of the phenyl-propanoid pathway and the necrotic reaction of hypersensitive tobacco to tobacco mosaic virus. *Phytochemistry*, **15**, 1353–1359.

Letcher, R., Widdowson, B., Deverall, B. and Mansfield, J. (1976) Identification and activity of wyerone acid as a phytoalexin in broad bean (*Vicia faba*) after infection by *Botrytis*. *Phytochemistry*, **9**, 249–252.

Lisker, N. and Kuć, J. (1977) Elicitors of terpenoid accumulation in potato tuber slices. *Phytopathology*, **67**, 1356–1359.

Lisker, N. and Kuć, J. (1978) Terpenoid accumulation and browning in potato sprouts inoculated with *Phytophthora infestans*. *Phytopathology*, **68**, 1284–1287.

Locci, R. and Kuć, J. (1967) Steroid glycoalkaloids as compounds produced by potato tubers under stress. *Phytopathology*, **57**, 1272–1273.

Lyon, G., Lund, B., Bayliss, C. and Wyatt, G. (1975) Resistance of potato tubers to *Erwinia carotovora* and formation of rishitin and phytuberin in infected tissue. *Physiol. Plant Pathol.*, **6**, 43–50.

Mass, M. R., Post, F. J. and Salunkhe, D. K. (1977) Production of steroid glycoalkaloids by *Phytophthora infestans* in complex and chemically defined media. *J. Food Safety*, **7**, 107–117.

McCance, D. and Drysdale, R. (1975) Production of tomatine and rishitin in tomato plants inoculated with *Fusarium oxysporum* f. sp. *lycopersici*. *Physiol. Plant Pathol.*, **7**, 221–230.

McKee, R. (1955) Host-parasite relationship in the dry-rot disease of potatoes. *Ann. appl. Biol.*, **43**, 147–148.

McKee, R. (1959) Factors affecting the toxicity of solanine and related alkaloids to *Fusarium caeruleum*. *J. Gen. Microbiol.*, **20**, 686–696.

Metlitskii, L., Ozeretskovskaya, O., Vasyukora, N., Davydova, M., Dorozhkin, N., Remneva, Z. and Ivanova, V. (1970) Potato resistance to *Phytophthora infestans* as related to leaf phytoalexin activity. *Acad. Nauk. SSSR Priklad. Biok i Microbiol.*, **6**, 568–573.

Mohanakumaran, N., Gilbert, J. and Buddenhagen, I. (1969) Relationship between tomatine and bacterial wilt resistance in tomato. *Phytopathology*, **59**, 14.

Molot, P. M., Mass, P., Conus, M., Ferriere, H. and Ricci, P. (1981) Relations between capsidiol concentrations, speed of fungal invasions, and level of induced resistance in cultivars of pepper (*Capsicum annuum*) susceptible or resistant to *Phytophthora capsici*. *Physiol. Plant Pathol.*, **18**, 379–389.

Müller, K. O. and Börger, H. (1940) Experimentelle Untersuchungen über die *Phytophthora* Resistenz der Kartoffel. *Arb. Biol. Versuchsanstalt fur Land U Forstwirtschaft* Berlin-Dahlem **23**, 189–231.

Murai, K., Abiko, A., Ono, M., Katsui, V. and Masamune, T. (1978) Structure revision and biogenetic relationships of aubergenone, a sesquiterpenoid phytoalexin of eggplants. *Chemistry Lett.*, 1209–1212.

Price, K. R., Howard, B. and Coxon, D. T. (1976) Stress metabolite production in potato tubers infected by *Phytophthora infestans*, *Fusarium avenaceum* and *Phoma exigua*. *Physiol. Plant Pathol.*, **9**, 189–197.

Ravisé, A. and Tanguy, J. (1973) Etude des reactions phénoliques de plantules de *Nicotiana* inoculées par des souches de *Phytophthora* de Bary. *Phytopathol. Z.*, **76**, 253–264.

Reuveni, M. and Cohen, Y. (1978) Growth retardation and changes in phenolic compounds, with special reference to scopoletin, in mildewed and ethylene-treated tobacco plants. *Physiol. Plant Pathol.*, **12**, 179–189.

Ride, J. P. (1980) The effect of induced lignification on the resistance of wheat cells to fungal degradation. *Physiol. Plant Pathol.*, **16**, 187–196.

Sato, N., Tomiyama, K., Katsui, N. and Masamune, T. (1968a) Isolation of rishitin from tubers of interspecific potato varieties containing different late blight genes. *Ann. Phytopathol. Soc. Japan*, **34**, 140–142.

Sato, N., Tomiyama, K. and Katsui, N. (1968b) Isolation of rishitin from tomato plants. *Ann. Phytopathol. Soc. Japan*, **34**, 344–345.

Shih, M. J. (1972) The accumulation of isoprenoids and phenols and its control as related to the interaction of potato (*Solanum tuberosum*) with *Phytophthora infestans*. Ph.D. Thesis, Purdue Univ., Lafayette, Indiana (U.S.A.).

Shih, M. and Kuć, J. (1973) Incorporation of ^{14}C from acetate and mevalonate into rishitin and steroid glycoalkaloids by potato slices inoculated with *Phytophthora infestans. Phytopathology*, **63**, 826–829.

Shih, M. and Kuć, J. (1974) α- and β-Solamarine in Kennebec *Solanum tuberosum* leaves and aged tuber slices. *Phytochemistry*, **13**, 997–1000.

Shih, M. J., Kuć, J. and Williams, E. B. (1973) Suppression of steroid glycoalkaloid accumulation as related to rishitin accumulation in potato tubers. *Phytopathology*, **63**, 821–826.

Stoessl, A., Robinson, J. R., Rock, G. L. and Ward, E. W. B. (1977a) Metabolism of capsidiol by sweet pepper tissue: some possible implications for phytoalexin studies. *Phytopathology*, **67**, 64–66.

Stoessl, A. and Stothers, J. B. (1980) 2-Epi and 15 dihydro-2-epilubimin: new stress compounds from potato. *Can. J. Chem.* **58**, 2069–2072.

Stoessl, A., Stothers, J. B. and Ward, E. W. B. (1974) Lubimin: A phytoalexin of several Solanaceae. Structure, revision and biogenetic relationships. *J. Chem. Soc. Chem. Comm.*, 709–710.

Stoessl, A., Stothers, J. B. and Ward, E. W. B. (1975) The structure of some stress metabolites from *Solanum melongena. Can. J. Chem.*, **53**, 3351–3358.

Stoessl, A., Stothers, J. B. and Ward, E. W. B. (1976) Sesquiterpenoid stress compounds of the Solanaceae. *Phytochemistry*, **15**, 855–872.

Stoessl, A., Stothers, J. B. and Ward, E. W. B. (1978) Biosynthetic studies of stress metabolites from potatoes: incorporation of sodium acetate-$^{13}C_2$ into 10 sesquiterpenes. *Can. J. Chem.*, **56**, 645–653.

Stoessl, A., Unwin, C. H. and Ward, E. W. B. (1972) Post-infectional inhibitors from plants. I. Capsidiol, an antifungal compound from *Capsicum frutescens. Phytopathol. Z.*, **74**, 141–152.

Stoessl, A., Unwin, C. H. and Ward, E. W. B. (1973) Postinfectional fungus inhibitors from plants: fungal oxidation of capsidiol in pepper fruit. *Phytopathology*, **63**, 1225–1231.

Stoessl, A., Ward, E. W. B. and Stothers, J. B. (1977b) "Biosynthetic relationships of sesquiterpenoidal stress compounds from the Solanaceae", in *Host Plant Resistance to Pests*, ed. P. A. Hedin, Amer. Chem. Soc., Washington, D.C., 61–77.

Takagi, Y., Fugimori, T., Kaneko, H. and Kato, K. (1979) Phytuberol from Japanese domestic tobacco, *Nicotiana tabacum* cv. Suifu. *Agric. Biol. Chem.*, **43**, 2395–2396.

Tanguy, J. and Martin, C. (1972) Phenolic compounds and the hypersensitivity reaction in *Nicotiana tabacum* infected with tobacco mosaic virus. *Phytochemistry*, **11**, 19–28.

Thomas, C. and Allen, E. (1970) An antifungal polyacetylene compound from *Phytophthora*-infected safflower. *Phytopathology*, **60**, 261–263.

Tjamos, E. C. and Smith, I. M. (1974) The role of phytoalexins in resistance of tomato to *Verticillium* wilt. *Physiol. Plant Pathol.*, **4**, 249–259.

Tomiyama, K. (1966) Double infection by an incompatible race of *Phytophthora infestans* of potato cell which has previously been infected by a compatible race. *Ann. Phytopathol. Soc. Japan*, **32**, 181–185.

Tomiyama, K., Sakuma, T., Ishizaka, N., Sato, N., Takasugi, M. and Katsui, T. (1968) A new antifungal substance isolated from potato tuber tissue infected by pathogens. *Phytopathology*, **58**, 115–116.

Uegaki, R., Fujimori, T., Kubo, S. and Kato, K. (1981) Sesquiterpenoid stress compounds from *Nicotiana* species. *Phytochemistry*, **20**, 1567–1568.

Vance, C. P., Sherwood, R. T. and Kirk, T. K. (1980) Lignification as a mechanism of disease resistance. *Ann. Rev. Phytopathology*, **18**, 259–288.

Varns, J. L. (1970) Biochemical response and its control in the Irish potato tuber. (*Solanum tuberosum*)–*Phytophthora infestans* interactions. Ph.D. Thesis, Purdue Univ., Lafayette, (U.S.A.).

Varns, J., Currier, W. W. and Kuć, J. (1971a) Specificity of rishitin and phytuberin accumulation by potato. *Phytopathology*, **61**, 968–971.

Varns, J. and Kuć, J. (1971) Suppression of rishitin and phytuberin accumulation and hypersensitive response in potato by compatible races of *Phytophthora infestans. Phytopathology*, **61**, 178–181.

Varns, J. and Kuć, J. (1972) "Suppression of the resistance response as an active mechanism for susceptibility in the potato–*Phytophthora infestans* interaction", in *Phytotoxins in Plant Diseases*, eds. R. K. S. Wood and A. Graniti, Academic Press, 465–468.

Varns, J., Kuć, J. and Williams, E. (1971*b*) Terpenoid accumulation as a biochemical response of the potato tuber to *Phytophthora infestans*. *Phytopathology*, **61**, 174–177.

Ward, E. W. B. (1976) Capsidiol production in pepper leaves in incompatible interactions with fungi. *Phytopathology*, **66**, 175–176.

Ward, E. W. B. and Stoessl, A. (1977) Phytoalexins from potatoes: evidence for the conversion of lubimin to 15-dihydrolubimin by fungi. *Phytopathology*, **67**, 468–471.

Ward, E. W. B., Stoessl, A. and Stothers, J. B. (1977) Metabolism of the sesquiterpenoid phytoalexins capsidiol and rishitin to their 13-hydroxy derivatives by plant cells. *Phytochemistry*, **16**, 2024–2025.

Ward, E. W. B., Unwin, C. H., Hill, J. and Stoessl, A. (1975) Sesquiterpenoid phytoalexins from fruits of eggplants. *Phytopathology*, **65**, 859–863.

Ward, E. W. B., Unwin, C. H., Rock, G. L. and Stoessl, A. (1976) Postinfectional inhibitors from plants. XXIII. Sesquiterpenoid phytoalexins from fruit capsules of *Datura stramonium. Can. J. Bot.*, **54**, 25–29.

Ward, E. W. B., Unwin, C. H. and Stoessl, A. (1973) Postinfectional inhibitors from plants. VI. Capsidiol production in pepper fruit infected with bacteria. *Phytopathology*, **63**, 1537–1538.

4 Phytoalexins from other families

D. T. COXON

Scope of review

For this review I have attempted to include all chemically characterized phytoalexins which have been reported in the abstract literature until mid-1980, excluding those from the Leguminosae and Solanaceae, which are dealt with in chapters 2 and 3 respectively in this volume. I will leave the arguments about how one defines a phytoalexin to other authors in this volume. For the purpose of this chapter my definition of a phytoalexin is any fungally induced compound synthesized from remote precursors which shows *in vitro* fungitoxicity such that it *could* exert an antifungal effect *in vivo* if present at the right time and in the right place in the plant tissue in which it is formed. This definition serves to exclude quite a large number of induced stress metabolites for which no antifungal activity has been demonstrated, but will include many compounds which may not play any significant part in plant defence mechanisms.

Not unexpectedly, the phytoalexins from the range of plant families covered in this chapter represent many different chemical classes. The majority of these phytoalexins are phenolic compounds but many different ring structures are represented. Since there is no obvious chemical or phyllogenetic classification which would prove useful I have arranged the individual plant families alphabetically. Table 4.1 lists the names of the phytoalexins which have been isolated from the 13 plant families which are considered in this review. Molecular formulae and physical data are presented in Table 4.2 (p. 126) and the chemical structures are given in the text.

Characterized phytoalexins

Amaryllidaceae

Three phenolic phytoalexins have been isolated from bulb scales of the daffodil (*Narcissus pseudonarcissus*) following inoculation with suspensions of conidia

1 7-Hydroxyflavan

2 7,4′-Dihydroxyflavan

3 7,4′-Dihydroxy-8-methylflavan

of *Botrytis cinerea*. The compounds, which are flavans, were characterized and synthesized (Coxon *et al.*, 1980) thereby proving their structures to be 7-hydroxyflavan (**1**), 7,4′-dihydroxyflavan (**2**) and 7,4′-dihydroxy-8-methylflavan (**3**). Each of the compounds was shown to be active in tests against germinated spores of *B. cinerea* in liquid culture. ED_{50} values against germ tube growth were 22, 65 and 32 $\mu g/ml$ for **1**, **2** and **3** respectively.

The occurrence of flavans, which are rather uncommon natural products, has been recently reviewed (Gottlieb, 1977) and it appears that this is the first report of flavans possessing antifungal activity and accumulating in response to fungal infection.

Chenopodiaceae

The fungus responsible for a leaf spot disease of sugar beet (*Beta vulgaris*) is *Cercospora beticola*. Infected leaves contain enhanced quantities of two antifungal flavonoid compounds which are present in only barely detectable quantities in apparently healthy tissue (Geigert *et al.*, 1973). Extracts from infected leaves of a highly resistant cultivar showed antifungal activity against *C. beticola*, whereas the extract from healthy leaves showed little or no activity. Also, extracts from non-infected or infected leaves of a highly susceptible variety showed no antifungal activity. The compounds betagarin (**4**) and betavulgarin (**5**) were isolated from leaves of infected plants of resistant cultivars (Johnson

4 Betagarin

5 Betavulgarin

Table 4.1 List of phytoalexins in alphabetical order of plant family

Family	Species	Phytoalexin	Reference
Amaryllidaceae	*Narcissus pseudonarcissus*	7-hydroxyflavan (**1**) 7,4′-dihydroxyflavan (**2**) 7,4′-dihydroxy-8-methylflavan (**3**)	Coxon *et al.*, 1980
Chenopodiaceae	*Beta vulgaris*	betagarin (**4**) betavulgarin (**5**)	Geigert *et al.*, 1973
Compositae	*Carthamus tinctorius*	safynol (**6**) dehydrosafynol (**7**)	Allen and Thomas, 1971a Allen and Thomas, 1971c
Convolvulaceae	*Ipomoea batatas*	ipomeamarone (**8**) dehydroipomeamarone (**9**) ipomeamaronol (**10**)	Kubota and Matsuura, 1953 Oguni and Uritani, 1974a Kato *et al.*, 1971
Euphorbiaceae	*Ricinus communis*	casbene (**11**)	Sitton and West, 1975
Gramineae	*Oryza sativa*	momilactone A (**12**) momilactone B (**13**)	Cartwright *et al.*, 1977
Linaceae	*Linum usitatissimum*	coniferyl alcohol (**14**) coniferyl aldehyde (**15**)	Keen and Littlefield, 1979
Malvaceae	*Gossypium barbadense* *Gossypium hirsutum*	hemigossypol (**16**) isohemigossypol (**17**) gossyvertin (**18**) 6-methoxyhemigossypol (**19**) 6-deoxyhemigossypol (**20**) gossypol (**21**) 6-methoxygossypol (**22**) 6,6′-dimethoxygossypol (**23**)	Zaki *et al.*, 1972a Sadykov *et al.*, 1974 Karimdzhanov *et al.*, 1976 Bell *et al.*, 1975 Bell, 1967 Stipanovic *et al.*, 1975b

Moraceae	Morus alba	moracin A (24)	Takasugi et al., 1978a
		moracin B (25)	Takasugi et al., 1978b
		moracin C (26)	Takasugi et al., 1979
		moracin D (27)	
		moracin E (28)	
		moracin F (29)	
		moracin G (30)	
		moracin H (31)	
		oxyresveratrol (32)	Takasugi et al., 1978c
		4'-prenyloxyresveratrol (33)	Takasugi et al., 1980
	Broussonetia papyrifera	broussonin A (34)	
		broussonin B (35)	
Orchidaceae	Orchis militaris	orchinol (36)	Hardegger et al., 1963a
	Loroglossum hircinum	loroglossol (37)	Ward et al., 1975
		hircinol (38)	Urech et al., 1963
Rutaceae	Citrus limon	xanthoxylin (39)	Hartmann and Nienhaus, 1974
Umbelliferae	Daucus carota	6-methoxymellein (40)	Condon and Kuć, 1962
	Pastinaca sativa	falcarinol (41)	Harding and Heale, 1980
		xanthotoxin (43)	Johnson et al., 1973
Vitaceae	Vitis vinifera	ε-viniferin (44)	Langcake and Pryce, 1977a
		α-viniferin (45)	Pryce and Langcake, 1977
		pterostilbene (46)	Langcake et al., 1979

et al., 1976) and were identified as 2′,5-dimethoxy-6,7-methylenedioxyflavanone
and 2′-hydroxy-5-methoxy-6,7-methylenedioxyisoflavone, respectively. Both
compounds were assayed for inhibition of mycelial growth against *C. beticola*
and *Monilinia fructicola*. The flavanone betagarin was of low activity whereas
betavulgarin, an isoflavone, had significantly greater activity. ED_{50} values for
inhibition of mycelial growth of *C. beticola* were more than $200 \mu g/ml$ for
betagarin and between 50 and $100 \mu g/ml$ for betavulgarin. Against *M. fructicola*
betagarin had an ED_{50} of $200 \mu g/ml$ whereas for betavulgarin the value was less
than $25 \mu g/ml$.

An HPLC assay was developed for analysis of these phytoalexins in extracts
from leaf tissue. Healthy leaf tissue of a highly resistant cultivar contained $13 \mu g$
of betagarin and $0.4 \mu g$ of betavulgarin/g of dried tissue. Infected leaf tissues
contained 344 and $47 \mu g$, respectively. In a highly susceptible cultivar, healthy
tissue contained $0.8 \mu g$ of betagarin and $1.9 \mu g$ of betavulgarin. Although
infected leaves of this susceptible cultivar had many more lesions than the
resistant cultivar they contained only $1.5 \mu g$ of betagarin and $12.7 \mu g$ of betavulgarin. Thin layer chromatograms of acetone extracts of lesions of various
cultivars indicated no apparent relationship between the amounts of betagarin
and the degree of resistance. However, infected resistant cultivars generally had
a higher concentration of betavulgarin than did more susceptible cultivars. The
authors concluded that betagarin plays a minor role, if any, in disease resistance
but that betavulgarin has many of the properties of a phytoalexin and may
contribute to the resistance of sugar beet to *C. beticola*. The structures of
betagarin and betavulgarin were deduced from MS and by IR, UV and NMR
spectroscopy. In particular, the NMR spectra allowed the aromatic substitution
patterns of both rings to be deduced. In order to confirm the structure of
betagarin it was synthesized in eight steps starting from 2-hydroxy-3-methoxy-
benzaldehyde.

Compositae

Two antifungal polyacetylenes have been isolated from the safflower (*Carthamus
tinctorius*) following infection with *Phytophthora drechsleri*, a fungus which
causes a root and stem-rot disease. Allen and Thomas (1971*a*) found that
extracts of hypocotyls and first internodes of the resistant cultivar Biggs infected
with *P. drechsleri* were highly fungitoxic to the pathogen compared with extracts
from non-infected hypocotyls. Initially they isolated the compound safynol (**6**)
which was the major antifungal compound present in these extracts. UV, IR
and MS indicated that safynol was identical with the known polyacetylene
3E,11E-3,11-tridecadiene-5,7,9-triyne-1,2-diol (Bohlmann and Herbst, 1959).
Safynol accumulation in infected hypocotyls following inoculation with *P.
drechsleri* was followed using resistant (Biggs) and susceptible (Nebraska-10)
cultivars of safflower (Allen and Thomas, 1971*b*). In the resistant cultivar the
concentration of safynol increased rapidly from $1.8 \mu g/g$ healthy tissue to

$$CH_3CH=CH(C\equiv C)_3CH=CH\underset{\underset{\displaystyle OH}{|}}{CH}CH_2OH$$

6 Safynol

$$CH_3CH=CH(C\equiv C)_4\underset{\underset{\displaystyle OH}{|}}{CH}CH_2OH$$

7 Dehydrosafynol

$26.7\,\mu g/g$ infected tissue 48 h after inoculation and reached $33.6\,\mu g/g$ in 96 h. In the susceptible cultivar safynol also accumulated rapidly from $1.0\,\mu g/g$ healthy tissue to $15.6\,\mu g/g$ infected tissue 24 h after inoculation but then decreased to $10\,\mu g/g$ at 96 h. Safynol was found to inhibit the linear growth of *P. drechsleri* by 50% at $12\,\mu g/ml$ and at $30\,\mu g/ml$ growth was prevented (Thomas and Allen, 1970). Ninety-six hours after inoculation the external cell layers of resistant tissues of Biggs safflower contained sufficient concentration of safynol to inhibit completely the growth of the pathogen *in vitro*. Since little or no enlargement of lesions occurred the authors concluded that safynol might play an important role in disease resistance.

Safynol may be estimated by optical density measurement at 269 nm of fractions obtained by thin layer chromatography of methanol extracts of inoculated tissue. Further investigation of these methanol extracts by thin layer chromatography led to the isolation of a second antifungal polyacetylene. This was identified as dehydrosafynol (7), (11E)-11-tridecene-3,5,7,9-tetrayne-1,2-diol (Allen and Thomas, 1971c). Healthy stems contained less than 2 ng dehydrosafynol and $0.83\,\mu g$ safynol/100 g tissue. Infected stems at 48 h after inoculation contained $2.96\,\mu g$ dehydrosafynol and $33.7\,\mu g$ safynol/100 g tissue. Dehydrosafynol, ED_{50} $1.7\,\mu g/ml$, was however several times more fungitoxic than safynol, ED_{50} $12\,\mu g/ml$. Both antifungal compounds were found in high concentrations in tissues exterior to the vascular ring.

The disease resistance of Biggs safflower stems was dependent on the period of illumination which the plants received (Thomas and Allen, 1971). Plants which received 0 or 8 h light every 24 h post-inoculation period showed susceptible or moderately susceptible disease reactions and accumulated low concentrations of safynol and dehydrosafynol. Plants receiving 16 or 24 h light per 24 h period were resistant to infection and contained high concentrations of both phytoalexins. The level of resistance, as affected by the post-inoculation light period, was related to the concentration of the two antifungal polyacetylene compounds, providing further evidence that these compounds might determine the type of reaction which occurred.

Nakada *et al.* (1977) have synthesized both enantiomeric forms of safynol and shown that natural safynol has the (R)-configuration. Both (R)- and (S)-safynols obtained synthetically were tested for inhibition of mycelial growth against a range of fungi including *P. drechsleri*. The results showed that the absolute configuration does not affect the antifungal activity.

Convolvulaceae

Ipomeamarone (**8**), an antifungal furanosesquiterpene, was first isolated nearly forty years ago from sweet potato (*Ipomoea batatas*) roots infected with *Ceratocystis fimbriata*. Since that time, numerous other furanoterpenoid metabolites with either unchanged or degraded sesquiterpene skeletons have been isolated from sweet potato roots damaged in various ways, such as by fungal infection or chemical treatment. Many papers have appeared over the last twenty years on stress metabolites of sweet potatoes, many of them emanating from the laboratories of Professor Uritani in Japan and Professor Wilson in the U.S.A. However, of all the stress metabolites known to be produced by sweet potato roots, only three, ipomeamarone (**8**), dehydroipomeamarone (**9**) and ipomeamaronol (**10**), have been described as phytoalexins. As with many stress metabolites

8 Ipomeamarone   **9** Dehydroipomeamarone

10 Ipomeamaronol

and phytoalexins the formation of ipomeamarone in sweet potato tissue is not a specific effect induced only by fungal infection. Treatment of sweet potato slices with various metabolic inhibitors including mercuric chloride, trichloracetic acid, monoiodoacetic acid, and 2,4-dinitrophenol, causes ipomeamarone production when used in concentrations high enough to induce death of the cells contacted by the agent (Uritani *et al.*, 1960). Ipomeamarone synthesis in root slices of resistant (Norin 1) and susceptible (Norin 2) varieties of sweet potato inoculated with *C. fimbriata* has been studied (Akazawa and Wada, 1961). After a lag phase of about 24 h synthesis occurred rapidly in both varieties and reached a maximum level about 72 h after inoculation. The rate of synthesis and the maximum concentration attained in the infected tissues was higher in the resistant than in the susceptible variety thus inferring a possible defence mechanism within the host tissue. The actual concentrations of ipomeamarone found in infected tissues 72 h after inoculation were 13.2 mg/g fresh tissue for the susceptible variety and 30.7 mg/g fresh tissue for the resistant variety. Thus compared to many phytoalexins very large amounts of ipomeamarone are produced in infected sweet potato tissue.

When sweet potato root tissue of the resistant variety Norin 1 was inoculated with *C. fimbriata* isolates from prune or coffee, fungal invasion was confined to 3 to 5 cell layers. In contrast, penetration by an isolate from sweet potato proceeded continuously into the inner cells (Hyodo *et al.*, 1969). Furanoterpenoids including ipomeamarone accumulated at higher concentrations in tissue infected by the sweet potato isolate than in tissue infected by the coffee or prune isolates. The conclusion reached was that the formation of furanoterpenoids in infected tissue was related to the degree of mycelial penetration and the resultant alteration of host metabolism, rather than a defence action of the host. Clearly these experiments raised doubts about the function of the compounds in resistance. However, more recent work would seem to support a defence role for induced furanoterpenoids (Kojima and Uritani, 1976). An infected tissue extract containing furanoterpenoids (mainly ipomeamarone, dehydroipomeamarone and ipomeamaronal) was prepared from Norin 1 sweet potato slices inoculated with a pathogenic sweet potato strain of *C. fimbriata.* Mycelial growth and spore germination of the sweet potato strain was only slightly inhibited when tested in the presence of the infected tissue extract, whereas non-pathogenic strains isolated from coffee, prune, oak, taro and almond were severely inhibited. Furanoterpenoid compounds appeared to be the main cause of the inhibitory activity. The depth of accumulation of furanoterpenoids and their concentration in samples from infected tissues was greater in tissue infected by the sweet potato strain than in tissues infected by the non-pathogenic strains but the concentration of furanoterpenoids in the tissues infected by all strains appeared high enough to inhibit fungal growth. A further significant observation was that when furano-terpenoid production of sweet potato root tissue was reduced by treatment with cycloheximide, the coffee strain of *C. fimbriata* behaved as a pathogenic fungus. Dehydroipomeamarone (**9**) (Oguni and Uritani, 1974*a*) would appear to be the immediate biosynthetic precursor of ipomeamarone as demonstrated by the incorporation of radioactive dehydroipomeamarone into ipomeamarone (Oguni and Uritani, 1974*b*). Ipomeamarone itself is a precursor of ipomeama-ronol (**10**) (Kato *et al.*, 1971; Yang *et al.*, 1971).

Euphorbiaceae

Casbene (**11**) is a diterpene hydrocarbon which has been isolated from cell-free extracts obtained from castor bean (*Ricinus communis*) seedlings germinated in the presence of certain fungal cultures (Sitton and West, 1975). Casbene has been described as a phytoalexin since the castor bean enzyme responsible for its

11 Casbene

production appears to be induced by fungal infection of the castor bean seedling and it has been shown that casbene has antifungal activity at concentrations of 10 μg/ml against *Aspergillus niger*, one of the fungal inducers. However, the formation of casbene has not been studied in a conventional host-parasite interaction and suggestions that it functions in disease resistance remain speculative. The formation of casbene from mevalonate in cell-free extracts of castor bean does seem to be a rather specialized response. At least four other diterpene hydrocarbons are also produced by cell-free extracts of castor bean incubated with mevalonate or geranylgeranyl pyrophosphate. Casbene production in cell free extracts is however greatly stimulated (20–40 times) by exposure of castor bean seedlings to *Aspergillus niger*, *Rhizopus stolonifer* or *Fusarium moniliforme*, whereas there is no consistent or striking stimulation of synthesis of any of the other diterpenes on exposure of the seedlings to fungi.

If casbene does act as a phytoalexin it is unique in a number of ways. Firstly, it is the first hydrocarbon reported to have such properties and one of the few diterpenoid substances to be described as phytoalexins (Cartwright *et al.*, 1977). Secondly, it is the product of a single enzyme-catalysed transformation of geranylgeranyl pyrophosphate (Dueber *et al.*, 1978) and its biosynthesis is therefore simple compared with most phytoalexins. The castor bean system would therefore seem an ideal one for the study of the enzymological and other processes associated with phytoalexin production.

Gramineae

Two diterpenoid compounds occur as phytoalexins in leaves of rice (*Oryza sativa*) infected by the rice blast disease fungus *Pyricularia oryzae*. The synthetic systemic fungicide 2,2-dichloro-3,3-dimethylcyclopropane carboxylic acid (WL28325) appears to function by activating this natural resistance mechanism in the rice plant (Cartwright *et al.*, 1977). Both the natural and WL28325 induced resistance of rice to infection by *P. oryzae* involve a hypersensitive type of browning reaction but the response is accentuated and accelerated by previous treatment of the plants through the roots with WL28325. Diffusion droplets from inoculation wounds on treated and untreated leaves were tested for inhibition of germ tube growth of *P. oryzae*. Inhibition of germ tube growth was very evident in diffusates collected from treated infected leaves. Diffusates collected from untreated infected leaves were considerably less inhibitory whilst those collected from uninfected leaves, either treated or untreated, showed no antifungal activity.

Experiments with ethanol extracts of infected leaves gave similar results and when the active extracts were chromatographed on thin layer plates and assayed for antifungal compounds using *Cladosporium cucumerinum*, two major anti-fungal zones were detected which were absent from extracts of uninfected leaves. Similar antifungal activity could be induced by ultraviolet irradiation of either leaves or dark-grown coleoptiles of rice. Under these conditions the production of antifungal compounds was not influenced by treatment of the plants with WL28325. The two major phytoalexins were isolated and identified as the 9-β-pimaradiene diterpenes, momilactone A (12) and momilactone B (13)

12 Momilactone A   13 Momilactone B

(Cartwright *et al.*, 1981). These had earlier been isolated as plant growth inhibitors from rice husks (Kato *et al.*, 1977). Concentrations of momilactones A and B giving 50% inhibition of germ tube growth of *P. oryzae* were 5 and 1 μg/ml, respectively.

Linaceae

Two simple phenylpropanoid phytoalexins, coniferyl alcohol (14) and coniferyl aldehyde (15), are believed to contribute to the resistance of flax (*Linum usitatissimum*) to the rust *Melampsora lini* (Keen and Littlefield, 1979). The two compounds accumulated more rapidly in incompatible than in compatible reactions. In addition their accumulation was more rapid in resistant geno-types typified by rapid restriction of fungus growth and small lesions than in those typified by later restriction of fungus growth and larger lesion areas.

14 Coniferyl alcohol   15 Coniferyl aldehyde

Incompatible infection sites, but not compatible ones, stained with phloro-glucinol, indicating the occurrence of coniferyl aldehyde and perhaps lignin in the incompatible sites. Initial detection of phloroglucinol staining was chrono-logically correlated with the cessation of fungal growth and inversely correlated with ultimate lesion size.

In a germ tube bioassay with *M. lini* inhibition of germination occurred with coniferyl aldehyde at 40 μg/ml and coniferyl alcohol at 140 μg/ml. The concen-trations of coniferyl aldehyde and coniferyl alcohol in tissue lesions were not measured but the accumulation of both compounds in facilitated diffusates from flax leaves of various genotypes was measured as a function of time. The overall evidence indicated that *de novo* production of coniferyl alcohol and coniferyl aldehyde may be the mechanism for the restriction of fungal growth in in-compatible flax-rust interactions.

Malvaceae

A whole range of gossypol-related naphthaldehyde compounds have been isolated from cotton plants (*Gossypium* spp.) infected with wilt disease caused by *Verticillium* spp. Gossypol (**21**) itself is a natural pigment found in certain tissues, particularly pigment glands, of healthy cotton plants. Early studies on induced antifungal compounds in cotton plants (Bell, 1967) suggested a possible role for gossypol and related compounds in disease resistance. When boll cavities or xylem vessels of excised stems of *Gossypium hirsutum* L. or *G. barbadense* L. were inoculated with *Verticillium albo-atrum* there was a marked accumulation of ether-soluble phenolic compounds in the tissue after 24 to 72 h. The major compound in this extract was identified as gossypol on the basis of R_f values, UV spectra and the formation of derivatives with aniline, 2,4-dinitro-phenylhydrazine and phloroglucinol. Some cotton cultivars have no pigment glands and have little or no free gossypol in their tissues. This fact however did not affect the induced formation of gossypol: inoculated bolls or stem sections of both glanded and glandless types produced similar amounts of gossypol. Gossypol synthesis could also be induced by inoculation of tissues with *Rhizopus nigricans* or by chemical treatment with cupric and mercuric ions and various metabolic inhibitors. Concentrations of purified gossypol giving 50% inhibition of spore germination against various fungi ranged from 20 to 100 μg/ml. Further studies (Bell, 1969) showed that *Verticillium* wilt resistance in *Gossypium* species was not simply due to quantitative differences in the amounts of phytoalexin formed in infected tissue or to the lack of phytoalexin synthesis in susceptible varieties. Both virulent and avirulent strains of *V. albo-atrum* induced phytoalexin synthesis in xylem vessels of either susceptible or resistant cottons. Furthermore, ED_{50} values for inhibition of spore germina-tion by either authentic gossypol or crude phytoalexin preparations were similar for all *V. albo-atrum* strains. Resistance in cotton varieties and species appeared to depend on the speed at which a fungistatic concentration of phytoalexin was formed at an infection site relative to the speed of fungal development at that site. This later study showed that in addition to gossypol one other major and four minor unidentified fungitoxic compounds were also present in xylem fluid from cotton inoculated with *V. albo-atrum*.

CHO OH structures:

16 Hemigossypol **17** Isohemigossypol **18** Gossyvertin

19 6-Methoxyhemigossypol **20** 6-Deoxyhemigossypol

Zaki *et al.* (1972*b*) failed to detect gossypol in extracts from *Verticillium*-inoculated cotton stems but isolated two other induced antifungal compounds which were given the trivial names hemigossypol and vergosin (Zaki *et al.*, 1972*a*). Hemigossypol (**16**) was correctly identified as 8-formyl-1,6,7-trihydroxy-5-isopropyl-3-methylnaphthalene but vergosin, which was thought to be 8-formyl-1-hydroxy-5-isopropyl-7-methoxy-3-methylnaphthalene has since been shown to be 3-hydroxy-5-isopropyl-4-methoxy-7-methyl-2H-naphtho[1,8-bc] furan or deoxy-6-methoxyhemigossypol (Stipanovic *et al.*, 1975*a*). Hemigossypol and vergosin were reported as the major antifungal compounds in cotton stems inoculated with *V. albo-atrum* (Zaki *et al.*, 1972*b*). There is now some doubt as to whether vergosin itself is antifungal since under the conditions of the bioassay used by Zaki *et al.* (1972*b*) it is readily autoxidized to another compound, 6-methoxy-hemigossypol (**19**) which also has antifungal activity. Thus it may be that vergosin is merely the biosynthetic precursor of the true phytoalexin 6-methoxyhemigossypol.

Recent investigations (Bell *et al.*, 1975) confirm that hemigossypol (**16**) is the major fungitoxic compound formed in *Verticillium* infected stele tissue of both *G. barbadense* and *G. hirsutum*. Hemigossypol (**19**), 6-methoxyhemigossypol and 6-deoxyhemigossypol (**20**) were found to be the major sesquiterpene aldehydes in infected stele tissue from a range of *Gossypium* species and other genera of the Malvaceae. Russian workers have isolated two other closely related compounds from stem tissues of cotton plants infected with *Verticillium dahliae*. Isohemigossypol (**17**) (Sadykov *et al.*, 1974) and gossyvertin (**18**) (Karimdzhanov *et al.*, 1975) were isolated from xylem extracts where very little gossypol was found.

It now appears therefore that gossypol itself is certainly not the most important contributor to the phytoalexin response of the cotton plant although it does have antifungal properties. 6-Methoxygossypol (**22**) and 6,6'-dimethoxy-gossypol (**23**) (Stipanovic *et al.*, 1975) also occur in the cotton plant and are fungitoxic, but their role in resistance is not established, although their presence in uninfected root tissue suggests that they may also function as preformed toxins.

21 Gossypol ($R^1 = R^2 = H$)
22 6-Methoxygossypol ($R^1 = CH_3, R^2 = H$)
23 6,6'-Dimethoxygossypol ($R^1 = R^2 = CH_3$)

Several terpenoid compounds including gossypol, 6-methoxygossypol, and 6,6'-dimethoxygossypol may be involved in the age-related disease resistance of cotton seedlings to *Rhizoctonia solani* (Hunter *et al.*, 1978). The resistance of cotton seedlings to *R. solani* increased between 5 and 12 days after planting in parallel with a corresponding increase, in uninfected hypocotyls, in the concentrations of deoxyhemigossypol, deoxy-6-methoxyhemigossypol, hemigossypol, 6-methoxyhemigossypol, gossypol, 6-methoxygossypol and 6,6'-dimethoxy-gossypol. Furthermore, the concentration of each compound in hypocotyls of either age increased even more during the first 48 hours after inoculation with *R. solani*. However, as the authors point out, the toxicity of the individual terpenoid compounds to *R. solani* must be thoroughly tested before it is known whether they play an active part in resistance of older tissues.

Moraceae

A very productive area of recent research has been the investigation of mulberry phytoalexins. At least twelve compounds have been described as phytoalexins from *Morus* spp. They include three structurally different types of phytoalexins: (1) the moracins; (2) closely related stilbene derivatives, and (3) the broussonins, 1,3-diphenylpropane derivatives.

The moracins (Takasugi *et al.*, 1978*a*, 1978*b*, 1979) were isolated from acetone extracts of cortex and phloem tissues of the shoots of mulberry (*Morus alba*) infected with *Fusarium solani* f. sp. *mori* and were not present in detectable quantities in uninfected tissue. Fractionation was assisted by bioassay against

24 Moracin A

25 Moracin B

26 Moracin C

27 Moracin D

28 Moracin E

29 Moracin F

30 Moracin G

31 Moracin H

Cochliobolus miyabeanus, a non-pathogen of mulberry. At least eight moracins are known, designated as moracins A (**24**) to H (**31**).

All the moracins showed activity against fungi which were both pathogenic and non-pathogenic to mulberry. Minimum concentration (μg/ml) required for complete inhibition of fungal growth of *F. solani* f. sp. *mori* were 224, moracin C; 112, moracin D; *ca.* 200, moracin E; *ca.* 1000, moracin F; *ca.* 500, moracin G and *ca.* 1000, moracin H.

32 Oxyresveratrol   33 4'-Prenyloxyresveratrol

Two stilbene-type antifungal compounds were isolated from acetone extracts of fungus-infected xylem tissues of mulberry shoots (Takasugi, 1978c); they were absent from corresponding extracts of healthy tissues. The compounds were identified as oxyresveratrol (32) and 4'-prenyloxyresveratrol (33). Oxyresveratrol (32) is a known constituent of heartwoods of several *Morus* species and was found in large amounts (1.7% of the dried tissues) in heartwood. However, oxyresveratrol was not detected in sapwood so its presence there as a preformed inhibitor must be ruled out. Minimum concentrations required for 100% inhibition of fungal growth of *F. solani* f. sp. *mori* were 285 to 560 and more than 224 μg/ml for oxyresveratrol and 4'-prenyloxyresveratrol respectively.

34 Broussonin A   35 Broussonin B

Broussonin A (34) and broussonin B (35) occur together with the known coumarin, marmesin, in cortex and phloem tissues of the shoots of paper mulberry (*Broussonetia papyrifera*) infected with *Fusarium solani* f. sp. *mori* (Takasugi *et al.*, 1980). Again, the antifungal compounds were absent from corresponding healthy tissues. Minimum concentrations required for preventing growth of *F. solani* f. sp. *mori* were *ca.* 230 μg/ml, for both compounds.

Orchidaceae

Three dihydrophenanthrene phytoalexins have been isolated from the infected tubers of orchid species. Orchinol (36), isolated originally from *Orchis militaris* infected with *Rhizoctonia repens*, has been identified as 2,4-dimethoxy-7-hydroxy-9,10-dihydrophenanthrene (Hardegger *et al.*, 1963a, 1963b). Loroglossol (37) and hircinol (38) were isolated from *Loroglossum hircinum* after infection with *Rhizoctonia versicolor* (Hardegger *et al.*, 1963a; Urech *et al.*, 1963) but loroglossol was regarded as biologically inactive. Loroglossol and hircinol form a common trimethoxydihydrophenanthrene derivative and were identified as 2,4-dimethoxy-5-hydroxy-9,10-dihydrophenanthrene and 4-methoxy-2,5-

36 Orchinol

37 Loroglossol

38 Hircinol

dihydroxy-9,10-dihydrophenanthrene, respectively (Letcher and Nhamo, 1973). All three orchid phytoalexins have since had their structures confirmed by synthesis (Steiner et al., 1974; Stoessl et al., 1974). Orchinol and hircinol both exhibit in vitro antifungal activity at concentrations in the range of 10^{-5} to 10^{-4} M (2.5 to 25 μg/ml) against a range of microorganisms (Gaumann, 1963; Gaumann et al., 1960) and more recently loroglossol has been shown to have a similar level of activity in vitro against Monilinia fructicola and Phytophthora infestans (Ward et al., 1975). Hircinol has also been identified in tubers of yam (Dioscorea rotundata) where its presence as a preformed antifungal compound may be important (Ogundana and Coxon, unpublished data).

Rutaceae

Phenolic compounds accumulated in Citrus spp. infected by Phytophthora citrophthora and Hendersonula toruloidea (Hartmann and Nienhaus, 1974; Musumeci and Oliveira, 1976). Two fungitoxic compounds not found in healthy bark were found in the bark of Citrus limon Burm. f. after infection.

39 Xanthoxylin

One of these compounds was identified as xanthoxylin (39) 2-hydroxy-4,6-dimethoxyacetophenone. Xanthoxylin was also found in the infected bark of other Citrus spp. i.e., C. limon, C. aurantifolia and C. medica, after inoculation with P. citrophthora, H. toruloidea or Diplodia natalensis. It was not detected in extracts of bark which had been mechanically or chemically wounded and it seems therefore to be a specific product of *Citrus* species after fungal infection. The concentration of xanthoxylin found in lesion tissues was 3 to 5 times higher than the concentration inhibiting the fungus *in vitro* ($ED_{50} = 150$ to $175 \mu g/ml$). The accumulation of xanthoxylin appears to be the result of the necrotic reaction and it may at least contribute to eliminating the pathogen after it has been stopped by other defence mechanisms.

Umbelliferae

One of the earliest reported phytoalexins was the compound 6-methoxymellein (**40**) which was isolated from carrot root slices after inoculation with *Ceratocystis fimbriata*, a fungus which causes black rot disease in sweet potato but is not pathogenic to carrots (Condon and Kuć, 1960). The induced production of 6-methoxymellein accounted for the resistance of the carrot tissue to infection. This conclusion was based on determination *in vitro* of its inhibitory effects on *C. fimbriata*. Further studies (Condon *et al.*, 1963) showed that production of 6-methoxymellein was not a specific response to fungal infection and that various chemical treatments would also induce its formation. Ethylene treatment of carrot slices was also found to be a very effective way of inducing 6-methoxymellein and other phenolic compounds (Chalutz *et al.*, 1969; Coxon *et al.*, 1973). It was also shown that common carrot storage pathogens and not just non-pathogens would induce 6-methoxymellein formation.

It has been claimed that 6-methoxymellein is important in the active defence of cold-stored carrots to *Botrytis cinerea* (Goodliffe and Heale, 1978). When carrots which had been stored at 4 to 6°C were wound-inoculated with mycelial discs of *B. cinerea* the proportion of roots in which the resulting infection remained localized, decreased as the pre-inoculation storage time increased.

40 6-Methoxymellein

$$CH_3(CH_2)_6 CH = CH CH_2 (C \equiv C)_2 \underset{\overset{|}{OH}}{CH} CH = CH_2$$

41 Falcarinol

$$CH_3(CH_2)_6 CH = CH \underset{\overset{|}{OH}}{CH} (C \equiv C)_2 \underset{\overset{|}{OH}}{CH} CH = CH_2$$

42 Falcarindiol

When tissue from resistant lesions was analysed 55 days after inoculation it contained much higher levels of 6-methoxymellein than either healthy tissue or tissue from the edges of spreading lesions. The level of 6-methoxymellein which had accumulated in resistant root tissue 55 days after inoculation declined with increasing pre-inoculation cold-storage time. In time-course experiments carried out at 20°C, 6-methoxymellein accumulated to fungitoxic concentrations in the resistant reaction but not in the susceptible one.

Artificially induced resistance to infection of carrot slices with *B. cinerea* by pretreatment with heat-killed conidia prior to inoculation with live spores leads to the production of additional antifungal compounds by the carrot tissue. 6-Methoxymellein (ED_{50} 104 μg/ml), *p*-hydroxybenzoic acid (ED_{50}, 607 μg/ml) and the polyacetylene falcarinol (**41**) (ED_{50}, 9.2 μg/ml) have been identified as inhibitors produced in this situation (Harding and Heale, 1980, 1981). The induced formation of falcarinol (**41**) is interesting since the related compound falcarindiol (**42**) has been reported as a preformed antifungal compound in carrot tissue and is highly active against *Mycocentrospora acerina* (Garrod *et al.*, 1978). However, concentrations of falcarindiol do not increase following infection and it cannot be regarded as a phytoalexin on the evidence available.

43 Xanthotoxin

Xanthotoxin (**43**) has been identified as a phytoalexin in parsnip (*Pastinaca sativa*) (Johnson *et al.*, 1973). Several non-pathogens of parsnip including *Ceratocystis fimbriata*, *Helminthosporium carbonum*, *Alternaria* spp. or *Colletotrichum lindemuthianum* caused xanthotoxin accumulation in inoculated root discs. Xanthotoxin prevented growth *in vitro* of *C. fimbriata* at a concentration of 10^{-3} M (216 μg/ml) and produced 50% inhibition of radial growth of the fungus at 10^{-4} M concentration (21.6 μg/ml).

Vitaceae

Among the more recently discovered phytoalexins are a new class isolated from grapevine (*Vitis vinifera*) and given the trivial name of viniferins (Langcake and Pryce, 1977*a*). The viniferins are oligomers of *trans*-resveratrol and at least four such compounds, designated as α-, β-, γ- and ε-viniferins, occur in UV-irradiated or *Botrytis cinerea*-infected detached vine leaves (Langcake and Pryce, 1977*b*). They are a dimer, ε-viniferin (**44**), a cyclic trimer, α-viniferin (**45**), a cyclic tetramer (β-viniferin) and an uncharacterized oligomer (γ-viniferin). *Trans*-resveratrol itself was also present in extracts from UV-irradiated or *B. cinerea* infected leaves. All the viniferins have antifungal activity against *B. cinerea*, *Cladosporium cucumerinum*, *Pyricularia oryzae* and *Plasmopara viticola*, α- and ε-viniferin generally being the most active. Concentrations causing 50% inhibition of spore germination (ED_{50}) of *Cladosporium cucumerinum* were 47, 150 and 37 µg/ml for α-, γ- and ε-viniferin, respectively. By comparison, the ED_{50} value for inhibition of spore germination of *C. cucumerinum* by *trans*-resveratrol was greater than 200 µg/ml. It is suggested that *trans*-resveratrol probably functions as a biosynthetic precursor of the viniferins, but due to its low antifungal activity it should not be considered as a phytoalexin (Langcake and Pryce, 1977*a*). Although not normal constituents of healthy leaves, *trans*-resveratrol and ε-viniferin are also present at high concentrations (more than 500 µg/g) in lignified stem tissue of grapevine.

In leaves infected with *B. cinerea*, α-viniferin was the major antifungal compound (50µg/g of leaf tissue) while ε-viniferin was present at 10 µg/g. In addition to α-viniferin (20 µg/g) and ε-viniferin (15 µg/g), UV-irradiated leaves contained β-viniferin (9 µg/g) and γ-viniferin (80 µg/g) together with other unidentified antifungal compounds. The structure of ε-viniferin (**44**), the simplest member of the class, was established largely from MS, UV and NMR spectra. The mass spectra of the pentamethyl and pentaacetyl derivatives of ε-viniferin (**44**) established its molecular formula as $C_{28}H_{22}O_6$ and comparison of its UV and NMR spectra with those of *trans*-resveratrol ($C_{14}H_{12}O_3$) indicated its

44 ε-Viniferin   45 α-Viniferin

46 Pterostilbene

dimeric nature. It has been proposed that the viniferins are synthesized by an oxidative process analogous to that involved in the formation of dimeric lignans. Support for this hypothesis was obtained when resveratrol was treated with horseradish peroxidase and hydrogen peroxide. The major product of this *in vitro* reaction, obtained in 40% yield, was an antifungal resveratrol dehydro-dimer with a dihydrobenzofuran structure analogous to that of ε-viniferin but with a different coupling orientation (Langcake and Pryce, 1977c). α-Viniferin (**45**) is the only other viniferin which has been extensively characterized (Pryce and Langcake, 1977). It has a non-symmetrical tricyclic structure analogous to that of the dimeric ε-viniferin. The detailed stereochemistry of α-viniferin was derived largely from proton NMR spectra of the compound and its methyl ether and acetate derivatives.

A methylated resveratrol derivative, *trans*-pterostilbene (**46**) (3,5-dimethoxy-4'-hydroxystilbene) has also been described as a phytoalexin from *Vitis vinifera* (Langcake *et al.*, 1979). *Trans*-pterostilbene is produced during infection of vine leaves by the downy mildew pathogen, *Plasmopara viticola*, and although only a minor component of the phytoalexin response of *V. vinifera*, its antifungal activity is relatively high by comparison with resveratrol and the viniferins.

Uncharacterized phytoalexins

There have been many claims for the existence of unidentified phytoalexins in other plant families and a few of these merit consideration.

Ginkgoaceae

The resistance of leaves of the primitive gymnosperm *Ginkgo biloba* to fungal penetration by *Botrytis allii* appears to be due to an induced resistance mechanism involving the production of an epidermal penetration inhibitor (Christensen, 1972). It was found that enzymically isolated leaf cuticle was penetrated by test fungi which were non-pathogenic to *Ginkgo*, and that leaf surface wax, obtained by dipping leaves in chloroform, enhanced spore germination *in vitro*, mycelial growth and the penetration of collodion membranes. However, fungi which germinated and grew on the surface of leaves did not penetrate the cuticle. A defence reaction was observed in the epidermis beneath appressoria, each reaction being characterized by a halo of small clear globules of an unidentified substance. Such reaction zones were absent in leaves rendered susceptible to fungal attack by freeze-drying or other treatments.

Leaves inoculated with *B. allii* and incubated for 10 days were extracted with acetone and the extract was found to contain a fungal inhibitor. This inhibitor was absent from extracts of healthy non-inoculated leaves and inoculated leaves rendered susceptible to penetration. When bioassayed *in vitro* the extract inhibited the penetration of collodion membranes by *B. allii* but did not inhibit spore germination or mycelial growth. This is the first report of a phytoalexin

Table 4.2 Physical properties of phytoalexins

Compound	Elemental composition (mol. wt.)	M.p. °C	Derivatives (M.p. °C)	λ_{max} nm (EtOH)	Reference
(1)	$C_{15}H_{14}O_2$ (226)	197–198		211, 285, 290sh	Coxon et al., 1980
(2)	$C_{15}H_{14}O_3$ (242)	132–135		226, 284, 290sh	Coxon et al., 1980
(3)	$C_{16}H_{16}O_3$ (256)	187–195		225, 279, 283sh	Coxon et al., 1980
(4)	$C_{18}H_{16}O_6$ (328)	147–150		242, 279, 335	Geigert et al., 1973
(5)	$C_{17}H_{12}O_6$ (312)	122.5–123.5		252, 285sh, 317	Geigert et al., 1973
(6)	$C_{13}H_{12}O_2$ (200)			215, 225, 235, 246, 255, 269, 290, 309, 330, 354	Allen and Thomas, 1971a
(7)	$C_{13}H_{10}O_2$ (198)			212, 222, 231, 240, 256, 270, 288, 305, 325, 349, 376	Allen and Thomas, 1971c
(8)	$C_{15}H_{22}O_3$ (250)		semicarbazone (133–134)		Akazawa, 1960
(9)	$C_{15}H_{20}O_3$ (248)			211	Oguni and Uritani, 1974a
(10)	$C_{15}H_{22}O_4$ (266)		3,5-dinitrobenzoate (81)		Kato et al., 1971
(11)	$C_{20}H_{32}$ (282)				Sitton and West, 1975
(12)	$C_{20}H_{26}O_3$ (314)	235–236			Kato et al., 1977
(13)	$C_{20}H_{26}O_4$ (330)	242d.			Kato et al., 1977
(14)	$C_{10}H_{12}O_3$ (180)	74		213, 265, 300sh	Keen and Littlefield, 1979
(15)	$C_{10}H_{10}O_3$ (178)	72–3		204, 250sh, 305sh, 340	Keen and Littlefield, 1979
(16)	$C_{15}H_{16}O_4$ (260)	159–163	triacetate	229, 279, 288sh, 298sh, 374	Bell et al., 1975
(17)	$C_{15}H_{16}O_4$ (260)	142–144		267, 365 ($CHCl_3$)	Sadykov et al., 1974
(18)	$C_{16}H_{18}O_4$ (274)	147–149		223, 266, 350, 385	Karimdzhanov et al., 1976
(19)	$C_{16}H_{18}O_4$ (274)	156–160		225, 268, 281sh, 352, 388	Bell et al., 1975

(20)	$C_{15}H_{16}O_3$ (244)	174–178.5		222, 259, 281sh, 336, 339	Bell et al., 1975
(21)	$C_{30}H_{30}O_8$ (518)	184–187		236, 283sh, 289, 376	Stipanovic et al., 1975b
(22)	$C_{31}H_{32}O_8$ (532)	146–149		235, 288, 369	Stipanovic et al., 1975b
(23)	$C_{32}H_{34}O_8$ (546)	181–184		231, 253, 287, 360, 390sh	Stipanovic et al., 1975b
(24)	$C_{16}H_{14}O_5$ (286)	83–85	diacetate (126–127)	217, 304, 313, 326	Takasugi et al., 1978a
(25)	$C_{16}H_{14}O_5$ (286)	184–185	diacetate (82–84) dimethyl ether (104–105)	218, 285, 294, 325, 337	Takasugi et al., 1978a
(26)	$C_{19}H_{18}O_4$ (310)	198–199	triacetate (156–157)	219, 287sh, 296sh, 319, 333	Takasugi et al., 1978b
(27)	$C_{19}H_{16}O_4$ (308)	130–131	diacetate (125–126)	219, 329, 342, 360	Takasugi et al., 1978b
(28)	$C_{19}H_{16}O_4$ (308)	184–185	diacetate (87–88)		Takasugi et al., 1979
(29)	$C_{16}H_{14}O_5$ (286)	188–189	dimethyl ether (101–103)	217, 282, 291, 321, 334	Takasugi et al., 1979
(30)	$C_{19}H_{16}O_4$ (308)	198–199			Takasugi et al., 1979
(31)	$C_{20}H_{18}O_5$ (338)	191–192		217, 306, 315, 329	Takasugi et al., 1979
(32)	$C_{14}H_{12}O_4$ (244)	199–200	tetraacetate (141–142)	218, 235sh, 290sh, 301, 330	Takasugi et al., 1978c
(33)	$C_{19}H_{20}O_4$ (312)	196–197	tetramethyl ether (90–91)	220, 240sh, 294sh, 303, 330	Takasugi et al., 1978c
(34)	$C_{16}H_{18}O_3$ (258)	101–101.5	diacetate	225, 280, 287sh	Takasugi et al., 1980
(35)	$C_{16}H_{18}O_3$ (258)	99.5–100	diacetate	225, 280, 287sh	Takasugi et al., 1980
(36)	$C_{16}H_{16}O_3$ (256)	127–128	methyl ether (86–87)	213, 280, 292, 300sh	Hardegger et al., 1963a
(37)	$C_{16}H_{16}O_3$ (256)	98–99.5	acetate (108–109)	214, 273, 293, 301	Urech et al., 1963
(38)	$C_{15}H_{14}O_3$ (242)	162.5–164	diacetate (126.5–127.5)	214, 274, 293, 302	Urech et al., 1963
(39)	$C_{10}H_{12}O_4$ (196)	80.5–81		288, 320	Hartmann and Nienhaus, 1974
(40)	$C_{11}H_{12}O_4$ (208)	75–76		217, 267, 302	Condon and Kuć, 1960
(41)	$C_{17}H_{24}O$ (244)			229, 240, 254	Harding and Heale, 1980
(43)	$C_{12}H_8O_4$ (216)	146–147		249, 262sh, 300, 340sh	Johnson et al., 1973
(44)	$C_{28}H_{22}O_6$ (454)	155–160	pentaacetate	224, 286sh, 310, 324, 345sh	Langcake and Pryce, 1977a
(45)	$C_{42}H_{30}O_9$ (678)		hexaacetate	225sh, 281sh, 286, 293sh	Pryce and Langcake, 1977
(46)	$C_{16}H_{16}O_3$ (256)	88–89	acetate (128)		Langcake et al., 1979

apparently acting selectively by inhibiting the process of fungal penetration, whilst allowing germination and growth to proceed.

Gramineae

One form of resistance of corn (*Zea mays*) to *Helminthosporium turcicum* is conditioned by a single dominant gene known as Ht-gene (Lim *et al.*, 1968). Evidence was obtained that this monogenic resistance is chemical in nature and due to phytoalexin accumulation. By examining leaf diffusates and extracts from both healthy and inoculated leaves from both susceptible and resistant cultivars of corn it was found that only the inoculated leaves of resistant cultivars contained active quantities of compounds which inhibited germination of *H. turcicum*. No substance inhibitory to spore germination was detected in non-infected corn or in inoculated susceptible corn. Thus it appears that the production of phytoalexin is conditioned by the Ht-gene and occurs in resistant corn only when host and pathogen interact. An inhibitory extract obtained from leaf diffusates was chromatographed and separated into two fractions (Lim *et al.*, 1970). These fractions were both UV absorbing (λ_{max} 327, 280; λ_{max} 270 nm), fluorescent under UV illumination and reacted with diazotized sulphanilic acid, suggesting that they were phenolic in nature. Their chemical structures were not further investigated.

A further study (Obi, 1979) showed the presence of four additional antifungal compounds in an extract from *H. turcicum* inoculated corn leaves but apart from their phenolic nature, little further evidence was obtained for their structures.

Piperaceae

When leaves of various *Peperomia* species were inoculated with avirulent strains of *Phytophthora nicotianae* var. *parasitica* hypersensitive lesions were formed resulting in an induced resistance to colonization by virulent strains of the same fungus (Siradhana *et al.*, 1969). Fungal inhibitors could not be extracted from hypersensitive lesions with either water or ethanol, but aqueous diffusates from hypersensitive tissue were inhibitory to fungi. Leaf disc diffusates induced by either avirulent and virulent isolates were both inhibitory when collected 72 h after inoculation. The inhibitory substances could not be extracted from the crude diffusates with organic solvents but could be separated by adsorbing them on an anion exchange resin and eluting with aqueous formic acid. The inhibitory substances were toxic to *P. nicotianae* var. *parasitica*, *Sclerotinia fructicola* and *Glomerella cingulata*, but not *Alternaria solani*. Inhibition of spore germination was demonstrated for 24 h diffusates from hypersensitive tissue only, but after 48 h or more, toxic substances diffused from both hypersensitive and rotted tissues. There was evidence from UV absorption of the anionic fractions that the compounds in the 48 h diffusates were different from those in the 24 h diffusates.

The anionic fractions from the 24 h diffusate absorbed in the UV spectrum at λ_{max} 199 and 232 nm.

The overall evidence presented in this work suggests a resistance mechanism in which phytoalexin(s) could be involved. However, the anionic nature of the crude substances obtained and their UV spectra suggest that if this is the case they are very different from the known classes of phytoalexins.

Rosaceae

Evidence for the production of phytoalexins in strawberry roots has been presented by Mussell and Staples (1971). Growth of *Phytophthora fragariae* on susceptible strawberry plants leads to a disease condition known as red core. In resistant cultivars inoculated with *P. fragariae* two fractions inhibitory to *Cladosporium cucumerinum* were produced. These substances were not detected in extracts from healthy roots but appeared in infected resistant cultivars within 48 h. Both components were extracted from inoculated resistant roots with ethanol and were subsequently shown to be soluble both in diethyl ether and 5 % sodium hydroxide solution. After separation by thin layer chromatography, both compounds were shown to be inhibitory to mycelial growth of *P. fragariae*. Inoculation of a susceptible strawberry cultivar resulted in the appearance of only one of the compounds but it was not detected until 5 to 8 days after infection. Further characterization of these compounds was not however carried out.

Concluding remarks

At present, available data are inadequate to allow generalization about the diversity of plant families which may utilize phytoalexin production as one method of disease resistance. Deverall (1977), for example, described his failure to detect phytoalexins either in wheat leaves undergoing a hypersensitive reaction to *Puccinia graminis* or in cucumber leaves challenged with *Colletotrichum* spp. More recently, however, Cartwright and Russell (1981) and Kuć and Caruso (1977) have reported preliminary results indicating that phytoalexins are indeed produced by these plants. Until thorough investigations have shown the contrary, it is therefore tempting to suggest that phytoalexins may be produced by all angiosperms, although they may not necessarily accumulate in large quantities. Further studies, particularly with Gramineae and gymnosperms, could clarify the possible ubiquitous nature of phytoalexins.

REFERENCES

Akazawa, T. (1960) Chromatographic isolation of pure ipomeamarone and reinvestigation of its chemical properties. *Arch. Biochem. Acta*, **90**, 82–89.

Akazawa, T. and Wada, K. (1961) Analytical study of ipomeamarone and chlorogenic acid alterations in sweet potato roots infected by *Ceratocystis fimbriata*. *Plant Physiol.*, **36**, 139–144.

Allen, E. H. and Thomas, C. A. (1971a) Trans-trans-3,11-tridecadiene-5,7,9-triyne-1,2-diol, an antifungal polyacetylene from diseased safflower (*Carthamus tinctorius*). *Phytochemistry*, **10**, 1579–1582.

Allen, E. H. and Thomas, C. A. (1971b) Time course of safynol accumulation in resistant and susceptible safflower infected with *Phytophthora drechsleri*. *Physiol. Plant Pathol.*, **1**, 235–240.

Allen, E. H. and Thomas, C. A. (1971c) A second antifungal polyacetylene from *Phytophthora*-infected safflower. *Phytopathology*, **61**, 1107–1109.

Bell, A. A. (1967) Formation of gossypol in infected or chemically irritated tissues of *Gossypium* species. *Phytopathology*, **57**, 759–764.

Bell, A. A. (1969) Phytoalexin production and *Verticillium* wilt resistance in cotton. *Phytopathology*, **59**, 1119–1127.

Bell, A. A., Stipanovic, R. D., Howell, C. R. and Fryxell, P. A. (1975) Antimicrobial terpenoids of *Gossypium*: Hemigossypol, 6-methoxyhemigossypol and 6-deoxyhemigossypol. *Phytochemistry*, **14**, 225–231.

Bohlmann, F. and Herbst, P. (1959) Polyacetylenverbindungen. XXV. Synthesen der Polyine aus *Centaurea ruthenica* L. *Chem. Ber.*, **92**, 1319–1328.

Cartwright, D., Langcake, P., Pryce, R. J., Leworthy, D. P. and Ride, J. P. (1977) Chemical activation of host defence mechanisms as a basis for crop protection. *Nature*, **267**, 511–513.

Cartwright, D. W., Langcake, P., Pryce, R. J., Leworthy, D. P. and Ride, J. P. (1981) Isolation and characterization of phytoalexins from rice as momilactones A and B. *Phytochemistry*, **20**, 535–537.

Cartwright, D. W. and Russell, G. E. (1981) Possible involvement of phytoalexins in durable resistance of winter wheat to yellow rust. *Trans. Br. mycol. Soc.*, **78**, 323–325.

Chalutz, E., DeVay, J. E. and Maxie, E. C. (1969) Ethylene-induced isocoumarin formation in carrot root tissue. *Plant Physiol.*, **44**, 235–241.

Christensen, T. G. (1972) A study of the resistance of *Gingko biloba* L. to fungi: phytoalexin production induced by *Botrytis allii* Munn. *Diss. Abstr. Int. B.*, **32**, 4340.

Condon, P. and Kuć, J. (1960) Isolation of a fungitoxic compound from carrot root tissue inoculated with *Ceratocystis fimbriata*. *Phytopathology*, **50**, 267–270.

Condon, P. and Kuć, J. (1962) Confirmation of the identity of a fungitoxic compound produced by carrot root tissue. *Phytopathology*, **52**, 182–183.

Condon, P., Kuć, J. and Draudt, H. N. (1963) Production of 3-methyl-6-methoxy-8-hydroxy-3,4-dihydroisocoumarin by carrot root tissue. *Phytopathology*, **53**, 1244–1250.

Coxon, D. T., Curtis, R. F., Price, K. R. and Levett, G. (1973) Abnormal metabolites produced by *Daucus carota* roots stored under conditions of stress. *Phytochemistry*, **12**, 1881–1885.

Coxon, D. T., O'Neill, T. M., Mansfield, J. W. and Porter, A. E. A. (1980) Identification of three hydroxyflavan phytoalexins from daffodil bulbs. *Phytochemistry*, **19**, 889–891.

Deverall, B. J. (1977) *Defence Mechanisms of Plants*, Cambridge University Press, Cambridge.

Dueber, M. T., Adolf, W. and West, C. A. (1978) Biosynthesis of the diterpene phytoalexin casbene. *Plant Physiol.*, **62**, 598–603.

Garrod, B., Lewis, B. G. and Coxon, D. T. (1978) *Cis*-heptadeca-1,9-diene-4,6-diyne-3,8-diol, an antifungal polyacetylene from carrot root tissue. *Physiol. Plant Pathol.*, **13**, 241–246.

Gaumann, E., Nuesch, J. and Rimpau, R. H. (1960) Weitere Untersuchungen uber die Chemischen Abwehrreaktionen der Orchideen. *Phytopath. Z.*, **38**, 274–308.

Gaumann, E. (1963) Weitere Untersuchungen uber die Chemische Infektabwehr der Orchideen. *Phytopath. Z.*, **49**, 211–232.

Geigert, J., Stermitz, F. R., Johnson, G., Maag, D. D. and Johnson, D. K. (1973) Two phytoalexins from sugarbeet (*Beta vulgaris*) leaves. *Tetrahedron*, **29**, 2703–2706.

Goodliffe, J. P. and Heale, J. B. (1978) The role of 6-methoxymellein in the resistance and susceptibility of carrot root tissue to the cold-storage pathogen *Botrytis cinerea*. *Physiol. Plant Pathol.*, **12**, 27–43.

Gottlieb, O. R. (1977) The flavonoids: Indispensable additions to a recent coverage. *Israel J. Chem.*, **16**, 45–51.

Hardegger, E., Schellenbaum, M. and Corrodi, H. (1963a) Welkstoffe und Antibiotika. Uber induzierte Abwehrstoffe bei Orchideen II. *Helv. Chim. Acta*, **46**, 1171–1180.

Hardegger, E., Biland, H. R. and Corrodi, H. (1963b) Synthese von 2,4-Dimethoxy-6-hydroxy-phenanthren und Konstitution des Orchinols. *Helv. Chim. Acta*, **46**, 1354–1360.

Harding, V. K. and Heale, J. B. (1980) Isolation and identification of the antifungal compounds accumulating in the induced resistance response of carrot slices to *Botrytis cinerea*. *Physiol. Plant Pathol.*, **17**, 277–289.

Harding, V. K. and Heale, J. B. (1981) The accumulation of inhibitory compounds in the induced resistance response of carrot root slices to *Botrytis cinerea*. *Physiol. Plant Pathol.*, **18**, 7–15.

Hartmann, G. and Nienhaus, F. (1974) The isolation of xanthoxylin from the bark of *Phytophthora*- and *Hendersonula*-infected *Citrus limon* and its fungitoxic effect. *Phytopath. Z.*, **81**, 97–113.

Hunter, R. E., Halloin, J. M., Veech, J. A. and Carter, W. W. (1978) Terpenoid accumulation in hypocotyls of cotton seedlings during ageing and after infection by *Rhizoctonia solani*. *Phytopathology*, **68**, 347–350.

Hyodo, H., Uritani, I. and Akai, S. (1969) Production of furanoterpenoids and other compounds in sweet potato root tissue in response to infection by various isolates of *Ceratocystis fimbriata*. *Phytopath. Z.*, **65**, 332–340.

Johnson, C., Brannon, D. R. and Kuć, J. (1973) Xanthotoxin: a phytoalexin of *Pastinaca sativa* root. *Phytochemistry*, **12**, 2961–2962.

Johnson, G., Maag, D. D., Johnson, D. K. and Thomas, R. D. (1976) The possible role of phytoalexins in the resistance of sugarbeet (*Beta vulgaris*) to *Cercospora beticola*. *Physiol. Plant Pathol.*, **8**, 225–230.

Karimdzhanov, A. K., Ismailov, A. I., Abdullaev, Z. S., Islambekov, Sh. Yu., Kamaev, F. G. and Sadykov, A. S. (1976) Structure of gossyvertin—a new phytoalexin of the cotton plant. *Khim. Prir. Soedin.*, 238–242.

Kato, N., Imaseki, H., Nakashima, N. and Uritani, I. (1971) Structure of a new sesquiterpenoid, ipomeamaronol, in diseased sweet potato root tissue. *Tetrahedron Lett.*, 843–846.

Kato, T., Tsunakawa, M., Sasaki, N., Aizawa, H., Fujita, K., Kitahara, Y. and Takahashi, N. (1977) Growth and germination inhibitors in rice husks. *Phytochemistry*, **16**, 45–48.

Keen, N. T. and Littlefield, L. J. (1979) The possible association of phytoalexins with resistant gene expression in flax to *Melampsora lini*. *Physiol. Plant Pathol.*, **14**, 265–280.

Kojima, M. and Uritani, I. (1976) Possible involvement of furano-terpenoid phytoalexins in establishing host-parasite specificity between sweet potato and various strains of *Ceratocystis fimbriata*. *Physiol. Plant Pathol.*, **18**, 97–111.

Kubota, T. and Matsuura, T. (1953) Chemical studies on the black rot disease of sweet potato V. Chemical constitution of ipomeamurone. *J. Chem. Soc. Japan*, **74**, 248–251.

Kuć, J. and Caruso, F. L. (1977) "Activated co-ordinated chemical defense against disease in plants", in *Host Plant Resistance to Pests*, ed. P. A. Hedin, American Chemical Society, Washington, D.C., 78–89.

Langcake, P. and Pryce, R. J. (1977a) A new class of phytoalexins from grapevines. *Experientia*, **33**, 151–152.

Langcake, P. and Pryce, R. J. (1977b) The production of resveratrol and the viniferins by grapevines in response to ultraviolet irradiation. *Phytochemistry*, **16**, 1193–1196.

Langcake, P. and Pryce, R. J. (1977c) Oxidative dimerisation of 4-hydroxystilbenes *in vitro*: production of a grapevine phytoalexin mimic. *J. Chem. Soc. Chem. Commun.*, 208–210.

Langcake, P., Cornford, C. A. and Pryce, R. J. (1979) Identification of pterostilbene as a phytoalexin from *Vitis vinifera* leaves. *Phytochemistry*, **18**, 1025–1027.

Letcher, R. M. and Nhamo, L. R. M. (1973) Structure of orchinol, loroglossol and hircinol. *J. Chem. Soc., Perkin Trans 1*, 1263–1265.

Lim, S. M., Paxton, J. D. and Hooker, A. L. (1968) Phytoalexin production in corn resistant to *Helminthosporium turcicum*. *Phytopathology*, **58**, 720–721.

Lim, S. M., Hooker, A. L. and Paxton, J. E. (1970) Isolation of phytoalexins from corn with monogenic resistance to *Helminthosporium turcicum*. *Phytopathology*, **60**, 1071–1075.

Mussell, H. W. and Staples, R. C. (1971) Phytoalexin-like compounds apparently involved in strawberry resistance to *Phytophthora fragariae*. *Phytopathology*, **61**, 515–517.

Musumeci, M. R. and Oliveira, A. R. (1976) Accumulation of phenols and phytoalexins in Citrus tissues following inoculation with *P. citrophthora*. *Summa Phytopathol.*, **2**, 27–31.

Nakada, H., Kobayashi, A. and Yamashita, K. (1977) Stereochemistry and biological activity of phytoalexin "safynol" from safflower. *Agric. Biol. Chem.*, **41**, 1761–1765.

Obi, I. U. (1979) Additional phytoalexin-like compounds in Ht-gene resistance of corn to *Helminthosporium turcicum*. *Ann. appl. Biol.*, **92**, 377–381.

Oguni, I. and Uritani, I. (1974a) Dehydroipomeamarone from infected *Ipomoea batatas* root tissue. *Phytochemistry*, **13**, 521–522.

Oguni, I. and Uritani, I. (1974b) Dehydroipomeamarone as an intermediate in the biosynthesis of ipomeamarone, a phytoalexin from the sweet potato root infected with *Ceratocystis fimbriata*. *Plant Physiol.*, **53**, 649–652.

Pryce, R. J. and Langcake, P. (1977) α-Viniferin: an antifungal resveratrol trimer from grapevines. *Phytochemistry*, **16**, 1452–1454.

Sadykov, A. S., Metlitskii, L. V., Karimdzhanov, A. K., Ismailov, A. I., Mukhamedova, R. A., Avazkhodzhaev, M. K. H. and Kamaev, F. G. (1974) Isohemigossypol—the phytoalexin of the cotton plant. *Dokl. Akad. Nauk. SSSR*, **218**, 1472–1475.

Siradhana, B. S., Schmitthenner, A. F. and Ellett, C. W. (1969) Formation of phytoalexin in *Peperomia* in relation to resistance to *Phytophthora nicotianae* var. *parasitica*. *Phytopathology*, **59**, 405–410.

Sitton, D. and West, C. A. (1975) Casbene: an antifungal diterpene produced in cell-free extracts of *Ricinus communis* seedlings. *Phytochemistry*, **14**, 1921–1925.

Steiner, K., Egli, C., Rigassi, N., Helali, S. E. and Hardegger, E. (1974) Welkstoffe und Antibiotika. Zur Synthese des Orchinols. *Helv. Chim. Acta.*, **57**, 1137–1141.

Stipanovic, R. D., Bell, A. A. and Howell, R. D. (1975a) Naphthofuran precursors of sesquiterpenoid aldehydes in diseased *Gossypium*. *Phytochemistry*, **14**, 1809–1811.

Stipanovic, R. D., Bell, A. A., Mace, M. E. and Howell, C. R. (1975b) Antimicrobial terpenoids of *Gossypium*: 6-methoxygossypol and 6,6′-dimethoxygossypol. *Phytochemistry*, **14**, 1077–1081.

Stoessl, A., Rock, G. L. and Fisch, M. H. (1974) An efficient synthesis of orchinol and other orchid phenanthrenes. *Chem. and Ind. (London)*, 703–704.

Takasugi, M., Nagao, S., Masamune, T., Shirata, A. and Takahashi, K. (1978a) Structure of moracin A and B, new phytoalexins from diseased mulberry. *Tetrahedron Lett.*, 797–798.

Takasugi, M., Nagao, S., Ueno, S., Masamune, T., Shirata, A. and Takahashi, K. (1978b) Moracin C and D, new phytoalexins from diseased mulberry. *Chem. Lett.*, 1239–1240.

Takasugi, M., Munoz, L., Masamune, T., Shirata, A. and Takahashi, K. (1978c) Stilbene phytoalexins from diseased mulberry. *Chem. Lett.*, 1241–1242.

Takasugi, M., Nagao, S., Masamune, T., Shirata, A. and Takahashi, K. (1979) Structures of moracins E, F, G and H, new phytoalexins from diseased mulberry. *Tetrahedron Lett.*, 4675–4678.

Takasugi, M., Anetai, M., Masamune, T., Shirata, A. and Takahashi, K. (1980) Broussonins A and B, new phytoalexins from diseased paper mulberry. *Chem. Lett.*, 339–340.

Thomas, C. A. and Allen, E. H. (1970) An antifungal polyacetylene compound from *Phytophthora* infected safflower. *Phytopathology*, **60**, 261–263.

Thomas, C. A. and Allen, E. H. (1971) Light and antifungal polyacetylene compounds in relation to resistance of safflower to *Phytophthora drechsleri*. *Phytopathology*, **61**, 1459–1461.

Urech, J., Fechtig, B., Nuesch, J. and Vischer, E. (1963) Hircinol, eine antifungisch wirksame Substanz aus Knollen von *Loroglossum hircinum* (L.) Rich. *Helv. Chim. Acta*, **46**, 2758–2766.

Uritani, I., Uritani, M. and Yamada, H. (1960) Similar metabolic alterations induced in sweet potato by poisonous chemicals and by *Ceratostomella fimbriata*. *Phytopathology*, **50**, 30–34.

Ward, E. W. B., Unwin, C. H. and Stoessl, A. (1975) Loroglossol: an orchid phytoalexin. *Phytopathology*, **65**, 632–633.

Yang, D. T. C., Wilson, B. J. and Harris, T. M. (1971) The structure of ipomeamaronol: a new toxic furanosesquiterpene from mouldy sweet potatoes. *Phytochemistry*, **10**, 1653–1654.

Zaki, A. I., Keen, N. T., Sims, J. J. and Erwin, D. C. (1972a) Vergosin and hemigossypol, antifungal compounds produced in cotton plants inoculated with *Verticillium albo-atrum*. *Phytopathology*, **62**, 1398–1401.

Zaki, A. I., Keen, N. T. and Erwin, D. C. (1972b) Implication of vergosin and hemigossypol in the resistance of cotton to *Verticillium albo-atrum*. *Phytopathology*, **62**, 1402–1406.

5 Biosynthesis of phytoalexins

A. STOESSL

Introduction

The term *biosynthesis* refers to the process by which a given compound is elaborated from less complex substances by living organisms. Usually, this comprises a lengthy sequence of enzyme mediated, endergonic reactions (Brown, 1972). Main aspects of interest are the participating molecular components (precursors and intermediates; enzymes) and the mechanisms (i.e. sequence of bonds made and broken) by which the reactions are effected. The expression *biogenesis* has a wider meaning, with less emphasis on details of mechanism, but with coverage extending to exergonic and catabolic reactions which are not properly described as biosynthetic. Nevertheless, in many situations the terms can be used interchangeably. Individual biosynthetic pathways, though essentially limitless in number, are variations on a relatively few main routes on which all natural products are generated. These biosynthetic highways form a rational basis for the arrangement of compounds in this chapter. Familiarity with these basic routes is assumed. For information on experimental strategies and techniques, readers may refer to, *inter alia*, the article by Brown (1972) as a concise general introduction, and that by Tanabe (1973) as an account of recent tracer methodologies which exploit the advantages offered by stable isotopes.

Phytoalexins by definition are secondary metabolites, i.e. compounds which are not obviously essential to growth and normal metabolism of the producing organism (Martin and Demain, 1980). They exhibit the range and complexity of structure typical of secondary metabolites of higher plants, in contradistinction to the frequently more esoteric compounds produced by microorganisms. It is especially noteworthy that different main biosynthetic routes are utilized in the construction of different phytoalexins. This diversity is inconsistent with a common evolutionary origin; it indicates that phytoalexins are not, in any strict sense, the agents of a single defence mechanism (Stoessl, 1977, 1980). In general, a given plant family produces phytoalexins belonging to not more than two or three, characteristic structural–biogenic groups. To a limited extent, the

133

converse is also true; i.e. specific classes of phytoalexins may be associated with specific plant families. Examples are pterocarpans and the Leguminosae (VanEtten and Pueppke, 1976) and rearranged eudesmanoids and the Solanaceae (Stoessl *et al.*, 1976, 1977). However, this is not an invariable rule and some structural classes, e.g. stilbenes, are represented by phytoalexins from a fairly wide spectrum of unrelated plant families.

Phytoalexins derived from shikimic acid

Compounds derived from the simple shikimic acid route

Only a few representatives of this large group of plant products have been considered to be phytoalexins. Coniferaldehyde (**1a**) and coniferyl alcohol (**1b**) in flax (Keen and Littlefield, 1979) probably arise by reduction of ferulic acid (**1c**) (Amrhein and Zenk, 1977), but no direct evidence is available. Radioactivity from glucose-^{14}C and phenylalanine-^{14}C was incorporated into benzoic acid (**2a**) produced in infected apples, or apples treated with fungal proteases (Swinburne and Brown, 1975), probably by a degradative process via cinnamic acid (**1d**) or an ester thereof. Such an origin has been demonstrated for 4-hydroxybenzoic acid (**2b**), a congener of **2a** in apples, which originates from

$$R_2\text{—}\underset{R_3}{\bigcirc}\text{—CH}\!=\!\text{CH—}R_1$$

1a, R_1 = CHO, R_2 = OH, R_3 = OMe

1b, R_1 = CH$_2$OH, R_2 = OH, R_3 = OMe

1c, R_1 = CO$_2$H, R_2 = OH, R_3 = OMe

1d, R_1 = CO$_2$H, R_2 = R_3 = H

1e, R_1 = CO·O , R_2 = R_3 = OH

1f, R_1 – CO$_2$H, R_2 = OH, R_3 = H

1g, R_1 = COSCoA, R_2 = OH, R_3 = H

1h, R_1=COAgm; R_2=OH, R_3=H

$$R_2\text{—}\bigcirc\text{—}R_1$$

2a, R_1 = CO$_2$H, R_2 = H

2b, R_1 = CO$_2$H, R_2 = OH

2c, R_1 = R_2 = OH

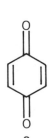

3

the action of fungal enzymes on the host's chlorogenic acid (**1e**) and related acids (Fawcett and Spencer, 1968). However, attempts to demonstrate that degradation to **2a** is effected by fungi in an analogous manner were unsuccessful (Swinburne, 1975). Benzoquinone (**3**) and its dihydro-derivative, hydroquinone (**2c**), both of which were recently claimed as phytoalexins of barley (Evans and Pluck, 1978), must also arise by degradation of cinnamic acid derivatives (Barz and Hösel, 1975) . Given these presumptive origins, it is a question. of terminology whether any of these and some other, similarly implicated compounds can be regarded as true phytoalexins.

A number of coumarins, whose biosynthesis from cinnamic acids is relatively well understood, at least under non-pathological conditions (Brown, 1978), occur as stress compounds, e.g. in potato (Clarke and Baines, 1976) and tobacco (Fritig *et al.*, 1972; Sequeira, 1969) but are not usually considered to be phytoalexins. An exception is xanthotoxin (**4**) which is regarded as a phytoalexin of parsnip (Johnson *et al.*, 1973). The biosynthesis of this compound in parsnip has been studied, but under non-pathological conditions and with results which could not be completely evaluated because poor incorporation precluded location of the labels by degradative studies (Brown, 1970). However, subsequent investigations, in several laboratories but with different plants and furocoumarins, leave little doubt that xanthotoxin, like other furocoumarins, is constructed by prenylation of umbelliferone (**5**), cyclization, loss of the C_3 side-chain from the prenyl residue, hydroxylation, and *O*-methylation (Brown,

Scheme 1   Presumptive biosynthesis of xanthotoxin

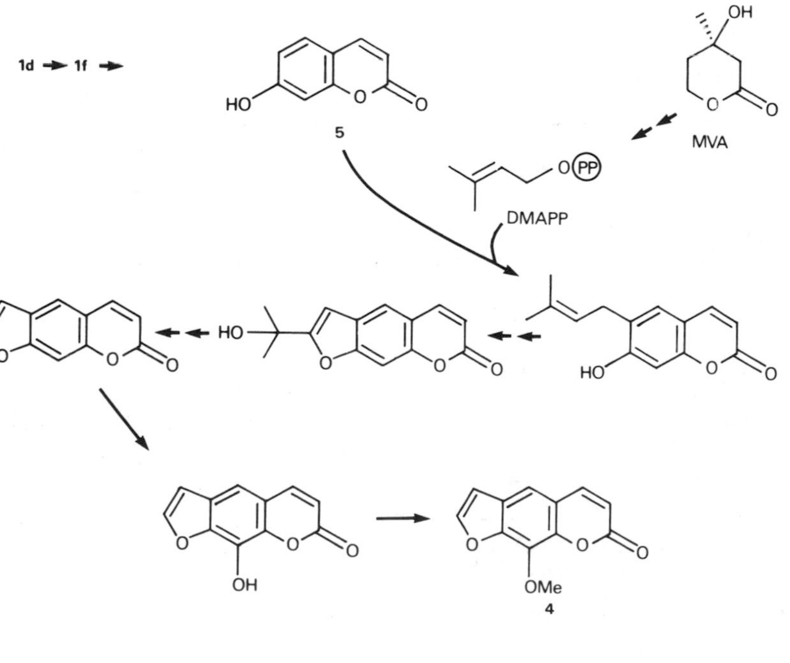

1978) as in Scheme 1, though not necessarily in that precise sequence. Prenylation of an aromatic nucleus, by dimethylallylpyrophosphate (DMAPP) derived from mevalonic acid (MVA) or its lactone is·a very common biosynthetic event which will be encountered repeatedly.

Compounds formed on the shikimic–polymalonic acid route

The intermediate formed by reaction of a cinnamic acid enzyme ester, e.g. *p*-coumaroyl coenzyme A (**1g**), with three units of malonyl coenzyme A can cyclize in different ways, depending on its conformation. This leads to two important series of compounds, each of which can be further sub-divided into

Scheme 2   Main pathways on the shikimic—polymalonic acid route

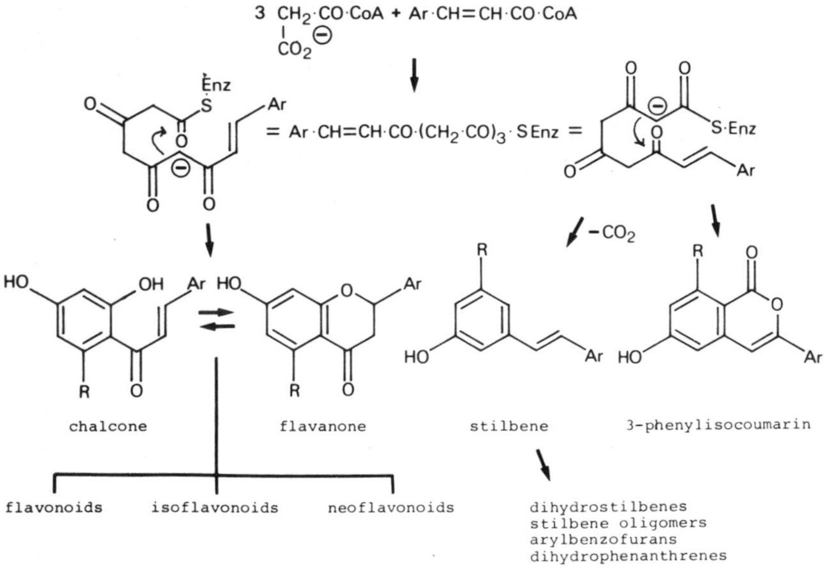

several branches (Scheme 2) but not all of these are pertinent to phytoalexin studies at present.

1. Flavonoids

Flavonoid biosynthesis, including enzymology, has been extensively studied and reviewed (e.g. Hahlbrock and Grisebach, 1975) and its broad outlines are well established. However, the position of chalcones in the general scheme of flavonoid biosynthesis is still a matter of active enquiry. Until the mid-1970's, it was widely believed that chalcones were the first products of the cyclization reaction, giving rise to flavanones by a second ring closure in a further step (Grisebach, 1962). These assumptions became less tenable when an experiment with an enzyme preparation devoid of chalcone-isomerase indicated that a

6a, $R_1 = R_4 = H$, $R_2 = R_3 = R_5 = OH$

6b, $R_1 = R_3 = OMe$, $R_2 = H$, $R_4R_5 = -OCH_2O-$

7

flavanone, naringenin (**6a**) and not the corresponding chalcone **7** was the primary reaction product obtainable from p-coumaroyl coenzyme A and malonyl coenzyme A (Kreuzaler and Hahlbrock, 1975). More recently, however, work in the same laboratory has shown that in fact the chalcone accompanies the flavanone, and that the flavanone is formed non-enzymically. The question of the role of the chalcone therefore appears to be still open (Heller *et al.*, 1980).

The presence or absence of oxygen functions at the 5-,7-,4'- and, in some cases 2'-positions (flavone–isoflavone numbering; note that the numbering is reversed for chalcones) is probably always determined before flavanone–chalcone formation. Other hydroxyl groups are introduced subsequently, at various stages of further elaboration. Flavones, dihydroflavonols and flavonols are formed from flavanones or chalcones by oxidative steps (e.g. Wong and Wilson, 1976); flavanols and flavandiols arise from the more highly oxidized compounds by reduction (e.g. Jacques and Haslam, 1974). Introduction of alkyl groups, either on to oxygen or carbon, is a general biosynthetic process, as already noted for the prenylation of coumarins, and is not limited to a particular stage in the biosynthetic sequence at or beyond flavanones. Several routes may be available to the same compound in a given plant, forming a so-called "metabolic grid" (Bu'Lock, 1965).

(i) Unrearranged flavonoids
Compounds from this class are essentially ubiquitous in higher plants. It is now well established, largely through the work of Barz (1977), that they are normally

subject to rapid metabolic turnover, with the probable implication that they perform essential functions in normal life processes. Numerous reports in the literature describe changes, increases or decreases, in the concentrations of such flavonoids following infection or other trauma. A substantial number of the compounds are known to have antifungal properties but only a few have been described as phytoalexins. These include the broussonins A (**8a**) and B (**8b**),

8a, R$_1$ = H, R$_2$ = Me

8b, R$_1$ = Me, R$_2$ = H

9a, R$_1$ = R$_2$ = H

9b, R$_1$ = OH, R$_2$ = H

9c, R$_1$ = OH, R$_2$ = Me

unusual deoxydihydrochalcones from the mulberry *Broussonetia papyrifera* (Takasugi *et al.*, 1980); the hydroxyflavans **9a** to **9c** from daffodil bulbs (Coxon *et al.*, 1980), which are among the very few phytoalexins known from monocotyledons; and the flavanone betagarin (**6b**) from beet, *Beta vulgaris* (Johnson *et al.*, 1976). The biosynthetic origins of these compounds, by presumably unexceptional processes, can be inferred in outline from their structures but no experimental studies have been reported.

(ii) Isoflavonoids

Of the compounds in this class, pterocarpans, isoflavans and coumestans tend to be associated with stress conditions, either as compounds formed as phytoalexins or as constituents of roots and heartwood, i.e. dead tissue or tissue with maximal exposure to microorganisms. In contrast, isoflavones and isoflavanones are well-known as normal, secondary plant metabolites, though their distribution is restricted to a few specific plant families (Wong, 1975). The formation of the isoflavonoid skeleton through a 1,2-aryl shift in a flavonoid precursor, suggested by the structure of the compounds, was first demonstrated by Grisebach and Doerr (1960). However, neither the precursors nor the precise mechanism are known with certainty (Hahlbrock and Grisebach, 1975). A mechanistically plausible pathway from a chalcone (Scheme 3) proposed by Pelter *et al.* (1971) is supported by experimental studies (e.g., Crombie *et al.*, 1973) and by the characteristic presence of O-substituents in the 4′- or, much less frequently, 2′-position of all known isoflavonoids. In Scheme 3, X$^+$ may be a proton or other positively charged species (see below). A similar pathway from flavones can also be envisaged. Other isoflavonoids are probably derived from

Scheme 3   A plausible pathway for the isoflavonoid rearrangement (after Pelter *et al.*, 1971)

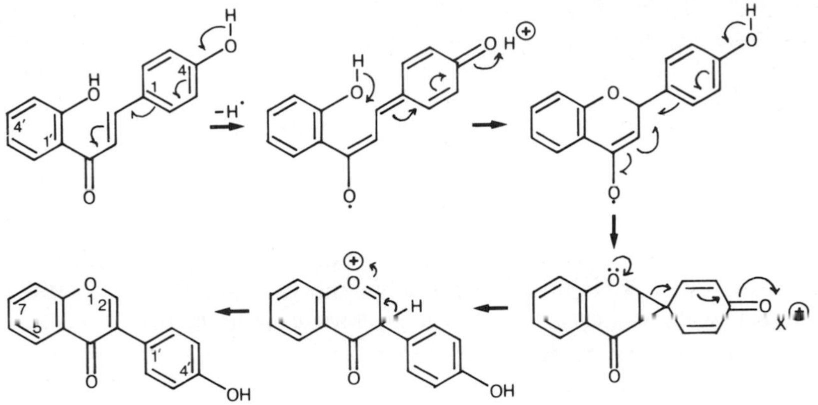

the isoflavones which are the first products of the aryl migration. A comprehensive scheme delineating the then probable relationships was proposed by Wong (1975). While generally still valid, the scheme needs modification in certain details, in terms of pathways more recently suggested by Martin and Dewick (1978), Dewick and Martin (1979) and Woodward (1980) (see below).

(*a*) *Isoflavones.*   Only three members of the class appear to have been reported as phytoalexins. Betavulgarin (**10a**) from beet (Johnson *et al.*, 1976) is somewhat unusual, both in the absence of 4′-oxygenation (but, as just noted, the 2′-hydroxyl will have served equally well in mediating the rearrangement of a flavonoid precursor), and as such appears to be the only known example of an isoflavonoid from the Chenopodiaceae. Its biosynthesis has not been investigated. Wighteone (**10b**) is a normal metabolite of *Lupinus* spp. but has also been described as a phytoalexin of *Glycine wightii* (Ingham *et al.*, 1977). An

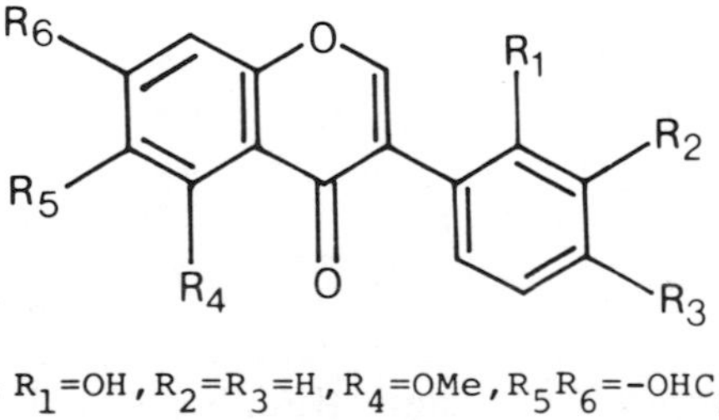

10 a, R_1=OH, R_2=R_3=H, R_4=OMe, R_5R_6=-OHC$_2$O-

10 b, R_1=R_2=H, R_3=R_4=R_6=OH, R_5=CH$_2$·CH=CMe$_2$

10 c, R_1=R_2=R_5=H, R_3=R_4=R_6=OH

10 d, R_1=R_3=R_4=R_6=OH, R_2=R_5=H

10 e, R_1=R_3=R_4=R_6=OH, R_2=CH$_2$·CH=CMe$_2$, R_5=H enzyme (a dimethylallylpyrophosphate:isoflavone dimethylallyl transferase) which catalysed the prenylation of genistein (**10c**) with DMAPP-^{14}C to give wighteone, has been isolated from *Lupinus albus* (Schröder *et al.*, 1979). Treatment of lupin hypocotyls with a fungal elicitor did not increase dimethyl-allyl transferase activity, in agreement with the synthesis of **10b** as a normal metabolite in this plant. 2′-Hydroxygenistein (**10d**), described as a phytoalexin of French and hyacinth bean (Ingham, 1977) is considered in the following subsection.

(*b*) *Isoflavanones.*   Several have been reported as phytoalexins, with kievitone (**11a**) (Bailey and Burden, 1973) probably the best known. As a class, iso-flavanones are generally accepted as being formed by reduction of isoflavones (Hahlbrock and Grisebach 1975; Martin and Dewick, 1978; Wong, 1975). Kievitone (**11a**), when induced by fungal inoculation of either *Lablab niger* (Ingham, 1977) or French bean (Woodward, 1979), is accompanied by much smaller amounts of the 5-hydroxyisoflavones genistein (**10c**) and 2′-hydroxy-genistein (**10d**), and the 5-hydroxyisoflavanone dalbergioidin (**11b**) (apart from

11a,   R=CH$_2$·CH=CMe$_2$
11b,   R=H

5-unsubstituted isoflavans and pterocarpans which are formed as major phyto-alexins). French bean, in addition, forms appreciable amounts of the 5-hydroxy-isoflavone phaseoluteone (**10e**). The co-occurrence and relative molar amounts of the 5-hydroxy compounds are consistent with the sequences, 4′,5′,7-tri-hydroxyflavanone → **10c** → **10d** → **11b** → **11a** and **10c** → **10d** → **10e** (Woodward, 1979); i.e. they conform to the generally accepted concepts.

(*c*) *Pterocarpans and isoflavans.*   Many of these compounds have been reported as phytoalexins but only those whose biosynthesis has been studied experimentally are discussed here. Pterocarpans and isoflavans are considered together because they are intimately connected biogenetically. They occur together in some plants, being interconvertible *in vivo* in at least some instances (see below). In general, they must either be in a precursor-product relationship or, as suggested by Dewick and Martin (1976), pterocarpans and isoflavans with equivalent substitution patterns may be derived from the same precursors. Both isoflavans and pterocarpans carry oxygen substituents at C-4′ (C-9 in pterocarpan numbering), characteristic of isoflavonoids, and at C-7 (C-3, pterocarpan numbering). The latter function is usually hydroxyl; when not, it is an ether group either known or likely to have arisen by alkylation of hydroxyl in a pterocarpan. This conforms with the suggestion of Cornia and Merlini (1975) that the C-7 hydroxyl of isoflavans participates in the oxidative ring closure to pterocarpans (see below). Finally, all plant isoflavans are oxygenated at the 2′-position, as required for interconversion with pterocarpans (Wong, 1975).

It is probable that all recoverable intermediates on the pathway from formononetin (**12a**) to the pterocarpan medicarpin (**13a**) and the isoflavans vestitol (**14a**) and sativan (**14b**) have been identified (Scheme 4), in studies which

Scheme 4 Probable pathway to pterocarpans and isoflavans (after Martin and Dewick, 1978)

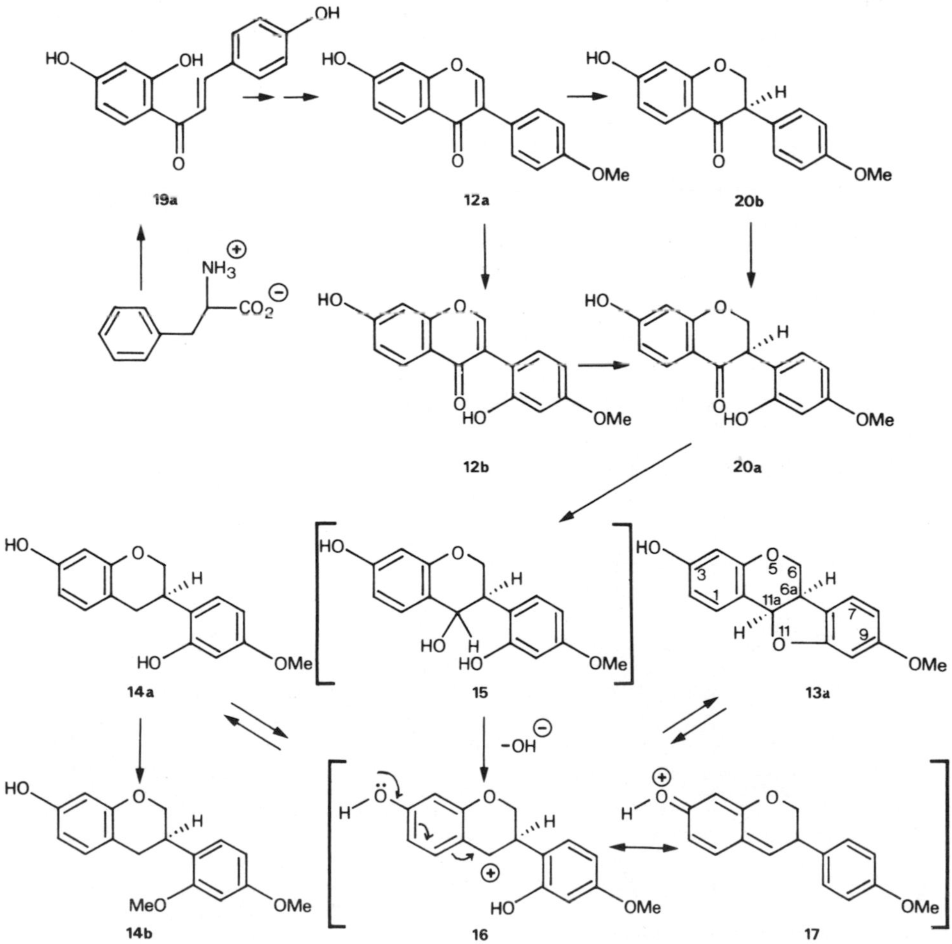

Scheme 5 Possible metabolic grid of isoflavans and pterocarpans

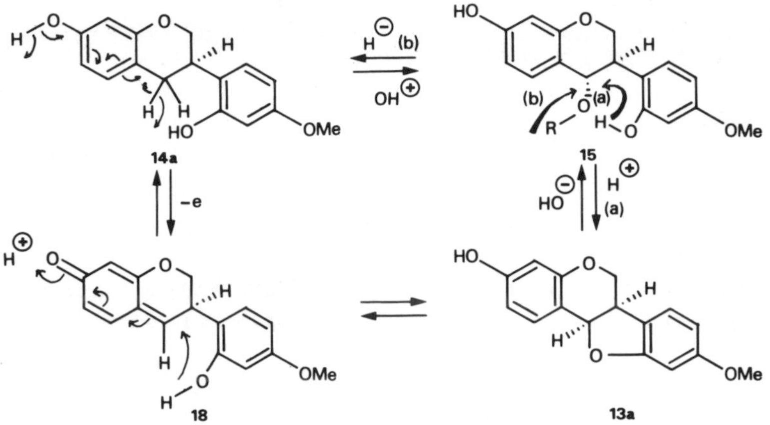

will be discussed in detail. However, the scheme will first be considered as a particular illustration of general principles. In the scheme, the 2′-hydroxy-isoflavan-4-ol **15** is a putative intermediate, no compounds of the type having been isolated from a natural source, presumably because of their high reactivity. Martin and Dewick (1978) suggested that **15** is itself converted to the carbonium ions **16** and **17** as the ultimate entities from which pterocarpans and isoflavans are formed.[1] An alternative possibility is that a quinonemethide, e.g. **18**, intervenes as suggested by Cornia and Merlini (1975) on the basis of the oxidative conversion of isoflavans into pterocarpans in the laboratory. Such an intervention can be accommodated in the overall scheme if a metabolic grid (Scheme 5) replaces the interplay of isoflavanol, carbonium ions and products in Scheme 4.

Allowing for a proper choice of substituents, essentially the same pathway as in Scheme 4 is almost certainly operative for other pterocarpans and isoflavans. An important result which emerges is that, since formononetin (**12a**) and genistein (**10c**) occur as normal metabolites of plants, the point of divergence from normal to stress metabolism is at the isoflavone or subsequent stages; i.e. it will be found either in reduction to the isoflavanone or the putative isoflavanol, or in the imposition of a metabolic block immediately beyond the pterocarpans and isoflavans. The latter case presupposes that the compounds are normal metabolites of the plant but normally do not accumulate to any but

[1] It seems both unnecessary and undesirable to invoke the carbonium ions. The 4-hydroxyl of the isoflavanol **15** would appear to be an ideal site of attachment for the enzyme(s) mediating hydride or phenoxide attack to furnish the isolated products by S_N2 type displacement; the enzyme may also be that effecting the reduction of the 2′-hydroxyisoflavanone **20a**, so that the isoflavanol may never exist in a free state. It seems improbable that an enzyme could stabilize the carbonium ions as postulated, since these at best are unlikely to be of finite existence; ion **17** in particular must be highly destabilized by the vicinal H-3, to give isoflavenes by an aromatization process.

minute traces because of rapid catabolism, as has been argued by Yoshikawa *et al.* (1979) for the soybean phytoalexins.

Medicarpin, vestitol and sativan. Scheme 4 is largely based on extensive studies by Dewick and co-workers. In these, isoliquiritigenin-CO-^{14}C (**19a**), formononetin-Me-^{14}C (**12a**), 2'-hydroxyformononetin-Me-^{14}C (**12c**) and ($\pm$)-2',7-dihydroxy-4-methoxyisoflavanone-^{14}C (**20a**) were all incorporated into medicarpin (**13a**) in Cu^{2+}-stimulated seedlings of *Trifolium pratense* (Dewick, 1975, 1977) and into **13a**, vestitol (**14a**) and sativan (**14b**) in similarly treated seedlings of *Medicago sativa* (Dewick and Martin, 1979a,b). Moreover, when administered to the latter plant, ($\pm$)-medicarpin was incorporated into **14a** and **14b**, while ($\pm$)-vestitol was incorporated into **13a** and **14b** in *M. sativa*, and into **13a** in *T. pratense* (Dewick and Martin, 1979a). The incorporation of **19a**

19a, R=H

19b, R=Me

21

22

into **12a** was demonstrated in *T. pratense* (Dewick, 1975). ($\pm$)-7-Hydroxy-4-methoxyisoflavanone (**20b**) also was incorporated into **13a** and **14a** in *M. sativa* (Dewick and Martin, 1979b), indicating that the critical 2'-hydroxylation occurs at either or both the isoflavone or isoflavanone stage. Throughout these researches incorporations were mostly excellent (dilution values < 100), occasionally extremely efficient, e.g. in some experiments, dilution values for the incorporation of **12b** and **20a** into **13a** were as low as 3, corresponding to a specific incorporation of better than 30%, despite being uncorrected for unused starting material.

Scheme 6   Stereochemistry of isoflavone reduction in medicarpin biosynthesis

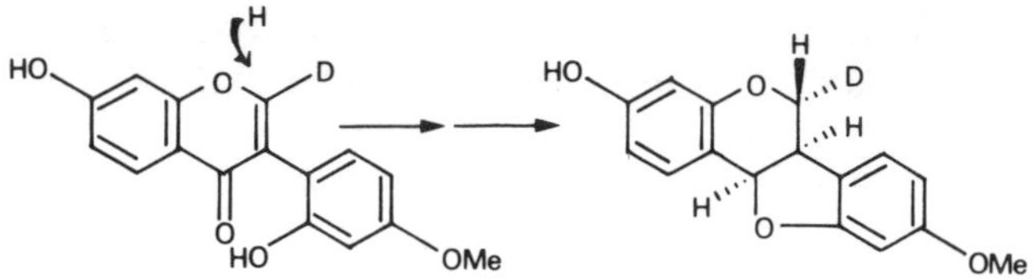

12c   13b

2'-Hydroxisoflavenes have been considered as possible precursors of ptero-carpans or intermediates of the pterocarpan ⇌ isoflavan interconversion. How-ever, in an elegant investigation, Martin and Dewick (1978) were able to disprove the latter role and to show that the former is an unlikely one. In separate experiments, they administered the doubly labelled compounds, (±)-medicarpin-6a-^{3}H-9-OMe-^{14}C and (±)-vestitol-3-^{3}H-4'-OMe-^{14}C to *M. sativa* seedlings pretreated with Cu^{2+} or UV light. The medicarpin (**13a**) was incorpor-ated into vestitol (**14a**) and sativan (**14b**) without significant change in the ^{3}H/^{14}C ratio. Analogous results were obtained for the incorporation of the labelled vestitol into **13a** and **14b**. Had the isoflavene been an intermediate, all the ^{3}H would have been lost, as was in fact observed, in the same experiments, for the incorporation of the two compounds into 9-O-methylcoumestrol (see below). In another experiment, 2',7'-dihydroxy-4'-methoxyisoflavene (**21**) was incorporated in only small amounts into **13a**, **14a** and **14b** under conditions in which other incorporations were excellent, e.g. of the isoflavene **21** into 9-O-methylcoumestrol, and of 2'-hydroxyformononetin (**12b**) into medicarpin and the isoflavans. The conclusion that 2'-hydroxisoflavenes are unlikely precursors of pterocarpans was also reached by van der Merwe *et al.* (1978) on chemical evidence, with steric strain indicated as a barrier to furan ring closure.

The stereochemistry of one of the two reductive steps between **12b** and **13a** was clarified by using the stable isotope deuterium ($= D = {}^2$H) as tracer (Dewick and Ward, 1977). The isoflavone, 96% labelled in the 2-position as in **12c**, was fed to fenugreek (*Trigonella foenum-graecum*) seedlings treated with Cu^{2+} or irradiated with UV light. Medicarpin labelled as in **13b** (0.6 atom ^{2}H by MS) was isolated and purified as the acetate. Both ^{1}H- and ^{2}H-NMR spectra confirmed the enrichment and located the label exclusively in the α, 6-R, equatorial position. This implies an overall *trans*-addition of hydrogen (Scheme 6).

The argument that pterocarpans and isoflavans are not formed sequentially but are derived via the same intermediate (Scheme 4) rests in part on the observation that in comparative experiments the incorporations of **13a** and **14a** into one another were no more efficient than the incorporations of other, more distant precursors (Dewick and Martin, 1979a). Support for a common inter-mediate was also adduced from experiments in which maximal incorporation of phenylalanine into **13a** and **14a** occurred within the first 6 hours of administration to *M. sativa*, whereas about twice that time was required for maximal incorporation into **14b**. However, these arguments, while suggestive, lack compulsion.[2] The exact nature of the relationship between pterocarpans and isoflavans, and the identity of their immediate precursors therefore remains uncertain.

23a, R_1=OH R_2=OMe R_3=H

23b, R_1=OH R_2=OMe R_3=OH

23c, R_1=H R_2=OH R_3=H

23d, R_1=H R_2=OH R_3=OH

24a, R=H

24b, R=OH

Adenosyl

$$\underset{\underset{(CH_2)_2 \cdot \underset{\underset{NH_2}{|}}{CH} \cdot CO_2H}{|}}{\overset{|}{S}} \oplus$$

25

Maackiain (= *inermin*). In some of the experiments with *T. pratense* just discussed (Dewick, 1975, 1977), the chalcone isoliquiritigenin (**19a**) and the isoflavone formononetin (**12a**) were found to be excellent precursors of maackiain (**22**) as well as of medicarpin (**13a**). In contrast, the medicarpin precursors **12b** and **20a** and (±)-medicarpin itself were incorporated into **22** only to an insignificant extent, with the implication that the route to the two phytoalexins branches at the formononetin stage. Starting from this consideration, Dewick and Ward (1978) synthesized and tested the CO-^{14}C labelled isoflavones calycosin (**23a**), Ψ-baptigenin (**24a**), and **24b** as likely precursors of **22**. All three

[2]They do not take into account the possible effects of factors such as compartmentalization and differential transport rates. In addition, much of the force of the first argument is dispelled by certain of the experimental data. Thus, the incorporation of (±)-vestitol (**14a**) into sativan (**14b**) which is almost certainly effected by methylation, in a formally single step should, by the terms of the argument, be exceptionally efficient. In fact, however, it is no more efficient than the incorporations of (±)-medicarpin (**13a**) into **14a** and of the more remote precursors **12b** and **20a** into **14b** (Dewick and Martin, 1979*a*).

compounds were very efficiently incorporated, with dilution values not exceeding 40. A fourth possible contender, **23b**, was too susceptible to oxidation to be tested. The very high incorporation rate observed for **24a** especially (dilution value 9), favours the route **12a**→**23a**→**24a**→**24b**→→**22** as the most likely or important pathway but, as the authors point out, this view must be tentative because transport efficiency, differential metabolism and other such factors may affect the course of the experiment. The true situation may well be a metabolic grid in which two or more alternative steps may be of comparable importance. These questions may eventually be resolved by enzymological studies. In the meantime, the data confirm that the pathway to maackiain (**22**) diverges from that to medicarpin (**13a**) at the formononetin (**12a**) stage by means of 3′-hydroxylation. Steps subsequent to the formation of **24b** are likely to be analogous to those on the medicarpin route.

A further important aspect of these investigations is that they support the conclusion from other studies (e.g. Crombie *et al.*, 1973) that methylation of the 4′-hydroxyl is, in at least some cases, an integral part of the rearrangement of chalcones or flavanones to 4′-methoxyisoflavonoids. This mechanistically attractive concept is readily accommodated in Scheme 3 if $X^{\oplus}$ is a methylating agent, e.g. S-adenosylmethionine (**25**). The evidence in the present case is that in comparative experiments in which **19a** and **12a** were excellent precursors of **22** and **13a** in *T. pratense*, the corresponding methoxychalcone **19b** and dihydroxy-isoflavone daidzein (**23c**) were incorporated into the two pterocarpans and formononetin only with greatly reduced efficiency, though still to a significant extent (Dewick, 1975), presumably on minor pathways. Very similar results were obtained for the incorporation of **19b** and, in part, **23c**, into **13a**, **14a** and **14b** in *M. sativa* (Dewick and Martin, 1979*b*).

Phaseollin (**26**) *and phaseollidin* (**27**).   In tracer studies of the biosynthesis of **26** in Cu^{2+} treated bean pods, radioactivity from daidzein-U-^{3}H (**23c**) was incorporated with satisfactory efficiency (Hess *et al.*, 1971). The result is relevant to a study by Woodward (1980) who identified a large number of isoflavonoids produced in *Phaseolus vulgaris* on inoculation with *Monilinia fructicola*, some of the compounds accruing in only very small amounts (daidzein (**23c**), 2′-hydroxy-daidzein (**23d**), the corresponding isoflavanones **28a** and 5-deoxykievitone (**28b**), 3,9-dihydroxypterocarpan (**30**), and 2′,4′,7-trihydroxyisoflavan (**31**)). Major products, apart from 5-oxyisoflavonoids (see isoflavonones) were phaseollidin (**27**) and phaseollin (**26**). 6a-Hydroxyphaseollin (**32**) and coumestrol were also detected. The compounds can be arranged in a sequence (Scheme 7) which constitutes a plausible biogenetic pathway to the phytoalexins **31** and **26**. Compound **29** is again a hypothetical intermediate which would account for both pterocarpan and isoflavan formation. The route to **30** and **31** in Scheme 7 is identical with that to **13a** and **14a** in Scheme 4 except for the substitution of 4′-OH for 4′-OMe in all compounds.

Pisatin.   In an early study of the biosynthesis of pisatin (**33a**) in Cu^{2+} treated pea pods, cinnamic acid was an excellent precursor, phenylalanine was incorpor-

Scheme 7 A plausible pathway to phaseollidin and phaseollin

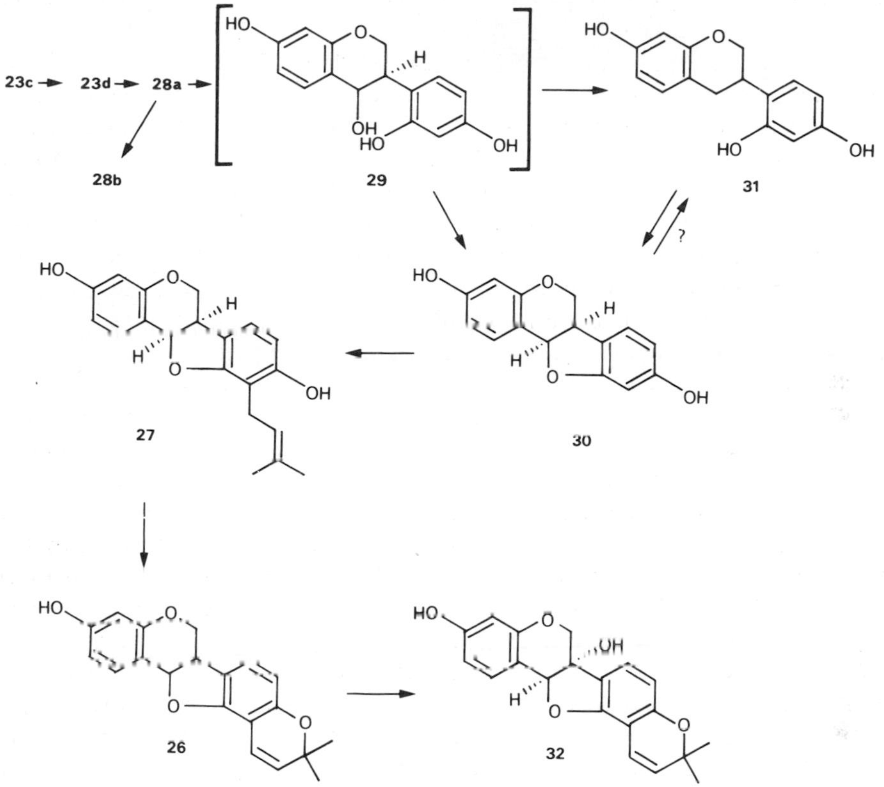

ated significantly though poorly but tyrosine (4-hydroxyphenylalanine) was not utilized (Hadwiger, 1965). These results indicate that the 4′-OH group associated with the flavone-isoflavone rearrangement is introduced at the cinnamic acid stage. Methionine-Me-^{14}C was also well incorporated in these experiments, presumably into the O-methyl and dioxymethylene carbons.

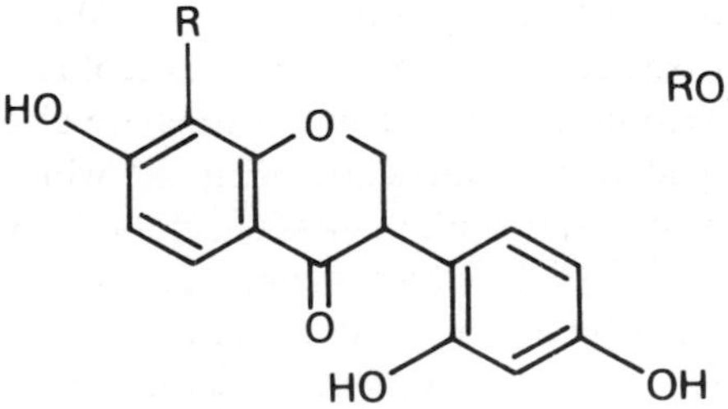

28a, R=H

28b, R=CH$_2$·CH=CMe$_2$

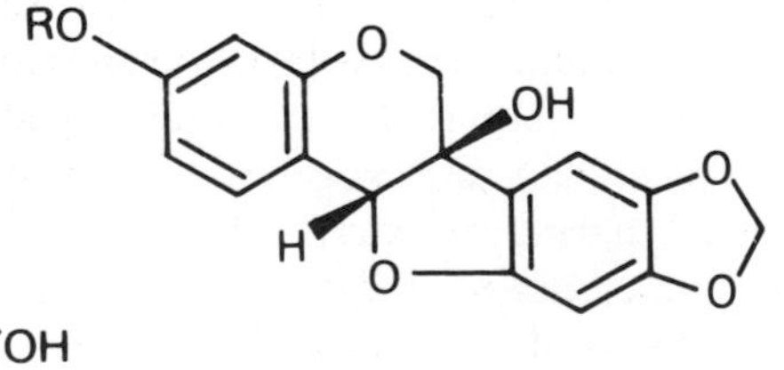

33a, R=Me

33b, R=H (stereochemistry presumed)

34

The formation of traces of maackiain (inermin) (**22**), in addition to much larger amounts of pisatin, in pea pods inoculated with *Monilinia fructicola* suggested that **22** might be an almost immediate precursor of **33a** (Stoessl, 1972). According to a preliminary report, the incorporation of **22** into **33a** has now been verified (Dewick and Banks, 1980).

Other compounds reported as precursors of **33a** were 6a-hydroxy-maackiain (**33b**, stereochemistry assumed), and the isoflavones **12a**, **23a**, **24a** and **24b**, all known as precursors of **22**. It is of particular significance that the pterocarpene, anhydropisatin (**34**) was also incorporated. (+)-Pisatin, although properly designated as 6aR, 11aR in terms of the Kahn-Ingold-Prelog nomenclature, on the reasonable though formally unproven assumption that its stereochemistry is as depicted in **33a** (Fuchs *et al.*, 1980; Sicherer and Sicherer-Roetman, 1980), is opposite in absolute configuration to 6aR, 11aR-(−)-maackiain (**22**), which is the form in which **22** was isolated from peas (Stoessl, 1972). If this is the enantiomer incorporated into **33a**, its hydroxylation must occur with configurational inversion at both chiral centres, e.g. *via* the corresponding isoflavene or the pterocarpene **34**, with epoxidation and epoxide cleavage as likely subsequent steps. The route *via* **34** is, of course, favoured by the observation (Dewick and Banks, 1980) that this compound can function as a precursor.

A different biosynthetic problem was investigated by application of the double labelling ^{13}C-NMR technique. This method depends on the magnetic interactions (spin–spin coupling) of *contiguous* ^{13}C atoms, revealed as characteristic splitting patterns in the ^{13}C-NMR spectrum. As a first approximation, such patterns will appear in products of biosynthetic studies only if the contiguous atoms are incorporated as an intact unit from the administered precursor. By these means, even simple, general precursors can be used to unravel subtleties in biosynthetic mechanisms which are not accessible otherwise. In the case of pisatin, sodium acetate-1,2-^{13}C$_2$ (90% ^{13}C), administered to pea pods pretreated with Cu^{2+}, was incorporated intact, with 2.3% enrichment at each of three sites in ring A (Stoessl and Stothers, 1979). Apart from minor, non-pertinent interactions, only the atoms linked by heavy bonds in representation **35a** were spin–spin coupled, each pair representing an intact acetate unit. This implies that the carbon chains of the putative *p*-coumaroyl-dioxohexanoate precursor **36** reacted in conformation **36a** (Scheme 8A, route a or b). Reaction from the second possible conformation **36b** by analogous routes would have

Scheme 8 Biosynthesis of pisatin

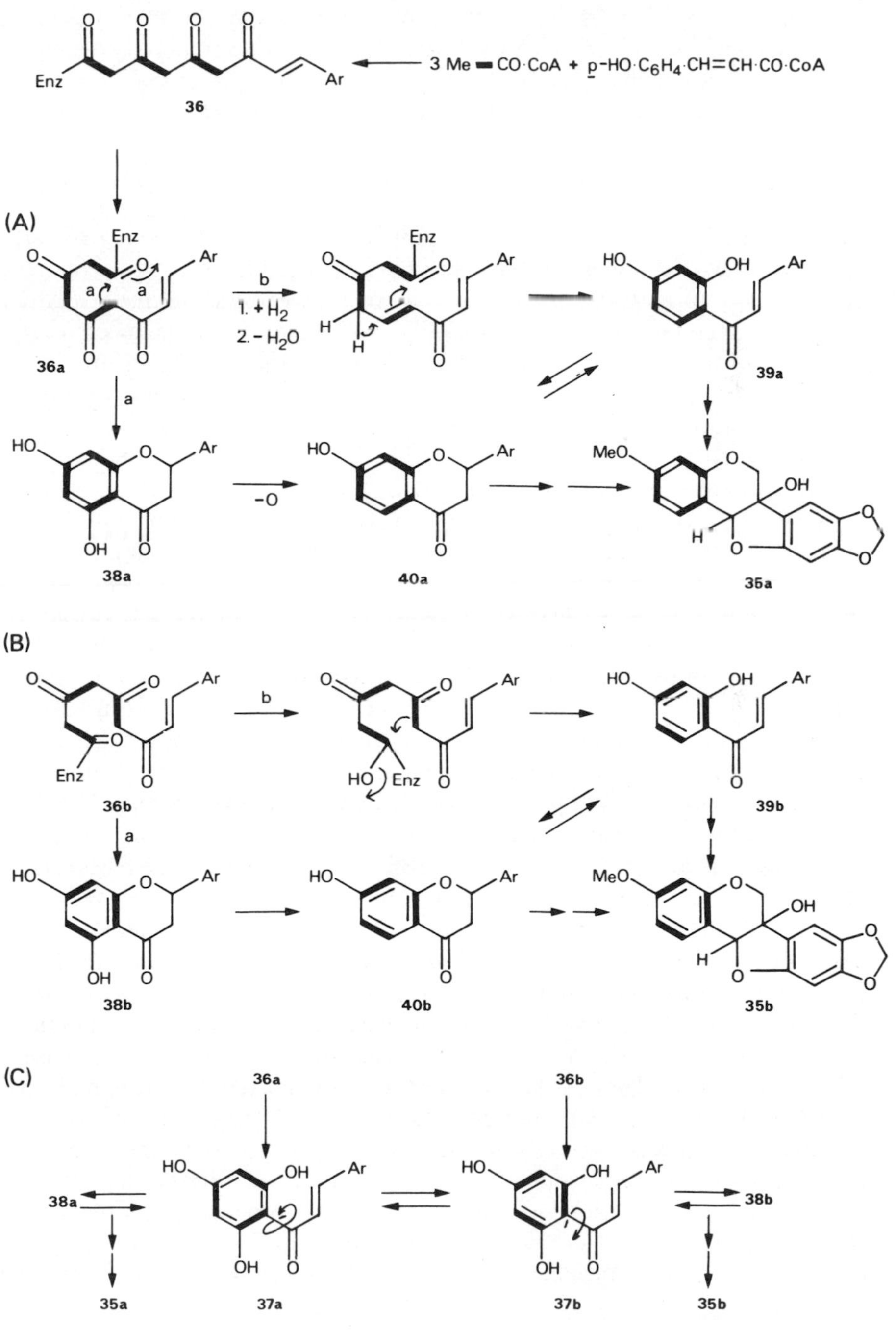

furnished pisatin labelled as in **35b** (Scheme 8B) but this was not observed. This route is also less favourable mechanistically. If the first product of the reaction is the chalcone **37a** (or **37b**) rather than the flavanone **38a** (or **38b**), or if these species are sufficiently long-lived to equilibrate, then the additional conclusion follows that the functionality of the eventual C-1 of pisatin (**35a**) is reduced *before* ring closure to the chalcone (**39a**) or flavanone (**40a**) is effected. The reason is that intervention of the symmetrical chalcones **37a** or **37b** as in Scheme 8C would provide a 1:1 mixture of **35a** and **35b**, contrary to the experimental result.

Glyceollins. Four compounds are known (glyceollins I to IV; **41a** to **41d** respectively) and a fifth (**41e**) has been tentatively identified. Some of these accrue in only small amounts and because their separation is difficult and, for many purposes, unnecessary, the name "glyceollin" frequently denotes mixtures in which **41a** usually predominates. The first biosynthetic study on the glyceollins (Keen *et al.*, 1972) was conducted with such a mixture and before the correct structure of the main component, glyceollin I (then regarded as 6a-hydroxyphaseollin and misnamed accordingly) was ascertained by Burden and Bailey (1975); however, this was without detriment to the significance of the study. It was shown that in soybean hypocotyls inoculated with an incompatible fungal strain, phenylalanine-U-^{14}C and, more importantly, isoliquiritigenin-CO-^{14}C (**19a**) were incorporated into glyceollin. Radioactivity was also incorporated into daidzein (**23c**), a normal soybean metabolite but one whose concentration nevertheless increased dramatically during the experiment, and into the coumestans coumestrol and sojagol (see below). Since **23c** is almost certainly an intermediate on the pathway to the glyceollins and coumestans, it is of interest that its concentration reached a plateau while maximal accumulation of glyceollin was taking place, but daidzein accumulation was resumed, and then maintained, just before glyceollin concentration began to decline.

The isolation of the 6a-hydroxypterocarpan **41f** from Cu^{2+}-treated soybean seedlings (Lyne and Mulheirn, 1978) opened the way to further biosynthetic studies. Other experiments had already indicated that MVA served as a precursor, as expected (Zähringer *et al.*, 1978). It could now be demonstrated that **41f** gave radioactive glyceollin IV (**41d**) when incubated with DMAPP-1-^{14}C in the presence of a dimethylallylpyrophosphate:trihydroxypterocarpan dimethylallyl transferase which had been isolated as a particulate fraction from fungal elicitor-treated soybean cotyledons (Zähringer *et al.*, 1979). Radioactive **41d** was also formed by a similar incubation of ^{3}H-labelled **41f** with unlabelled DMAPP. A minor product which was formed in both experiments was tentatively identified as **41e** (glyceollin V?). Dimethylallyl transferase activity was not observed in cotyledons which had been wounded but not treated with elicitor, and no transfer reaction was detected with isopentenylpyrophosphate (IPP)-1-^{14}C (Zähringer *et al.*, 1979). A plausible pathway is shown in Scheme 9, where the route from phenylalanine to **41f** via **23c** is analogous to that in Scheme 4.

Scheme 9

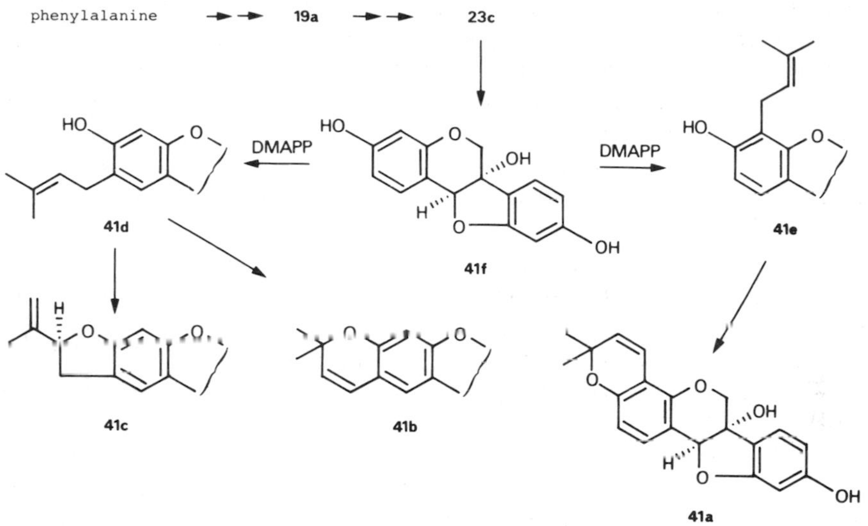

(d) *Coumestans.*   The status of these compounds as phytoalexins is uncertain. There are many reports of their accumulation after infection but they have usually been found to be non-fungitoxic. However, antibacterial and nemato-cidal properties have been reported and, in any case, the compounds are too closely linked to other isoflavonoids to be ignored. Coumestrol (**42a**), the simplest coumestan, has been the subject of numerous biosynthetic studies but discussion here must be limited to particularly salient points from the older work and to the most recent investigations.

Identified coumestrol precursors include isoliquiritigenin (**19a**)-4′-glucoside and daidzein (**23c**) in *M. sativa* studied in the absence of stress inducers; and **19a**, **23c**, **23d** and **28c** both in unstressed seedlings and cell suspension cultures of *Phaseolus aureus* (Berlin *et al.*, 1972; Dewick *et al.*, 1970; and references there cited). In fungus-infected soybean hypocotyls, coumestrol and its isoprenylated derivative sojagol **42b** commenced to accumulate at an appreciable rate only at about the time when concentrations of glyceollin began to decline (Keen *et al.*, 1972).

The accumulated incorporation data, biogenetic considerations and the known susceptibility of pterocarp-6a-enes to autoxidation to coumestans led to an attractive but largely hypothetical proposal in which a pterocarp-6a-ene was a crucial intermediate to coumestrol (Berlin *et al.*, 1972). This proposal, an alternative to one or two earlier schemes, found general favour for several years. Recently, however, Martin and Dewick (1978) have provided evidence that the biosynthesis of coumestans may proceed without pterocarpene intermediates. They showed that in Cu^{2+}-treated *M. sativa*, the pterocarpan **13a** is well

42a, R_1 = R_2 = H

42b, R_1-R_2 = -$CMe_2 \cdot CH_2 \cdot CH_2$-

42c, R_1 = Me, R_2 = H

incorporated into 9-O-methylcoumestrol **42c** (dilution value 25) but that incorporations of the isoflavene **21** and the phenylcoumarin **43** were spectacularly more efficient (dilution values 2.1 and 3.8). The isoflavone **12b** was also an excellent precursor. The authors therefore proposed a route (**12b**→**20a**→**15**→ **21**→**43**→**42c**) which diverges from that in Scheme 4 at the hypothetical intermediate isoflavan-4-ol **15** or the derived carbonium ions. Pterocarpan **13a** can then be incorporated into **42c** *via* the isoflavene **21**. An analogous pathway to coumestrol (**42a**) was suggested (Dewick and Martin, 1979*b*).

(*e*) *2-Arylbenzofurans.* Vignafuran (**44**) is one of several 2-arylbenzofuran phytoalexins from the Leguminosae. Its structure, and its co-occurrence with several isoflavonoids, e.g kievitone (**11a**), phaseollin (**26**) and medicarpin (**13a**) in *Vigna unguiculata* suggest a possible biogenesis from 9-O-methylcoumestrol (**42c**) by decarboxylation. This supposition was tested by Martin and Dewick (1979) who administered the putative, labelled precursors isoliquiritigenin (**19a**), daidzein (**23c**), formononetin (**12a**) and 2′-hydroxyformononetin (**12b**) to seedlings irradiated with UV-light, but no incorporation into vignafuran was observed. Phenylalanine-U-^{14}C, -1-^{14}C and -2-^{14}C, however, were incorporated. The location of the labels, determined by degradation of **44** into pertinent fragments, corresponded to an isoflavonoid character for the carbon skeleton, and hence, loss of C-3 of phenylalanine. This excluded an origin of **44** on either

44

the lignan or the stilbene pathway to arylbenzofurans. However, contrary to expectation, all radioactivity from the phenyl ring of the phenylalanine was located in the 2-aryl substituent of **44**, not in its benzofuran moiety. These results cannot be reconciled with a direct derivation of **44** from **42c**. Alternative origins, e.g. from the 9-O-methylcoumestrol precursor **43** can be discussed (Martin and Dewick, 1979) but are largely speculative.

2. Stilbene-derived compounds

(i) Stilbenes

Stilbenes are found in a wide range of unrelated plant families, frequently but not invariably under conditions indicative of stress, e.g. as heartwood constituents, as glycosides in roots, or as compounds formed on wounding or infection. Hillis and Ishikura (1969) have discussed the possibility that infection (and presumably other stresses) diverts normal metabolism to stilbene formation by blocking flavonoid biosynthesis. Biosynthetically, stilbenes are relatively simple products of the condensation of cinnamoyl or p-coumaroyl coenzyme A with three molecules of acetate (from malonyl coenzyme A), followed by cyclization from a characteristic conformation, and decarboxylation (Scheme 2). Further hydroxylation and/or methylation are subsequent processes (Rupprich et al., 1980). Of several stilbenes regarded as phytoalexins, two have been the subjects of biosynthetic studies.

Pinosylvin (**45a**) and its monomethylether, well-known as metabolites of *Pinus* and *Eucalyptus* spp., have been linked with defence reactions to fungi (e.g. Shain, 1967); they also occur normally in flowers of *Alnus* spp. (Suga et al., 1972). The specific incorporation of acetic acid and phenylalanine in the expected pattern in wounded tissue of *P. resinosa* was demonstrated by Rudloff and Jorgensen (1963). The reported isolation of an enzyme from *Eucalyptus* leaves, which catalysed the conversion of cinnamoyl-triacetic acid into **45a**, could not be replicated in later experiments (Hillis and Yazaki, 1971). However, a pinosylvin synthase, which brings about the formation of **45a** from cinnamoyl coenzyme A and malonyl coenzyme A, has been described by Schoeppner and

45a, $R_1 = R_2 = H$

45b, $R_1 = OH$, $R_2 = H$

45c, $R_1 = R_2 = OH$

46   48

Kindl (1979). This enzyme was normally present in the roots of the plant but was found in green needles only after induction with UV light.

Resveratrol (**45b**), a metabolite of *Pinus*, *Picea* and other plants, has been discussed as a phytoalexin of *Arachis hypogaea* (Ingham, 1976), *Trifolium* spp. (Ingham, 1978) and the Vitaceae (Langcake and Pryce, 1976). An enzyme preparation from rhubarb (*Rheum raphonticum*) rhizomes catalyses its formation from *p*-coumaroyl coenzyme A (Rupprich *et al.*, 1980). In UV-irradiated *Vitis vinifera*, phenylalanine was a good but tyrosine a poor precursor (Langcake and Pryce, 1977*a*), indicating that the 4'-hydroxyl is probably introduced at the cinnamic acid stage.

(ii) Stilbenoid benzofurans

Oxidative cyclization of stilbenes can lead to 2-arylbenzofurans which are biogenetically distinct from those discussed above. In some cases, the different origin may be manifest in the substitution patterns. An unambiguous distinction is always possible in principle by labelling experiments, since C-1 of the phenylpropanoid precursors will be incorporated into the furan ring of iso-flavonoid but into the chiefly acetate/malonate derived benzenoid ring of stilbenoid benzofurans (Scheme 2). At present, the only phytoalexins which are members of the class, as indicated by their characteristic substitution patterns, are the moracins from mulberry, e.g. moracin F (**46**) (Takasugi *et al.*, 1979). The assignment of these compounds to the stilbenoid pathway is strongly supported by their co-occurrence, in mulberry infected with *Fusarium solani* f. sp. *mori*, with similarly substituted stilbenes, e.g. **45c** (Takasugi *et al.*, 1978).

(iii) Dihydrophenanthrenes

The compounds of interest from this group are the orchid phytoalexins orchinol (**47a**), loroglossol (**47b**) and hircinol (**47c**) (Fisch *et al.*, 1973). A fourth compound, cypripedin (**48**) from the lady slipper *Cypripedium calceolus* (Schmalle and Hausen, 1979) is structurally and biogenetically pertinent although it was not isolated as a phytoalexin. Experimental biosynthetic studies have not been reported but nevertheless there is considerable justification for believing that the compounds are formed by intramolecular oxidative phenol coupling of dihydro-stilbene precursors, followed by a dienone-phenol rearrangement. The concept

Scheme 10 Putative biosynthetic routes to orchinol

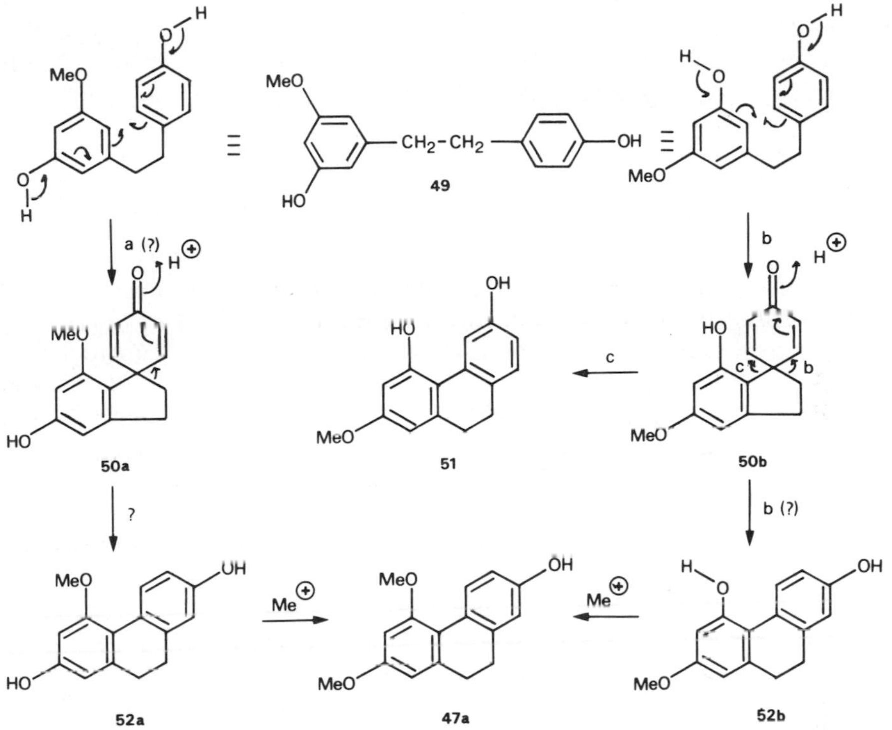

is a modification of one originally proposed by Birch (1966). Two of several equivalent pathways which differ only in minor details are shown in Scheme 10 for **47a**; Scheme 11 depicts one such route to **47b** and **c**. Whatever the details of these routes, the basic concept which is common to them is supported by the co-occurrence of dihydrostilbenes, phenanthrenes and dihydrophenanthrenes with appropriate oxygenation patterns in the Combretaceae (e.g. Letcher and Nhamo, 1972), Dioscoreaceae (e.g. Hashimoto and Tajima, 1978) and *Cassia garrettiana* (Hata *et al.*, 1979). However, the most impressive evidence for the concept comes from the co-occurrence, in *Cannabis sativa*, of the dihydrostilbene **49**, the spirodienone **50b** and the dihydrophenanthrene **51** in addition to other,

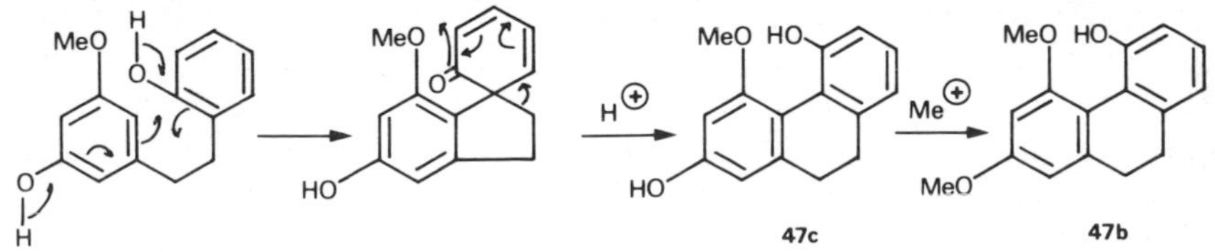

Scheme 11 Putative biosynthetic route to hircinol and loroglossol

closely related dihydrostilbenes and spiranes (Crombie *et al.*, 1979). Of these, **49** and **50b** are identical with postulated intermediates in Scheme 10 while **51** is an isomer of one of the two possible monodemethylorchinols (**52a**) and (**52b**). The biomimetic intramolecular phenol coupling of **49** to **50a** and **50b** has been accomplished with oxidizing agents in the laboratory (El-Feraly *et al.*, 1979).

(iv) Stilbene oligomers

ε-Viniferin (**53**) and α-viniferin (**54**), the major phytoalexins of *Vitis vinifera*, are formed under a variety of stress conditions together with the monomeric stilbene resveratrol (**45b**). As expected, radiolabelled phenylalanine, acetic and malonic acid were good precursors of **45b** in UV-irradiated vine leaves but tyrosine was incorporated only poorly (Langcake and Pryce, 1977*a*), indicating once again that the 4′-hydroxyl is probably introduced at the cinnamic acid stage (cf. pisatin). The specificity of the incorporations was confirmed by degradative experiments. In further work, oxidation of **45b** with the classic laboratory reagents, H_2O_2 and horseradish peroxidase (Freudenberg, 1965), gave structural analogues of **53** but not **53** or **54** (Langcake and Pryce, 1977*b*). Attempts to observe the direct incorporation of **45b** into the viniferins in vine leaves were also unsuccessful. Nevertheless, the sequence **45b**→**53**→**54** (Scheme 12 or equivalent) was supported by time-course studies with inoculated intact vines (Langcake and Pryce, 1977*a*). An analogy for the route is provided by the

Scheme 12   Probable route to viniferins

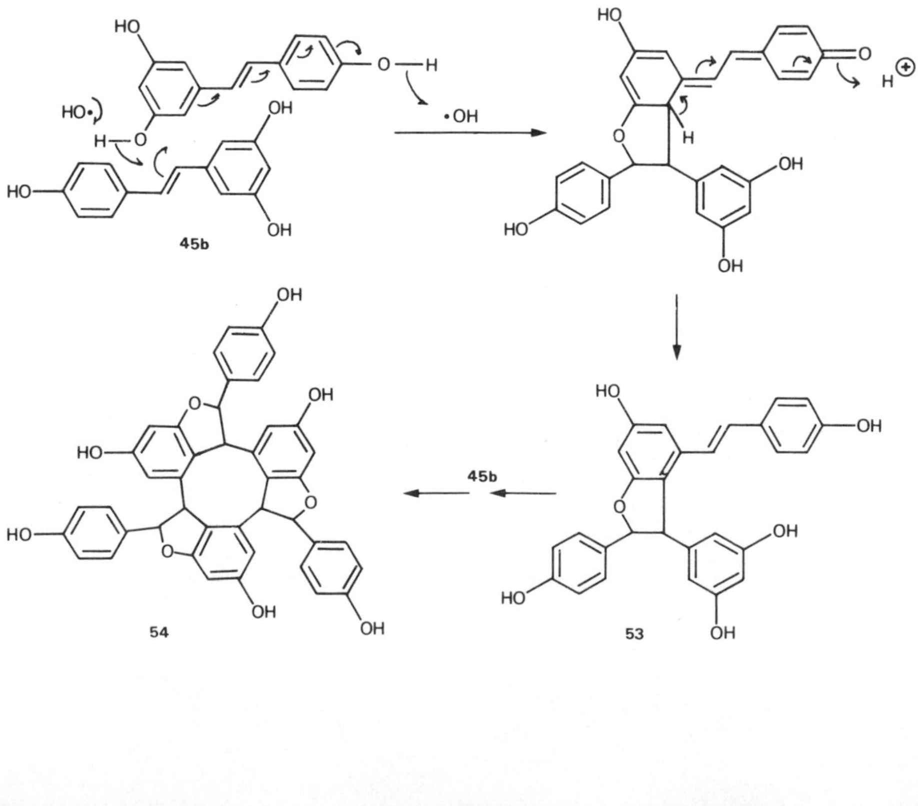

55a, $R_1 = CH_2OH$, $R_2 = OMe$

55b, $R_1 = CONH \cdot (CH_2)_4 \cdot NH \cdot C=NH \cdot NH_2$, $R_2 = H$

long-established, oxidative dimerization of phenylpropanoids, e.g. of coniferyl alcohol (**1b**) to dehydrodiconiferyl alcohol (**55a**) (Freudenberg, 1965), and that of coumaroylagmatine (**1h**) to give hordatine A (**55b**), one of the antifungal factors in barley (Stoessl, 1967). It is of interest that the last compound and its glucosides, while present in substantial amounts in healthy tissue, increase by a factor of 6 on infection (Smith and Best, 1978).

Phytoalexins derived from acetate-polymalonate

The condensation of acetyl coenzyme A with successive molecules of malonyl coenzyme A, in concert with decarboxylation, may be followed at each step by reduction, dehydration and, usually, hydrogenation affording long-chain fatty acids which may then undergo further modifications (fatty acid route); or alternatively, the carbonyl functions from successive acetate units may remain mainly intact until a chain of 4 or more units is built up, to enter into reactions which are typical of active methylene compounds—the polyketide route (Turner, 1971).

1. Polyacetylenes

Acetylenes are numerically probably the most important of the secondary metabolites formed on the fatty acid route in higher plants; it is from this class that the only known fatty acid-derived phytoalexins are drawn. At present, these are limited to wyerone (**56a**) and related metabolites from broad bean (*Vicia faba*) (Mansfield *et al.*, 1980) and *Lens culinaris* (Robeson, 1978) as well as to long-chain acetylenic compounds from *Carthamus tinctorius* (Allen and Thomas, 1971). Only the first of these two groups has been the subject of biosynthetic studies. Cain and Porter (1979), who recognized that conditions initiating *de novo* synthesis were of prime importance, achieved good incorporation of acetate-1-^{14}C, malonate-2-^{14}C and oleate-n9,10-[3]H (**57**) into **56a** in *V. faba* infected with *Botrytis cinerea*. The incorporation of **57**, an accepted general precursor of polyacetylenes (Simpson, 1977), was more efficient than that of

56a, R_1=Me, R_2R_3=O, R_4= —C=CEt
                                         H   H

56b, R_1=Me, R_2=OH, R_3=H, R_4= —C=CEt
                                         H   H

56c, R_1=H, R_2R_3=O, R_4= —C=CEt
                                   H   H

56d, R_1=Me, R_2R_3=O, R_4= —CH—CHEt
                                        \O/

malonate and acetate, in keeping with its role as an advanced precursor. Details of the necessarily complex sequence from **57** to **56a** are still mostly unknown. Hargreaves *et al.* (1977) have studied the time course of the accumulation of **56a** and its congeners; they interpret their results tentatively as favouring the sequences wyerol (**56b**)→wyerone (**56a**)→wyerone acid (**56c**) and **56a**→wyerone epoxide (**56d**).

2. Polyketides

The polyketide route, the main pathway to secondary metabolites in micro-organisms, is utilized only relatively rarely by higher plants. The only example of a polyketide phytoalexin is 6-methoxymellein (**58a**) from carrot, in which it is induced by infection or other stresses, together with the chromone eugenin (**59a**) and traces of the demethyl compounds **58b** and **59b** (Coxon *et al.*, 1973). The polyketide origin of the carbon skeletons of the compounds was demonstrated by the efficient incorporation of acetate-1-^{14}C and malonate-2-^{14}C into **58a** in carrot slices treated with either ethrel (a source of ethylene) or *Ceratocystis fimbriata* (Jaworski and Kuć, 1974); and into both **58a** and **59a** in a similar experiment employing ethylene gas as inducer (Sarkar and Phan, 1975). The latter authors also confirmed that **59b** precedes **59a** on the biosynthetic pathway by direct incorporation and trapping experiments. Finally, a ^{13}C-NMR study showed unequivocally that acetate-1,2-^{13}C$_2$ was incorporated into **58a** in the expected manner, Scheme 13 (Stoessl and Stothers, 1978). However, **59a**, isolated in the same experiment, was labelled to an equal extent in both possible

Scheme 13 Biosynthesis of 6-methoxymellein and eugenin

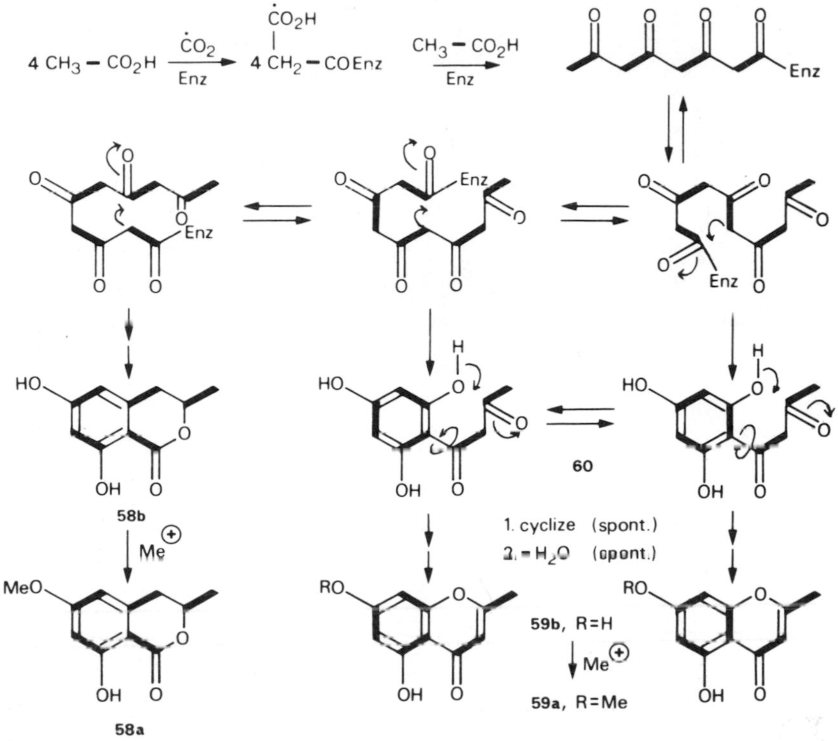

modes. This implies that it was derived from a symmetrical monocyclic intermediate, probably **60**, which was free to rotate about the bond indicated. It follows further that the cyclization of the intermediate to the chromone was probably spontaneous and not enzyme-mediated.

Phytoalexins with mevalonoid skeletons

Mevalonoid compounds (terpenes, sterols, carotenes) are ultimately derived from acetate as outlined in Scheme 14. This pathway is firmly established, in remarkable stereochemical and mechanistic detail (see Clayton, 1965; Turner, 1971; Cornforth, 1976). Further elaboration of the parent compounds shown in the scheme, often by complex rearrangements, leads to an extremely large array of different structural classes (for an overview, see Devon and Scott, 1972). Only a few of these need to be considered further here.

1. Monoterpenes

Monoterpenes, derived from geranylpyrophosphate (GPP), are ubiquitous in higher plants, sometimes accruing in large quantities, for example, as essential

Scheme 14 Derivation of mevalonoid compounds

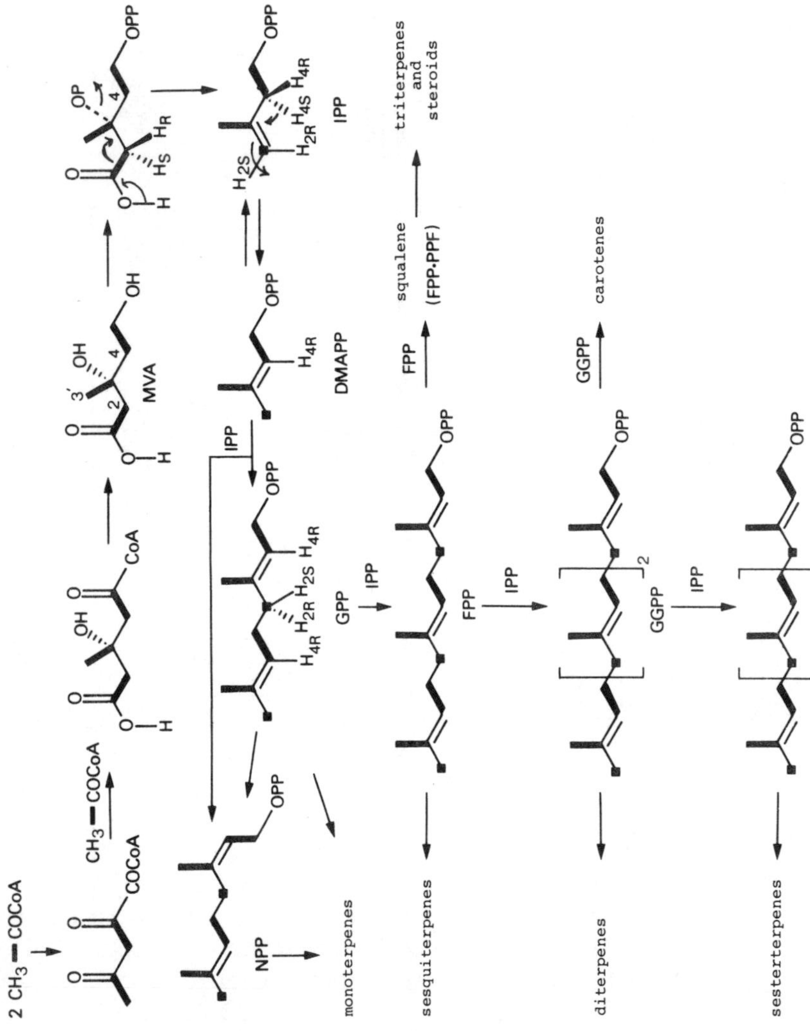

oils in pine exudates. Their antimicrobial properties and possible role in resistance have been reviewed by Hare (1966) and Kuć and Lisker (1978) *inter alia*. The compounds frequently accumulate as a consequence of infection or other stress. Nevertheless, they are not usually considered to be phytoalexins, though the reasons for this are unclear. A typical case which illustrates the nature of the compounds is the formation of myrcene (61) and Δ^3-carene (62) in fungus-infected wounds of *Abies grandis* (Russell and Berryman, 1976). The not to be underestimated intricacies of monoterpene biosynthesis have been explored in numerous, often very elegant studies; these have been admirably reviewed (Charlwood and Banthorpe, 1978). A pertinent example is the biosynthesis of 62 in pine, in the absence of stress conditions, which was investigated by the administration of MVA-2-^{14}C and careful location of the incorporated label by degradation experiments. As is the rule for monoterpenes,

Scheme 15   Probable biosynthesis of Δ^3-carene

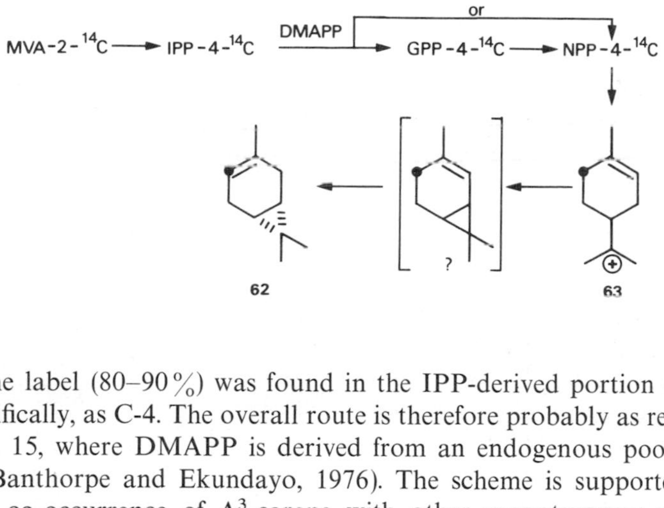

most of the label (80–90%) was found in the IPP-derived portion of 62 and more specifically, as C-4. The overall route is therefore probably as represented in Scheme 15, where DMAPP is derived from an endogenous pool and not labelled (Banthorpe and Ekundayo, 1976). The scheme is supported by the invariable co-occurrence of Δ^3-carene with other monoterpenes which are generally accepted as derived from the α-terpinyl ion 63 or equivalent. The observed losses of the *pro*-2S hydrogen, and of one of the two 5-hydrogens, when appropriately tritiated specimens of MVA were used as precursors, are consistent with this pathway but do not define it unambiguously.

2. Sesquiterpenes

Although they occur in numerous, often very different skeletal forms, all sesquiterpenes are probably derivatives of farnesyl pyrophosphate (FPP). Biosynthetic studies extending into the mid-1970's have been comprehensively reviewed (Cordell, 1976); to a large extent, results have vindicated earlier, mostly speculative concepts (cf. Devon and Scott, 1972; Parker *et al.*, 1967).

64a, R =

64b, R =

64c, R =

64d, R =

(i) Acarbocyclic sesquiterpenes

The 9-oxy, 9-oxo and 9,10-dehydro-11-oxy derivatives (**64b–d**) of nerolidol (**64a**), the allylic isomer of farnesol (**65**), are formed as stress compounds in eggplant (Stoessl *et al.*, 1975) together with two bicyclic compounds considered below. Their biogenetic derivation is self-evident in outline but has not been investigated experimentally. The compounds are relevant to the next paragraph.

Ipomeamarone (**66a**) and its congeners from sweet potato (*Ipomoea batatas*) have been extensively studied as stress compounds in Japan (Uritani, 1978) and,

65

66a, R =   67a

66b, R =   67b

mainly as hazardous mammalian toxins, in the United States (Burka and Wilson, 1976). In early biosynthetic studies Uritani and coworkers observed the incorporation into **66a** of ^{14}C-labelled MVA, acetate and acetate precursors. The pathway was defined more clearly when farnesol **(65)** and dehydro-ipomeamarone **(66b)** were shown to be efficient precursors of **66a** (4.8 and 21 % incorporation, dilution values 38 and 27 respectively) in experiments in which fungal infection was used to induce its formation (Oguni and Uritani, 1974). The specific nature of the incorporations was not confirmed by experimental location of the introduced labels but is supported by the low dilution values and by trapping experiments. Hydroxydehydromyoporone **(67b)** has also been claimed as a precursor of **66a** but only on the evidence of a time-course study (Inoue and Uritani, 1979) and without direct observation of its incorporation. The relationship is therefore tentative at best and is rendered even more doubtful by the unambiguously demonstrated conversion of ipomeamarone **(66a)** into hydroxymyoporone **(67a)** (1.1 % incorporation, dilution value 3.5) (Burka and Kuhnert, 1977). A sequence of oxidation and cyclization steps which might lead from **65** to **66a** has been discussed by Burka and Wilson (1976) but with an illustrative rather than definitive intent.

(ii) Carbocyclic sesquiterpenes

(a) *Gossypol and congeners.* Gossypol **(68)**, a dimeric sesquiterpene of the cadalane class, is a normal metabolite of cotton and other, related plants but, on account of certain properties, is often regarded as a phytoalexin. Its biosynthesis has been the subject of elegant studies which have recently been reviewed (Heinstein *et al.*, 1979). In the earlier work, it was shown that acetate-1-^{14}C,-2-^{14}C and MVA-2-^{14}C were incorporated into **68** in excised cotton roots in labelling patterns consistent with the mevalonoid route to cadalanes. However, in order to define the pathway with greater precision, GPP, neryl pyrophosphate (NPP) and the four possible stereoisomers of farnesyl pyrophosphate (*t,t*-, *c,t*-, *t,c*- and *c,c*-FPP), all synthesized in the laboratory, were tested as precursors in comparative experiments (Heinstein *et al.*, 1970). Possible problems arising from transport, permeability, pool size and like factors were obviated or minimized by the use of cell-free cotton root homogenates as incubation medium. Contrary to the then widely held view which favoured GPP and *t,t*-FPP as general intermediates of sesquiterpene biosynthesis, *c,c*-FPP was by far the best precursor, followed by NPP and *t,c*-FPP. This result was also in accord with the specific incorporation of labels, carefully ascertained by the appropriate degradative experiments (Scheme 16). The inference that **68** is formed from *c,c*-FPP via NPP is supported by the subsequent isolation of a prenyltransferase complex which is able to synthesize all four isomers of FPP from IPP. The structure of the immediate product of the cyclization of *c,c*-FPP is not yet known. Further intermediates between this compound and gossypol almost certainly include desoxyhemigossypol **(69)** and hemigossypol **(70)**, in this order.

Scheme 16

A. Percent incorporation of FPP-2-^{14}C isomers into gossypol (after Heinstein *et al.*, 1970).

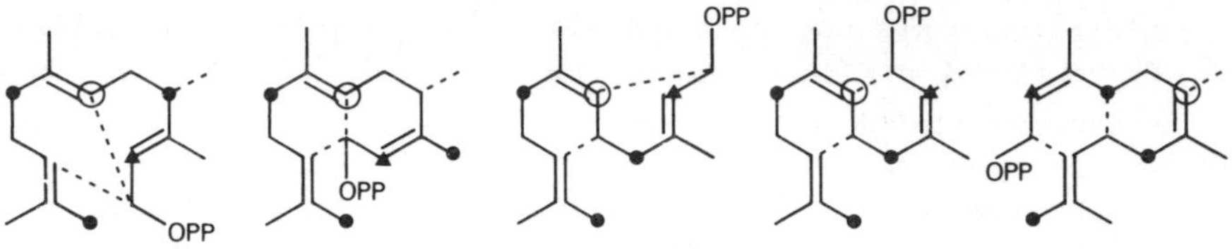

FPP- isomer:	t,t	c,t	t,c	c,c (conformer a)	and/or	c,c (conformer b)
incorporation observed, %:	0.15	0.16	7.64		38.9	

▲ label in FPP-2-^{14}C   --- additional C—C bonds in 68

B. Degradation of gossypol and observed labelling patterns.

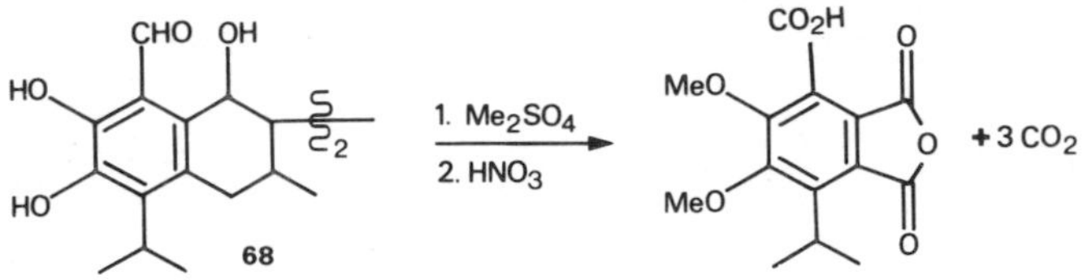

label from	MVA-2-^{14}C	●	3/3	0	t,t;c,t
	GPP-or NPP-2-^{14}C	○	1/1	0	c,c(b)
	FPP-2-^{14}C	▲	0	1/1	t,t;c,t;c,c(b)

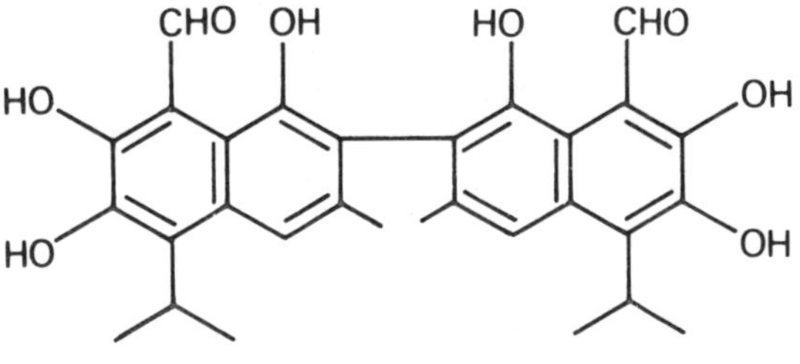

patterns excluded:

68

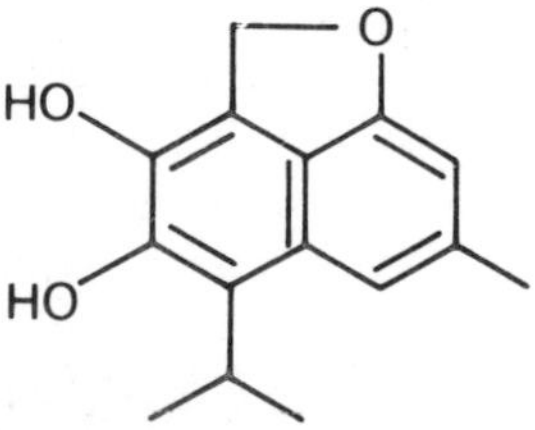

69

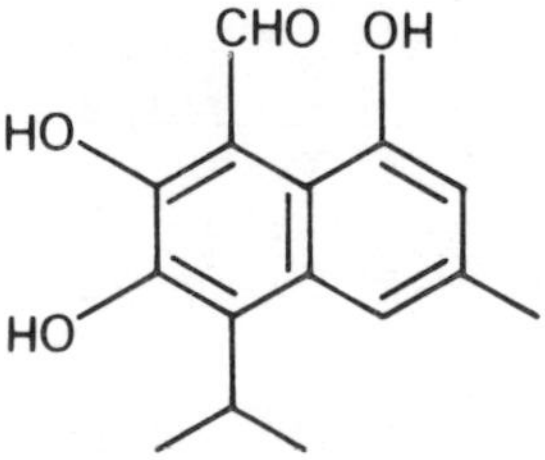

70

Both of these compounds have been isolated from cotton tissues. Desoxyhemi-gossypol (69) in air oxidizes spontaneously to 70 (Stipanovic et al., 1975) and the latter is converted to gossypol (68) by the action of peroxidase (Veech et al., 1976). The operation of this sequence in vivo is indicated by a recent time-course study (Bell and Stipanovic, 1978).

(b) Eudesmanoids. All the compounds in this sub-section are stress meta-bolites of the Solanaceae. Rishitin (71a), the first member of the group to be isolated (Tomiyama et al., 1968), was also the first whose mevalonoid origin was formally demonstrated by the highly efficient incorporation of acetate-1-^{14}C and MVA-2-^{14}C (Shih and Kuć, 1973). More than 30 other sesquiterpenoidal stress compounds, many of them with the properties of phytoalexins, are now known from the potato (Solanum tuberosum and related spp.) eggplant (S. melongena), tomato (Lycopersicon esculentum), sweet pepper (Capsicum annuum),

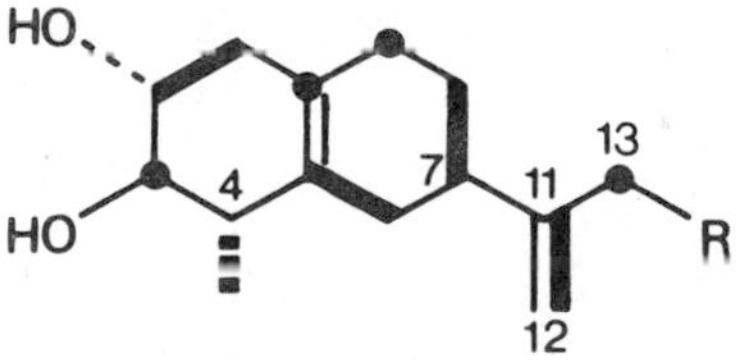

71a, R=H

71b, R=OH

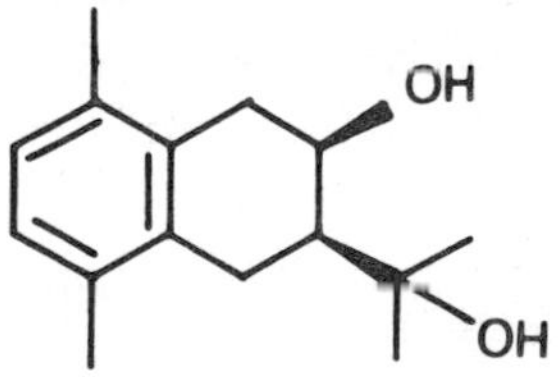

73

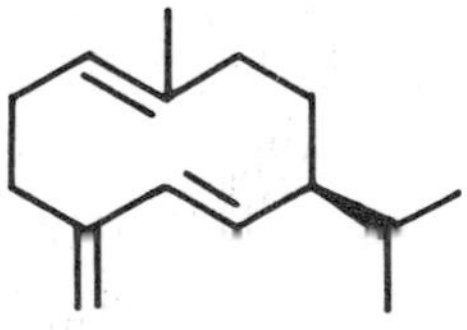

75

tobacco (Nicotiana spp.) and from Datura stramonium. Several have been found in more than one of these species. Because of the intervention of different molecular rearrangements in their biosynthesis, the close biogenetic relationships between the compounds is not always immediately apparent on a superficial inspection of their structures. However, it could be shown that the multicyclic compounds[3] of the group can be regarded as eudesmanes or as formally derived

Scheme 17   Biosynthesis of 2.3-dihydroxygermacrene A and aubergenone

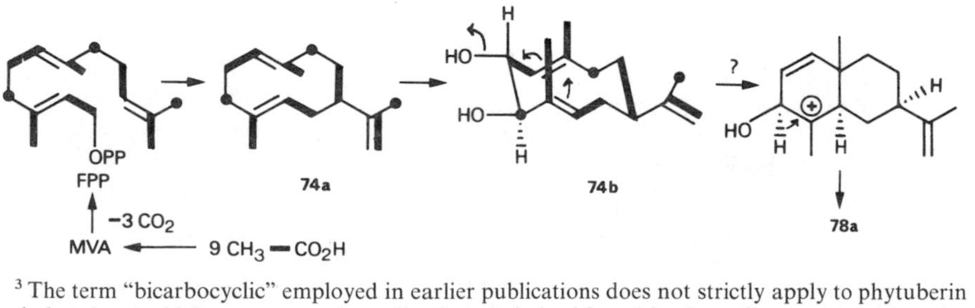

OPP

FPP

74a

74b

78a

$\uparrow$ $-3\,CO_2$

MVA $\longleftarrow$ 9 CH$_3$$\longrightarrow$CO$_2$H

[3] The term "bicarbocyclic" employed in earlier publications does not strictly apply to phytuberin and phytuberol which are monocarbocyclic, though derived from bicarbocyclic precursors. In order to circumvent this difficulty, the admittedly less definitive expression "multicyclic" is now substituted, though with some reluctance.

from a eudesmane cation **72** by unexceptional rearrangements (Stoessl, 1977; Stoessl *et al.*, 1976). Two additional rules which appeared to be observed were that with the sole exception of rishitinol (**73**), C-5 of the eudesmane skeleton served as a migration terminus and, again with the exception of **73**, oxygenation of ring carbons was confined to C-1 to C-4. The more than 20 compounds discovered since this pattern was first noted in 1975 have conformed to it, with only two further exceptions to the oxygenation rule. In addition, the principal features of several of the then postulated rearrangements are now firmly supported by direct experimental evidence. On the other hand, at least one rearrangement has turned out to be more complex than anticipated. At present, the compounds can be classified in seven different groups. Before discussing these individually, their common derivation must be given brief further consideration.

It is commonly accepted (Parker *et al.*, 1967; Cordell, 1976) that the general precursor of bicarbocyclic sesquiterpenes is a germacrene, e.g. in the simplest form, germacrene A (**74a**), which is derived by straightforward cyclization of *t,t*-FPP. While germacrene A itself is very rare as a natural product, germacrene D (**75**) has recently been found with great frequency.[4] Neither **74a** nor **75** has as yet been reported from the Solanaceae but 2,3-dihydroxygermacrene A (**74b**) occurs as a stress compound in *D. stramonium*. Its derivation from acetate-1,2-$^{13}C_2$ in the expected pattern (Scheme 17), presumably *via* MVA and FPP, has been confirmed by the ^{13}C-NMR method (Birnbaum *et al.*, 1976). The structure

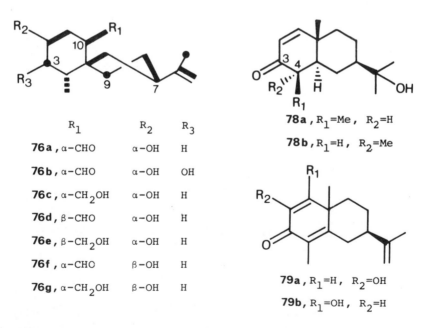

	R_1	R_2	R_3
76a, α-CHO	α-OH	H
76b, α-CHO	α-OH	OH
76c, α-CH$_2$OH	α-OH	H
76d, β-CHO	α-OH	H
76e, β-CH$_2$OH	α-OH	H
76f, α-CHO	β-OH	H
76g, α-CH$_2$OH	β-OH	H

78a, R_1=Me, R_2=H
78b, R_1=H, R_2=Me

79a, R_1=H, R_2=OH
79b, R_1=OH, R_2=H

[4] This suggests that the compound might well be a biosynthetic shunt.

of the compound at the time suggested that it might serve as a direct precursor to its bicyclic congeners lubimin (**76a**), 3-hydroxylubimin (**76b**) and capsidiol (**77a**), as well as for rishitin (**71a**) from potato and aubergenone (**78a**) from eggplant (Stoessl *et al.*, 1976). However, for reasons which will emerge presently, such a comprehensive function for **74b** now appears unlikely although a precursor relationship to one or two of the compounds remains a distinct possibility.

Unrearranged eudesmanes. Aubergenone from eggplant was initially formulated as **78b** but as a result of synthetic studies (Kelly *et al.*, 1978; Murai *et al.*, 1978*a*) is now known with certainty to be the thermodynamically *less* stable 4-epimer **78a**. This is in excellent accord with the predicted biosynthesis of aubergenone from the thermodynamically favoured C,C-conformation (Sutherland, 1974) of **74b** or other, equivalent germacrene (Scheme 17). The original biosynthetic proposal (Stoessl *et al.*, 1976) envisaged this but tacitly assumed that **78a**, the immediate product, would be epimerized to **78b**. Evidently, however, although this epimerization is readily accomplished in the laboratory (Kelly *et al.*, 1978), it does not take place *in vivo*. An alternative, similarly plausible biogenetic scheme, with T,T-germacrene A-1(10)-epoxide as point of departure, has been proposed by Murai *et al.* (1978*a*).

Another unrearranged eudesmane is 2-keto-α-cyperone (**79a**) from tobacco which, again, was formerly incorrectly regarded as 1-keto-α-cyperone (**79b**) (Murai *et al.*, 1978*b*).

4-epi-Eremophilanes. The biosynthesis of capsidiol (**77a**) in infected sweet pepper was the subject of a careful investigation by Baker *et al.* (1975). This study is noteworthy also as the first application of the powerful double-labelling ¹³C-NMR technique to biosynthetic investigations of phytoalexins, and also as the first instance in which the long-predicted methyl migration, the characteristic feature of the eudesmane to eremophilane rearrangement, has been

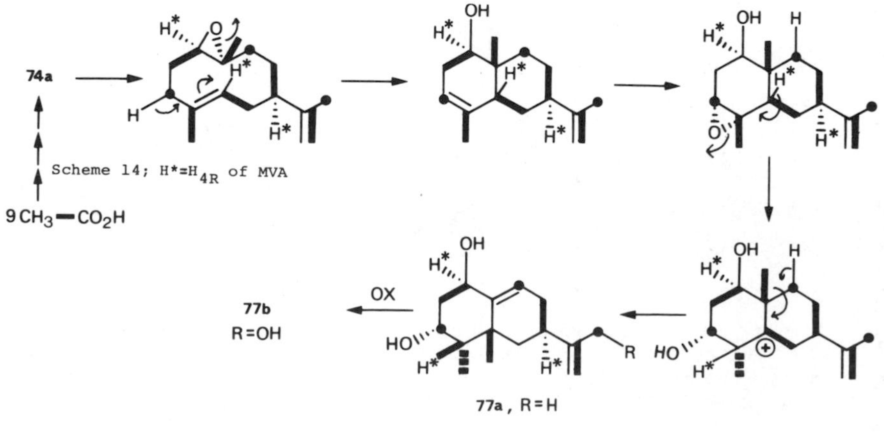

Scheme 18 Biosynthesis of capsidiol and hydroxycapsidiol

demonstrated experimentally. The precursor fed was acetate-1,2-^{13}C$_2$ which was specifically incorporated into **77a** to the extent of about 5% at each site. The ^{13}C-NMR spectrum of the enriched **77a** showed the pattern of singlet and paired doublet absorptions expected for a compound derived from acetate *via* farnesol (Scheme 14) except for the distinguishing feature that C-10 and C-15 (C-7 and C-14 in farnesol numbering) gave rise to singlet rather than doublet absorption; that is, these two atoms had become detached during the bio-synthetic process, precisely as predicted (Scheme 18).[5] A second important feature of Scheme 18 is that it predicts a 1,2-hydride shift from C-5 to C-4. This has now also been verified by direct observation (Hoyano *et al.*, 1980). MVA-4,4-^{2}H$_2$, when administered to sweet pepper infected with *Monilinia fructicola*, was incorporated into **77a** with retention of three deuterium atoms. Two of these, as determined by direct observation of the ^{2}H-NMR spectrum, were located at the normal 1α and 7α positions but the third, again exactly as predicted by Scheme 18, was found at C-4β.

Two other compounds are known from the 4-*epi*-eremophilane class. One is capsenone which is formed from capsidiol (**77a**) by fungal agency and therefore is not pertinent to these discussions. The second is 13-hydroxycapsidiol (**77b**), a transformation product of **77a** in healthy pepper tissue. The oxygenation of the C$_3$ sidechain of mono- and bicyclic sesquiterpenes is an extremely common process (cf. Devon and Scott, 1972); in this instance, it could be shown that it proceeds by direct attack on C-13 and without intervention of either an allylic rearrangement or epoxide formation. This conclusion follows in a straight-forward manner from the observation that capsidiol-^{13}C$_x$, biosynthesized from doubly labelled acetate as by Baker *et al.* (1975), was converted into the hydroxy derivative (dilution value, 1.0) with complete retention of the labelling pattern; i.e. doublet absorption in the side-chain continued to be exclusively associated with C-11 and C-12 (Stothers *et al.*, 1978).

It is intriguing, and presumably of physiological as well as biogenetic significance that, although very many eremophilanes are now known (Pinder, 1977) capsidiol and its two derivatives are still the only examples with a 4,5-*trans* configuration.

Scheme 19

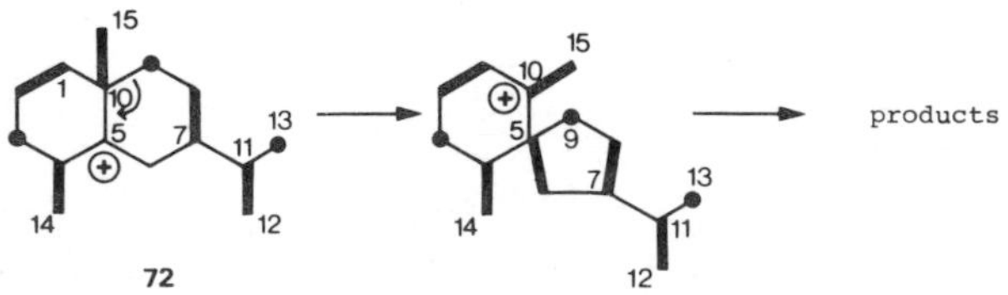

$$72 \longrightarrow \longrightarrow \text{products}$$

---

[5] A different route was considered at one time as a possible though less likely alternative (Stoessl *et al.*, 1976). This postulated the α-cyperone derivative from tobacco, with structure **79b**, as a crucial intermediate. However, since this compound is now known to be **79a** (preceding subsection), the rationale for the route has vanished.

Vetispiranes. The compounds comprising this class fall into several groups differentiated by stereochemistry and/or oxidation pattern. All appear to be derivable from eudesmane cations with the same *planar* projection **72**, by the same bond migration (Scheme 19). According to the scheme, the eudesmane to vetispirane rearrangement does not entail the cleavage of any of the six acetate residues carried over intact from MVA (cf. Scheme 14), irrespective of the stereochemistry or oxidation pattern of the vetispirane produced. This prediction has been verified for seven of the compounds. Acetate-1,2-$^{13}C_2$ was incorporated into lubimin (**76a**) and hydroxylubimin (**76b**) in the expected manner in fungus-inoculated *Datura stramonium* (Birnbaum *et al.*, 1976), with specific enrichments of the order of 7%. In analogous incorporation experiments with potatoes inoculated with either *Monilinia fructicola* or *Glomerella cingulata*, similar enrichments, with identical incorporation patterns, were realized for what may be called the "normal" series **76a**, **76b**, 15-dihydrolubimin (**76c**), and isolubimin (**80**), for the 10-*epi*-series 10-*epi*lubimin (**76d**) and 15-dihydro-10-*epi*lubimin (**76e**); and for solavetivone (**81a**). Three other enriched compounds isolated from these experiments are discussed below (Stoessl *et al.*, 1978). For each of these compounds, the incorporation pattern indicated in the structural representation was ascertained without ambiguity by the ^{13}C-NMR spectrum. It is not known whether **76a** and **76d** are connected through a relatively trivial epimerization or whether they arise independently, from different conformations of a germacrene precursor *via* different stereoisomeric forms of cation **72** or biological equivalent. It is also not known whether **76e** is in a product, precursor or equilibrium relationship to **76d**. In the case of **76c**, there can be no doubt that most of the compound which accumulates in the potato–*G. cingulata* interaction results from the reduction of **76a** by fungal

80

81a, R=H

81b, R=OH

82

enzymes (Ward and Stoessl, 1977). On the other hand, the isolation of trace amounts of **76c** from the potato–*M. fructicola* interaction (Stoessl *et al.*, 1978) seems to indicate that fungal reduction of **76a** is not the sole cause of its formation. The situation is complicated further by the recent isolation of members of a 2-*epi*-series, **76f** and **76g** (Stoessl and Stothers, 1980). It is not at all improbable that some or all of these compounds are in a dynamic equilibrium.

The available evidence concerning these matters is inconclusive. Kalan and Osman (1976) treated uninfected potato slices with solavetivone (**81a**) and found that isolubimin (**80**),[6] followed by lubimin (**76a**) and rishitin (**71a**), were formed on incubation. In a second report (Kalan *et al.*, 1976), an additional compound described as hydroxymethyllubimin, but which clearly was 15-dihydrolubimin (**76c**), was found in potato slices incubated with **81a**. When **76c** itself was applied to such slices, **76a** and **71a** were formed. The authors interpret these results in terms of the sequence, **81a**→**80**→**76c**→**76a**→**76b**→**71a**. Unfortunately, however, since unlabelled **81a** and **76a** were used in these experiments, the possibility remains that they were not precursors of the other compounds but rather stimuli which induced their *de novo* synthesis.

In any event, in direct contrast to these results, Stoessl and Stothers (1981), after administering **76c**-$^{13}C_x$ to healthy potato slices, isolated a mixture of isolubimin-$^{13}C_x$ (**80**), derived from **76c**-$^{13}C_x$ without detectable dilution, and lubimin (**76a**) which was completely unlabelled within the limits of experimental error ($\pm 0.3\%$). Unmetabolized **76c**-$^{13}C_x$ was similarly recovered without detectable dilution, clearly indicating that there was no significant interconversion, in either direction, between **76c** and **76a**.

However, it seems that at least part of the apparent conflict between the results from the two groups may arise from differences in experimental conditions, because the formation of **71a** from **76b**, as postulated by Kalan and Osman (1976) has been verified in the laboratories of a third group (see below).

Solanoscone (**82**), a metabolite from air-cured Burley tobacco, is probably formed directly from solavetivone (**81a**) by attack of one double bond on the other, a reaction which has been achieved in the laboratory by photochemical means (Fujimori *et al.*, 1978). Compounds derived from solavetivone by hydroxylation of the side chain are discussed in the following sub-section.

Noreudesmanes. In the experiments with inoculated potatoes described above (Stoessl *et al.*, 1978), acetate-1,2-$^{13}C_x$ was incorporated into rishitin in the expected manner (**71a**, heavy bonds and dots). However, although a direct derivation of **71a** from a germacrene via the eudesmane cation **72** in an oxidized form can be readily devised (Stoessl *et al.*, 1976), this has been shown to be an oversimplification. Administration of hydroxylubimin-^{14}C (**76b**) to healthy

---

[6] There is some doubt whether the compound described as isolubimin by Kalan and Osman (1976) belonged to the normal or the 10-*epi*-series (see Katsui *et al.*, 1978, but contrast Murai *et al.*, 1980). Spectra have been reported only in part, and no direct comparisons appear to have been made.

potato slices led to about 12% incorporation into **71a**, which was carefully crystallized to constant radioactivity as the bis-3,5-dinitrobenzoate. The efficient incorporation and very low dilution value leave no doubt that incorporation was specific and not the consequence of prior fragmentation. Moreover, the result was confirmed by dilution experiments in which addition of cold **76b** significantly decreased the incorporation of acetate-2-^{14}C into **71a** (Sato *et al.*, 1978).

83a , R=H

83b , R=OH

84a

84b

85

87

Glutinosone (**83a**) from tobacco (Burden *et al.*, 1975) and hydroxyglutinosone (**83b**) from potato (Murai *et al.*, 1980) are evidently derivable from **71a** by oxidation of the 2-OH and consequent migration of the 5,10 double bond into conjugation, possibly assisted by H_2O attack in the case of **83b**. Other compounds clearly related to rishitin are **84a** and **84b** which were isolated in trace amounts from tobacco (Demole and Enggist, 1978). Rishitin-M-1 (**71b**) and its 11,12-dihydro derivative M-2 are formed from **71a** in healthy potato slices (Murai *et al.*, 1977). The oxidation of **71a** to **71b** may be analogous to that of **77a** to **77b**, i.e. without participation of the 11,12 double bond, but this is not necessarily the case. In tobacco, a glucoside of 11,13 (or 11,12?)-dihydroxy-solavetivone (**85**) co-occurs with a glucoside of 13-hydroxysolavetivone (**81b**) (Anderson *et al.*, 1977).

Rishitinol (**73**) from potato (Katsui *et al.*, 1972) may be the product of a dienol-benzene rearrangement (with methyl migration) as suggested by Oliveira *et al.* (1976) who isolated its 8-epimer, together with several related compounds, from the trunk wood of *Emmotum nitens* in Brazil. More recently, rishitinol itself has also been isolated from the same source (Prof. A. Braga de Oliveira, private communication).

Scheme 20 Biosynthesis of phytuberin

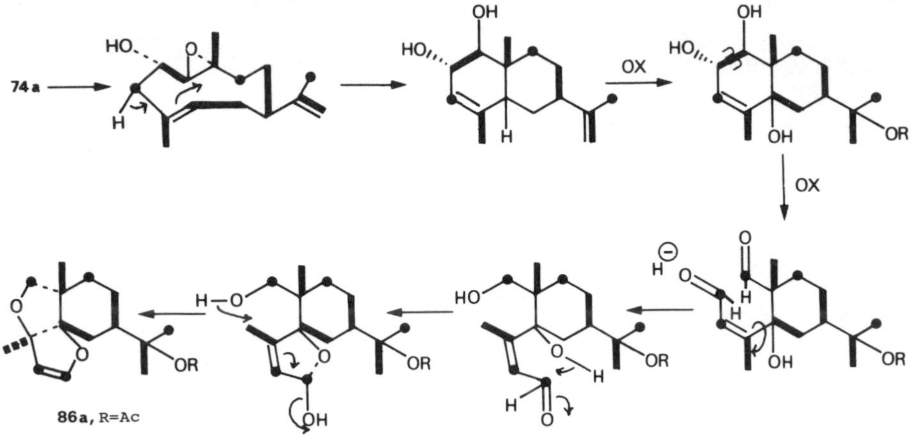

86a, R=Ac
86b, R=H

Phytuberin (**86a**). A plausible pathway to phytuberin and its desacetyl derivative phytuberol (**86b**) is delineated in Scheme 20. The salient feature of the route, the cleavage of the C-1/C-2 bond, has been verified by the incorporation of acetate-1,2-$^{13}C_2$, in the experiments with potatoes described above (Stoessl *et al.*, 1978), which gave both compounds clearly labelled in the expected, unusual pattern indicated.

Rishitinone (**87**) is a "normal" eremophilane recently reported from potato with few details accessible (Katsui *et al.*, 1980). It represents yet another family of sesquiterpenoids from the Solanaceae.

3. Diterpenes

Diterpenes are derived from geranylgeranylpyrophosphate or GGPP (Scheme 14). Although they are numerous, only very few have been reported as phytoalexins or stress compounds.

(i) Casbene (88)

This somewhat atypical phytoalexin (Sitton and West, 1975) is produced from MVA-2-^{14}C and GGPP-2-^{14}C by cell-free extracts from castor bean (*Ricinus communis*) seedlings. The amounts produced are greatly increased when the seedlings have been exposed to any one of several fungi or elicitors extracted from them. A partially purified and characterized casbene synthetase has been isolated from seedlings challenged with *Rhizopus stolonifer* but was not obtained in the absence of this treatment (Dueber *et al.*, 1978). Mechanistically, **88** can be envisaged to be formed by a straightforward cyclization of GGPP, followed by neutralization of the cationic charge by cyclopropane ring closure and loss of a proton (Robinson and West, 1970). As these authors pointed out, the cyclization of GGPP to **88** may represent a deflection from its normal elaboration to

88

89

90a

90b

gibberellins via copalylpyrophosphate (**89**). Alternatively, or additionally, since casbene is thought to be itself a precursor to other medium ring sized diterpenes from the Euphorbiaceae (Crombie *et al.*, 1980), its accumulation may perhaps result from a metabolic block in these pathways.

(ii) Momilactones

Momilactones A (**90a**) and B (**90b**) were isolated as phytoalexins of rice, again under somewhat atypical conditions (Cartwright *et al.*, 1977). They occur normally but only in minute amounts in rice seed where they appear to function as dormancy factors. No studies on their biosynthesis have been reported. However, they too are antipodally related to **89** and one may speculate, once again, that their biosynthesis represents a diversion from that of the gibberellins.

Concluding remarks

A remarkable feature of the biosynthetic experiments reviewed in this chapter is that in most instances, the incorporation of precursors into phytoalexins was realized with facility and high efficiency. This is in sharp contrast to similar studies with higher plants in a normal, i.e. healthy state, which are notorious for low, sometimes even abysmally low incorporation rates. Possible reasons for the difference are that stress conditions may ensure that precursors are added at a time of very active synthesis, and that the concomitant cellular disorganization may mitigate problems of transport and compartmentation. Stress conditions would therefore appear to offer considerable tactical advantages which might frequently be exploitable for biosynthetic studies in general.

The biosynthetic studies reviewed in this chapter are also of specific phyto-pathological interest because they bear directly on fundamental problems concerning the agents and mechanisms which bring about phytoalexin formation. Satisfactory solutions of these interconnected questions clearly require precise definition, in molecular terms, of the loci at which normal metabolism is diverted to stress compound accumulation. Operative mechanisms might include the imposition or removal of metabolic blocks relatively late on a normal metabolic route, or greatly stimulated synthesis of early, general biosynthetic precursors. Any of these alternatives could lead to the accumulation of compounds which otherwise occur only in trace amounts, as transient intermediates of normal metabolism; or to the deflection of such intermediates, or other normal metabolites, into normally suppressed metabolic routes. Only little information is as yet available on these matters. That relating to the loci which may regulate pterocarpan and isoflavan accumulation has been discussed above (pp. 141–2, 146). Other work, reviewed elsewhere in this volume, has documented the *de novo* synthesis and greatly increased activity of enzymes, e.g. phenylalanine ammonia lyase, which mediate the formation of general bio-synthetic precursors. Elicitors of phytoalexins, both exogenous and the probably more directly and specifically implicated constitutive elicitors, are the subject of chapter 9.

Acknowledgement

It is a pleasure to thank Gerald Lambert for preparing the structural diagrams.

REFERENCES

Allen, E. H. and Thomas, C. A. (1971) A second antifungal polyacetylene compound from *Phytophthora*-infected safflower. *Phytopathology*, 61, 1107–1109.

Amrhein, N. and Zenk, M. H. (1977) Metabolism of phenylpropanoid compounds. *Physiol. Veg.*, 15, 251–260.

Anderson, R. C., Gunn, D. M., Murray-Rust, J., Murray-Rust, P. and Roberts, J. S. (1977) Vetispirane sesquiterpene glucosides from flue-cured Virginia tobacco: structure, absolute stereo-chemistry and synthesis. X-Ray structure of the *p*-bromobenzenesulphonate of one of the derived aglucones. *J. Chem. Soc. Chem. Commun.*, 27–28.

Bailey, J. A. and Burden, R. S. (1973) Biochemical changes and phytoalexin accumulation in *Phaseolus vulgaris* following cellular browning caused by Tobacco Necrosis Virus. *Physiol. Plant Pathol.*, 3, 171–177.

Baker, F. C., Brooks, C. J. W. and Hutchinson, S. A. (1975) Biosynthesis of capsidiol in sweet peppers (*Capsicum frutescens*) infected with fungi: evidence for methyl group migration from ^{13}C Nuclear Magnetic Resonance spectroscopy. *J. Chem. Soc. Chem. Commun.*, 293–294.

Banthorpe, D. V. and Ekundayo, O. (1976) Biosynthesis of (+)-car-3-ene in *Pinus* species. *Phytochemistry*, 15, 109–112.

Barz, W. (1977) Degradation of polyphenols in plants and plant cell suspension cultures. *Physiol. Veg.*, **15**, 261–277.

Barz, W. and Hösel, W. (1975) "Metabolism of flavonoids", in *The Flavonoids*, eds. J. B. Harborne, T. J. Mabry and H. Mabry, Chapman and Hall, London, 916–969.

Bell, A. A. and Stipanovic, R. D. (1978) Biochemistry of disease and pest resistance in cotton. *Mycopathologia*, **65**, 91–106.

Berlin, J., Dewick, P. M., Barz, W. and Grisebach, H. (1972) Biosynthesis of coumestrol in *Phaseolus aureus*. *Phytochemistry*, **11**, 1689–1693.

Birch, A. J. (1966) Some natural antifungal agents. *Chem. Ind.*, 1173–1176.

Birnbaum, G. I., Huber, C. P., Post, M. L., Stothers, J. B., Robinson, J. R., Stoessl, A. and Ward, E. W. B. (1976) Sesquiterpenoid stress compounds of *Datura stramonium*: biosynthesis of the three major metabolites from [1,2-^{13}C] acetate and the X-ray structure of 3-hydroxylubimin. *J. Chem. Soc. Chem. Commun.*, 330–331.

Brown, S. A. (1970) Biosynthesis of furanocoumarins in parsnips. *Phytochemistry*, **9**, 2471–2475.

Brown, S. A. (1972) Methodology. *Biosynthesis*, **1**, 1–40.

Brown, S. A. (1978) Biochemistry of the coumarins. *Recent Adv. Phytochem.*, **12**, 249–285.

Bu'Lock, J. D. (1965) *The Biosynthesis of Natural Products.* McGraw-Hill, London, New York, Toronto, Sydney.

Burden, R. S. and Bailey, J. A. (1975) Structure of the phytoalexin from soybean. *Phytochemistry*, **14**, 1389–1390.

Burden, R. S., Bailey, J. A. and Vincent, G. G. (1975) Glutinosone, a new antifungal sesquiterpene from *Nicotiana glutinosa* infected with tobacco mosaic virus. *Phytochemistry*, **14**, 221–223.

Burka, L. T. and Kuhnert, L. (1977) Biosynthesis of furanosesquiterpenoid stress metabolites in sweet potatoes (*Ipomoea batatas*). Oxidation of ipomeamarone to 4-hydroxymyoporone. *Phytochemistry*, **16**, 2022–2023.

Burka, L. T. and Wilson, B. J. (1976) Toxic furanosesquiterpenoids from mold-damaged sweet potatoes (*Ipomoea batatas*). *Adv. Chem. Ser.*, **149**, 387–399.

Cain, R. O. and Porter, A. E. A. (1979) Biosynthesis of the phytoalexin wyerone in *Vicia faba*. *Phytochemistry*, **18**, 322–323.

Cartwright, D., Langcake, P., Pryce, R. J., Leworthy, D. P. and Ride, J. P. (1977) Chemical activation of host defence mechanisms as a basis for crop protection. *Nature*, **267**, 511–513.

Charlwood, B. V. and Banthorpe, D. V. (1978) "The biosynthesis of monoterpenes" in *Progress in Phytochemistry*, eds. L. Reinhold, J. B. Harborne and T. Swain, Vol. 5, Pergamon Press, Oxford, New York, Toronto, Sydney, Paris, Frankfurt, 65–125.

Clarke, D. D. and Baines, P. S. (1976) Host control of scopolin accumulation in infected potato tissue. *Physiol. Plant Pathol.*, **9**, 199–203.

Clayton, R. B. (1965) Biosynthesis of sterols, steroids, and terpenoids. *Quart. Rev. Chem. Soc.*, **19**, 168–200, 201–230.

Cordell, G. A. (1976) Biosynthesis of sesquiterpenes. *Chem. Rev.*, **76**, 425–460.

Cornforth, J. W. (1976) Asymmetry and enzyme action. *Science*, **193**, 121–125.

Cornia, M. and Merlini, M. (1975) A possible chemical analogy for pterocarpan biosynthesis. *J. Chem. Soc. Chem. Commun.*, 428–429.

Coxon, D. T., Curtis, R. F., Price, K. R. and Levett, G. (1973) Abnormal metabolites produced by *Daucus carota* roots stored under conditions of stress. *Phytochemistry*, **12**, 1881–1885.

Coxon, D. T., O'Neill, T. M., Mansfield, J. W. and Porter, A. E. A. (1980) Identification of three hydroxyflavan phytoalexins from daffodil bulbs. *Phytochemistry*, **19**, 889–891.

Crombie, L., Crombie, W. M. L. and Jamieson, S. V. (1979) Isolation of cannabispiradienone and cannabidihydrophenanthrene. Biosynthetic relationships between the spirans and dihydrostilbenes of Thailand cannabis. *Tetrahedron Lett.*, 661–664.

Crombie, L., Kneen, G., Pattenden, G. and Whybrow, D. (1980) Total synthesis of the macrocyclic diterpene (−)-casbene, the putative biogenetic precursor of lathyrane, tigliane, ingenane, and related terpenoid structures. *J. Chem. Soc., Perkin Trans.*, **1**, 1711–1717.

Crombie, L., Dewick, P. M. and Whiting, D. A. (1973) Biosynthesis of rotenoids. Chalcone, isoflavone, and rotenoid stages in the formation of amorphigenin by *Amorpha fructicosa* seedlings. *J. Chem. Soc., Perkin Trans. 1*, 1285–1294.

Demole, E. and Enggist, P. (1978) A chemical study of Virginia Tobacco flavour (*Nicotiana tabacum* L.). II. Isolation and synthesis of cis-2-isopropenyl-8-methyl-1,2,3,4-tetrahydro-1-naphthalenol and 3-isopropenyl-5-methyl-1,2-dihydronaphthalene. *Helv. Chim. Acta*, **61**, 1335–1341.

Devon, T. K. and Scott, A. I. (1972) *Handbook of Naturally Occurring Compounds. Vol. II. Terpenes.* Academic Press, New York and London.

Dewick, P. M. (1975) Pterocarpan biosynthesis: chalcone and isoflavone precursors of demethyl-homopterocarpin and maackiain in *Trifolium pratense. Phytochemistry*, **14**, 979–982.

Dewick, P. M. (1977) Biosynthesis of pterocarpan phytoalexins in *Trifolium pratense. Phytochemistry*, **16**, 93–97.

Dewick, P. M. and Banks, S. W. (1980) Biogenetic relationships amongst the isoflavonoids. *Planta Medica*, **39**, v/13.

Dewick, P. M. and Martin, M. (1976) Biosynthesis of isoflavonoid phytoalexins in *Medicago sativa*: the biosynthetic relationship between pterocarpans and 2′-hydroxyisoflavans. *J. Chem. Soc. Chem. Commun.*, 637–638.

Dewick, P. M. and Martin, M. (1979a) Biosynthesis of pterocarpan and isoflavan phytoalexins in *Medicago sativa*: the biochemical interconversion of pterocarpans and 2′-hydroxyisoflavans. *Phytochemistry*, **18**, 591–596.

Dewick, P. M. and Martin, M. (1979b) Biosynthesis of pterocarpan, isoflavan and coumestan metabolites of *Medicago sativa*: chalcone, isoflavone and isoflavanone precursors. *Phytochemistry*, **18**, 597–602.

Dewick, P. M. and Ward, D. (1977) Stereochemistry of isoflavone reduction during pterocarpan biosynthesis: an investigation using deuterium nuclear magnetic resonance spectroscopy. *J. Chem. Soc. Chem. Commun.*, 338–339.

Dewick, P. M. and Ward, D. (1978) Isoflavone precursors of the pterocarpan phytoalexin maackiain in *Trifolium pratense. Phytochemistry*, **17**, 1751–1754.

Dewick, P. M., Barz, W. and Grisebach, H. (1970) Biosynthesis of coumestrol in *Phaseolus aureus. Phytochemistry*, **9**, 775–783.

Dueber, M. T., Adolf, W. and West, C. A. (1978) Biosynthesis of the diterpene phytoalexin casbene. Partial purification and characterization of casbene synthetase from *Ricinus communis. Plant Physiol.*, **62**, 598–603.

El Feraly, F. S., Chan, Y. M., El-Sohly, M. and Turner, C. E. (1979) Biomimetic synthesis of cannabispiran. *Experientia*, **35**, 1131–1132.

Evans, R. L. and Pluck, D. J. (1978) Phytoalexins produced in barley in response to the halo spot fungus, *Selenophoma donacis. Ann. appl. Biol.*, **89**, 332–336.

Fawcett, C. H. and Spencer, D. M. (1968) *Sclerotinia fructigena* infection and chlorogenic acid content in relation to antifungal compounds in apple fruits. *Ann. appl. Biol.*, **61**, 245–253.

Fisch, M. H., Flick, B. H. and Arditti, J. (1973) Structure and antifungal activity of hircinol, loroglossol and orchinol. *Phytochemistry*, **12**, 437–441.

Freudenberg, K. (1965) Lignin: its constitution and formation from *p*-hydroxycinnamyl alcohols. *Science*, **148**, 595–600.

Fritig, B., Hirth, L. and Ourisson, G. (1972) Biosynthesis of phenolic compounds in healthy and diseased tobacco plants and tissue cultures. *Hoppe-Seyler's Z. Physiol. Chem.*, **353**, 134–135.

Fuchs, A., De Vries, F. W. and Platero Sanz, M. (1980) The mechanism of pisatin degradation by *Fusarium oxysporum* f. sp. *pisi. Physio!. Plant Pathol.*, **16**, 119–133.

Fujimori, T., Kasuga, R., Kaneko, H., Sakamura, S., Noguchi, M., Furusaki, A., Hashiba, N. and Matsumoto, T. (1978) Solanoscone: a novel sesquiterpene ketone from *Nicotiana tabacum.* X-ray structure determination of the corresponding oxime. *J. Chem. Soc. Chem. Commun.*, 563–564.

Grisebach, H. (1962) Die Biosynthese der Flavonoide. *Planta Medica*, **10**, 385–397.

Grisebach, H. (1975) Enzymologie der Flavonoid-biosynthese. *Ber. Dtsch. Bot. Ges.*, **88**, 61–69.

Grisebach, H. and Doerr, N. (1960) Zur Biogenese der Isoflavone. II. Mitt.: Über den Mechanismus der Umlagerung. *Z. Naturforsch. Teil B*, **15b**, 284–286.

Hadwiger, L. A. (1965) The biosynthesis of pisatin. *Phytochemistry*, **5**, 523–525.

Hahlbrock, K. and Grisebach, H. (1975) "Biosynthesis of flavonoids" in *The Flavonoids*, eds. J. B. Harborne, T. J. Mabry and H. Mabry, Chapman and Hall, London, 866–915.

Hargreaves, J. A., Mansfield, J. W. and Rossall, S. (1977) Changes in phytoalexin concentrations in tissues of the broad bean plant (*Vicia faba* L.) following inoculation with species of *Botrytis. Physiol. Plant Pathol.*, **11**, 227–242.

Hashimoto, T. and Tajima, M. (1978) Structures and synthesis of the growth inhibitors batatasins IV and V, and their physiological activities. *Phytochemistry*, **17**, 1179–1184.

Hare, R. C. (1966) Physiology of resistance to fungal diseases in plants. *Bot. Rev.*, **32**, 95–137.

Hata, K., Baba, K. and Kozawa, M. (1979) Chemical studies on the heartwood of *Cassia garrettiama* Craib. II. Nonanthraquinonic constituents. *Chem. Pharm. Bull.*, **27**, 984–989.

Heinstein, P. F., Herman, D. L., Tove, S. B. and Smith, F. H. (1970) Biosynthesis of gossypol. Incorporation of mevalonate-2-^{14}C and isoprenyl pyrophosphates. *J. Biol. Chem.*, **245**, 4658–4665.

Heinstein, P., Widmaier, R., Wegner, P. and Howe, J. (1979) Biosynthesis of gossypol. *Recent Adv. Phytochem.*, **12**, 313–333.

Heller, W., Gardiner, S. E. and Hahlbrock, K. (1980) Zum Reaktionsmechanismus der Chalcon-Synthase aus Belichteten Zellsuspensionskulturen von *Petroselinum hortense*. *Hoppe-Seyler's Z. Physiol. Chem.*, **361**, 265–266.

Hess, S. L., Hadwiger, L. A. and Schwochau, M. E. (1971) Studies on biosynthesis of phaseollin in excised pods of *Phaseolus vulgaris*. *Phytopathology*, **61**, 79–82.

Hillis, W. E. and Ishikura, N. (1969) An enzyme from *Eucalyptus* which converts cinnamoyl triacetic acid into pinosylvin. *Phytochemistry*, **8**, 1079–1088.

Hillis, W. E. and Yazaki, Y. (1971) The biosynthesis of stilbenes in eucalypt leaves. *Phytochemistry*, **10**, 1051–1054.

Hoyano, Y., Stoessl, A. and Stothers, J. B. (1980) Biosynthesis of the antifungal sesquiterpene capsidiol. Confirmation of a hydride shift by ^{2}H magnetic resonance. *Can. J. Chem.*, **58**, 2069–2072.

Ingham, J. L. (1976) 3,5,4'-Trihydroxystilbene as a phytoalexin from groundnuts (*Arachis hypogaea*). *Phytochemistry*, **15**, 1791–1793.

Ingham, J. L. (1977) Phytoalexins of hyacinth bean (*Lablab niger*). *Z. Naturforsch. Teil C*, **32c**, 1018–1020.

Ingham, J. L. (1978) Isoflavonoid and stilbene phytoalexins of the genus *Trifolium*. *Biochem. Syst. Ecol.*, **6**, 217–223.

Ingham, J. L., Keen, N. T. and Hymowitz, T. (1977) A new isoflavone phytoalexin from fungus-inoculated stems of *Glycine wightii*. *Phytochemistry*, **16**, 1943–1946.

Inoue, H. and Uritani, I. (1979) Biosynthetic correlation of various phytoalexins in sweet potato root tissue infected by *Ceratocystis fimbriata*. *Plant Cell Physiol.*, **20**, 1307–1314.

Jacques, D. and Haslam, E. (1974) Biosynthesis of plant proanthocyanidins. *J. Chem. Soc. Chem. Commun.*, 231–232.

Jaworski, J. G. and Kuć, J. (1974) Effect of ethrel and *Ceratocystis fimbriata* on the synthesis of fatty acids and 6-methoxy mellein in carrot root. *Plant Physiol.*, **53**, 331–336.

Johnson, C., Brannon, D. R. and Kuć, J. (1973) Xanthotoxin: a phytoalexin of *Pastinaca sativa* root. *Phytochemistry*, **12**, 2961–2962.

Johnson, G., Maag, D. D., Johnson, D. K. and Thomas, R. D. (1976) The possible role of phytoalexins in the resistance of sugarbeet (*Beta vulgaris*) to *Cercospora beticola*. *Physiol. Plant Pathol.*, **8**, 225–230.

Kalan, E. B. and Osman, S. F. (1976) Isolubimin: a possible precursor of lubimin in infected potato slices. *Phytochemistry*, **15**, 775–776.

Kalan, E. B., Patterson, J. M. and Schwartz, D. P. (1976) Metabolism of isolubimin and hydroxymethyllubimin by potato tuber slices. *Plant Physiol. Suppl.*, **57**, 91.

Katsui, N., Matsunaga, A., Imaizumi, K., Masamune, T. and Tomiyama, K. (1972) The structure and synthesis of rishitinol: a sesquiterpene alcohol from diseased potato tubers. *Bull. Chem. Soc. Jpn.*, **45**, 2871–2877.

Katsui, N., Yagihashi, F., Murai, A. and Masamune, T. (1978) Structure of oxyglutinosone and epioxylubimin, stress metabolites from diseased potato tubers. *Chemistry Lett.*, 1205–1206.

Katsui, N., Yagihashi, F., Murai, A. and Masamune, T. (1980) Isolation of a new sesquiterpene, rishitinone from potatoes and its structure. *Chemical Abstracts*, **93**, 46857.

Keen, N. T. and Littlefield, L. J. (1979) The possible association of phytoalexins and resistance gene expression in flax to *Melampsora lini*. *Physiol. Plant Pathol.*, **14**, 265–280.

Keen, N. T., Zaki, A. I. and Sims, J. J. (1972) Biosynthesis of hydroxyphaseollin and related isoflavonoids in disease-resistant soybean hypocotyls. *Phytochemistry*, **11**, 1031–1039.

Kelly, R. B., Alward, S. J., Murty, K. S. and Stothers, J. B. (1978) A revised structure for aubergenone, a sesquiterpenoid related to eudesmane:synthesis of 4-*epi*-aubergenone. *Can. J. Chem.*, **56**, 2508–2512.

Kreuzaler, F. and Hahlbrock, K. (1975) Enzymic synthesis of an aromatic ring from acetate units. Partial purification and some properties of flavanone synthase from cell-suspension cultures of *Petroselinum hortense*. *Eur. J. Biochem.*, **56**, 205–213.

Kuć, J. and Lisker, N. (1978) "Terpenoids and their role in wounded and infected plant storage tissue", in *Biochemistry of Wounded Plant Tissues*, ed. G. Kahl, de Gruyter, Berlin, 203–242.

Langcake, P. and Pryce, R. J. (1976) The production of resveratrol by *Vitis vinifera* and other members of the Vitaceae as a response to infection or injury. *Physiol. Plant Pathol.*, **9**, 77–86.

Langcake, P. and Pryce, R. J. (1977a) The production of resveratrol and the viniferins by grapevines in response to ultraviolet irradiation. *Phytochemistry*, **16**, 1193–1196.

Langcake, P. and Pryce, R. J. (1977b) Oxidative dimerisation of 4-hydroxystilbenes *in vitro*: production of a grapevine phytoalexin mimic. *J. Chem. Soc. Chem. Commun.*, 208–210.

Letcher, R. M. and Nhamo, L. R. M. (1972) Chemical constituents of the Combretaceae.Part III. Substituted phenanthrenes, 9,10-dihydrophenanthrenes, and bibenzyls from the heartwood of *Combretum psidioides*. *J. Chem. Soc., Perkin Trans. 1*, 2941–2946.

Lyne, R. L. and Mulheirn, L. J. (1978) Minor pterocarpinoids of soybean. *Tetrahedron Lett.*, 3127–3128.

Mansfield, J. W., Porter, A. E. A. and Smallman, R. V. (1980) Dihydrowyerone derivatives as components of the furanoacetylenic phytoalexin response of tissues of *Vicia faba*. *Phytochemistry*, **19**, 1057–1061.

Martin, J. F. and Demain, A. L. (1980) Control of antibiotic biosynthesis. *Microbiol. Rev.*, **44**, 230–251.

Martin, M. and Dewick, P. M. (1978) Role of an isoflav-3-ene in the biosynthesis of pterocarpan, isoflavan and coumestan metabolites of *Medicago sativa*. *Tetrahedron Lett.*, 2341–2344.

Martin, M. and Dewick, P. M. (1979) Biosynthesis of the 2-aryl-benzofuran phytoalexin vignafuran in *Vigna unguiculata*. *Phytochemistry*, **18**, 1309–1317.

Murai, A., Katsui, N., Yagihashi, F., Masamune, T., Ishiguri, Y. and Tomiyama, K. (1977) Structure of rishitin-M-1 and M-2, metabolites of rishitin in healthy potato tuber tissues. *J. Chem. Soc. Chem. Commun.*, 670–671.

Murai, K., Abiko, A., Ono, M., Katsui, N. and Masamune, T. (1978a) Structure revision and biogenetic relationship of aubergenone, a sesquiterpenoid phytoalexin of eggplants. *Chemistry Lett.*, 1209–1212.

Murai, A., Ono, M. and Masamune, T. (1978b). Structure revision of "1-keto-α-cyperone", a sesquiterpene isolated from tobacco. *Chemistry Lett.*, 1005–1006.

Murai, A., Taketsuru, H., Yagihashi, F., Katsui, N. and Masamune, T. (1980) The structure and configuration of (+)-glutinosone and (+)-oxyglutinosone. *Bull. Chem. Soc. Jpn.*, **53**, 1045–1048.

Oguni, I. and Uritani, I. (1974) Dehydroipomeamarone as an intermediate in the biosynthesis of ipomeamarone, a phytoalexin from sweet potato root infected with *Ceratocystis fimbriata*. *Plant Physiol.*, **53**, 649–652.

Oliveira, A. B. de, Oliveira, G. G. de, Liberalli, C. T. M., Gottlieb, O. R. and Magalhaes, M. T. (1976) Structure and absolute configuration of the sesquiterpenoid emmotins. *Phytochemistry*, **15**, 1267–127.

Parker, W., Roberts, J. S. and Ramage, R. (1967) Sesquiterpene biogenesis. *Quart. Rev. Chem. Soc.*, **21**, 331–363.

Pelter, A., Bradshaw, J. and Warren, R. F. (1971) Oxidation experiments with flavonoids. *Phytochemistry*, **10**, 835–850.

Pinder, A. R. (1977) The chemistry of the eremophilane and related sesquiterpenes. *Fortschr. Chem. Org. Naturst.*, **34**, 81–186.

Robeson, D. J. (1978) Furanoacetylene and isoflavonoid phytoalexins in *Lens culinaris*. *Phytochemistry*, **17**, 807–808.

Robinson, D. R. and West, C. A. (1970) Biosynthesis of cyclic diterpenes in extracts of seedlings of *Ricinus communis*. L. II. Conversion of geranylgeranylpyrophosphate into diterpene hydrocarbons and partial purification of the cyclization enzymes. *Biochemistry*, **9**, 80–89.

Rudloff, E. v. and Jorgensen, E. (1963) The biosynthesis of pinosylvin in the sapwood of *Pinus resinosa*. Ait. *Phytochemistry*, **2**, 297–304.

Rupprich, N., Hildebrand, H. and Kindl, H. (1980) Substrate specificity *in vivo* and *in vitro* in the formation of stilbenes, biosynthesis of rhaponticin. *Arch. Biochem. Biophys.*, **200**, 72–78.

Russell, C. E. and Berryman, A. A. (1976) Host resistance to the fir Engraver Beetle. 1. Monoterpene composition of *Abies grandis* pitch blisters and fungus-infected wounds. *Can. J. Bot.*, **54**, 14–18.

Sarkar, S. K. and Phan, C. T. (1975) The biosynthesis of 8-hydroxy-6-methoxy-3-methyl-3,4-dihydroisocoumarin and 5-hydroxy-7-methoxy-2-methylchromone in carrot root tissues treated with ethylene. *Physiol. Plant.*, **33**, 108–112.

Sato, K., Ishiguri, Y., Doke, N., Tomiyama, K., Yagihashi, F., Murai, A., Katsui, N. and Masamune, T. (1978) Biosynthesis of the sesquiterpenoid phytoalexin rishitin from acetate via oxylubimin in potato. *Phytochemistry*, **17**, 1901–1902.

Schmalle, H. and Hausen, B. M. (1979) A new sensitizing quinone from lady slipper (*Cypripedium calceolus*). *Naturwissenschaften*, **66**, 527–528.

Schoeppner, A. and Kindl, H. (1979) Stilbene synthase (pinosylvine synthase) and its induction by ultraviolet light. *FEBS Lett.*, **108**, 349–352.

Schröder, G., Zähringer, U., Heller, W., Ebel, J. and Grisebach, H. (1979) Biosynthesis of antifungal isoflavonoids in *Lupinus albus*. Enzymatic prenylation of genistein and 2'-hydroxy-genistein. *Arch. Biochem. Biophys.*, **194**, 635–636.

Sequeira, L. (1969) Synthesis of scopolin and scopoletin in tobacco plants infected with *Pseudomonas solanacearum*. *Phytopathology*, **59**, 473–478.

Shain, L. (1967) Resistance of sapwood in stems of loblolly pine to infection by *Fomes annosus*. *Phytopathology*, **57**, 1034–1045.

Shih, M. and Kuć, J. (1973) Incorporation of ^{14}C from acetate and mevalonate into rishitin and steroid glycoalkaloids by potato tuber slices inoculated with *Phytophthora infestans*. *Phytopathology*, **63**, 826–829.

Sicherer, C. A. X. G. F. and Sicherer-Roetman, A. (1980) The *R, S* nomenclature of pterocarpan stereochemistry. *Phytochemistry*, **19**, 485–486.

Simpson, T. J. (1977) Biosynthesis of polyketides. *Biosynthesis*, **5**, 1–33.

Sitton, D. and West, C. A. (1975) Casbene, an anti-fungal diterpene produced in cell-free extracts of *Ricinus communis* seedlings. *Phytochemistry*, **14**, 1921–1925.

Smith, T. A. and Best, G. R. (1978) Distribution of the hordatines in barley. *Phytochemistry*, **17**, 1093–1098.

Stipanovic, R. D., Bell, A. A. and Howell, C. R. (1975) Napthofuran precursors of sesquiterpenoid aldehydes in diseased *Gossypium*. *Phytochemistry*, **14**, 1809–1811.

Stoessl, A. (1967) The antifungal factors in barley. IV. Isolation, structure and synthesis of the hordatines. *Can. J. Chem.*, **45**, 1745–1760.

Stoessl, A. (1972) Inermin associated with pisatin in peas inoculated with the fungus *Monilinia fructicola*. *Can. J. Biochem.*, **50**, 107–108.

Stoessl, A. (1977) "Biogenetic relations between some bicyclic sesquiterpenoidal stress compounds of the Solanaceae", in *Current Topics in Plant Pathology*. (Proc. Conf. 1975; ed. Z. Kiraly), Publ. House Hung. Acad. Sci., Budapest, 61–72.

Stoessl, A. (1980) Phytoalexins—a biogenetic perspective. *Phytopathol. Z.*, **99**, 251–272.

Stoessl, A. and Stothers, J. B. (1978) A carbon-13 biosynthetic study of stress metabolites from carrot roots: eugenin and 6-methoxymellein. *Can. J. Bot.*, **56**, 2589–2593.

Stoessl, A. and Stothers, J. B. S. (1979) The incorporation of $[1,2\text{-}^{13}C_2]$ acetate into pisatin to establish the biosynthesis of its polyketide moiety. *Z. Naturforsch. Teil C*, **34c**, 87–89.

Stoessl, A. and Stothers, J. B. S. (1980) 2-Epi- and 15-dihydro-2-epilubimin: new stress compounds from the potato. *Can. J. Chem.*, **58**, 2069–2072.

Stoessl, A. and Stothers, J. B. S. (1981) A carbon-13 biosynthetic study of stress metabolites from potatoes: the origin of isolubimin. *Can. J. Bot.*, **59**, 637–639.

Stoessl, A., Stothers, J. B. S. and Ward, E. W. B. (1975) The structures of some stress metabolites from *Solanum melongena*. *Can. J. Chem.*, **53**, 3351–3358.

Stoessl, A., Stothers, J. B. and Ward, E. W. B. (1976) Sesquiterpenoid stress compounds of the Solanaceae. *Phytochemistry*, **15**, 855–872.

Stoessl, A., Ward, E. W. B. and Stothers, J. B. (1977) "Biosynthetic relationships of sesquiterpenoidal stress compounds from the Solanaceae", in *Host Plant Resistance to Pests*, (ACS Symposium Series 62; ed. P. A. Hedin), American Chemical Society, Washington, 61–77.

Stoessl, A., Stothers, J. B. and Ward, E. W. B. (1978) Biosynthetic studies of stress metabolites from potatoes: incorporation of sodium acetate-$^{13}C_2$ into 10 sesquiterpenes. *Can. J. Chem.*, **56**, 645–653.

Stothers, J. B., Stoessl, A. and Ward, E. W. B. (1978) A ^{13}C-NMR study of the biological oxidation of capsidiol. *Z. Naturforsch. Teil C*, **33c**, 149–150.

Suga, T., Iwata, N. and Asakawa, Y. (1972) Chemical constituents of the male flower of *Alnus pendula* (Betulaceae). *Bull. Chem. Soc. Jpn.*, **45**, 2058–2060.

Sutherland, J. K. (1974) Regio- and stereo-specificity in the cyclisation of medium ring 1,5-dienes. *Tetrahedron*, **30**, 1651–1660.

Swinburne, T. R. (1975) Microbial proteases as elicitors of benzoic acid accumulation in apples. *Phytopathol. Z.*, **82**, 152–162.

Swinburne, T. R. and Brown, A. E. (1975) The biosynthesis of benzoic acid in Bramley's seedlings infected by *Nectria galligena* Bres. *Physiol. Plant Pathol.*, **6**, 259–264.

Takasugi, M., Munõz, L., Masamune, T., Shirata, A. and Takahashi, K. (1978) Stilbene phytoalexins from diseased mulberry. *Chemistry Lett.*, 1241–1242.

Takasugi, M., Nagao, S., Masamune, T., Shirata, A. and Takahashi, K. (1979) Structures of moracins E, F, G and H, new phytoalexins from diseased mulberry. *Tetrahedron Lett.*, 4675–4678.

Takasugi, M., Anetai, M., Masamune, T., Shirata, A. and Takahashi, K. (1980) Broussonins A and B, new phytoalexins from diseased paper mulberry. *Chemistry Lett.*, 339–340.

Tanabe, M. (1973) Stable isotopes in biosynthetic studies. *Biosynthesis*, **2**, 241–299.

Tomiyama, K., Sakuma, T., Ishizaka, N., Sato, N., Katsui, N., Takasugi, M. and Masamune, T. (1968) New antifungal substance isolated from resistant potato tuber tissue infected by pathogens. *Phytopathology*, **58**, 115–116.

Turner, W. B. (1971) *Fungal Metabolites*, Academic Press, London, 74–83.

Uritani, I. (1978) "The biochemistry of host response to infection", in *Progress in Phytochemistry*, eds. L. Reinhold, J. B. Harborne and T. Swain, Vol. 5, Pergamon Press, Oxford, New York, Toronto, Sydney, Paris, Frankfurt, 29–63.

Van Der Merwe, P. J., Rall, G. J. H. and Roux, D. G. (1978) Novel isoflavan-pterocarpan inter-conversions: some structural requirements for cyclization. *J. Chem. Soc. Chem. Commun.*, 224–225.

VanEtten, H. D. and Pueppke, S. G. (1976) "Isoflavonoid phytoalexins", in *Biochemical Aspects of Plant-Parasitic Relationships*, eds. J. Friend and D. R. Threlfall, Academic Press, London, 239–289.

Veech, J. A., Stipanovic, R. D. and Bell, A. A. (1976) Peroxidative conversion of hemigossypol to gossypol. A revised structure for isohemigossypol. *J. Chem. Soc. Chem. Commun.*, 144–145, 528.

Ward, E. W. B. and Stoessl, A. (1977) Phytoalexins from potatoes: evidence for the conversion of lubimin to 15-dihydrolubimin by fungi. *Phytopathology*, **67**, 468–471.

Wong, E. (1975) "The isoflavonoids", in *The Flavonoids*, eds. J. B. Harborne, T. J. Mabry and H. Mabry, Chapman and Hall, London, 743–800.

Wong, E. and Wilson, J. M. (1976) Products of the peroxidase-catalyzed oxidation of 4,2′,4′-trihydroxychalcone. *Phytochemistry*, **15**, 1325–1332.

Woodward, M. D. (1979) Phaseoluteone and other 5-hydroxy-isoflavonoids from *Phaseolus vulgaris*. *Phytochemistry*, **18**, 363–365.

Woodward, M. D. (1980) Phaseollin formation and metabolism in *Phaseolus vulgaris*. *Phytochemistry*, **19**, 921–927.

Yoshikawa, M., Yamauchi, K. and Masago, M. (1979) Biosynthesis and biodegradation of glyceollin by soybean hypocotyls infected with *Phytophthora megasperma* var. *sojae*. *Physiol. Plant Pathol.*, **14**, 157–169.

Zähringer, U., Ebel, J. and Grisebach, H. (1978) Induction of phytoalexin synthesis in soybean. Elicitor-induced increase in enzyme activities of flavonoid biosynthesis and incorporation of mevalonate into glyceollin. *Arch. Biochem. Biophys.*, **188**, 450–455.

Zähringer, U., Ebel, J., Mulheirn, L. J., Lyne, R. L. and Grisebach, H. (1979) Induction of phytoalexin synthesis in soybean. Dimethylallylpyrophosphate: trihydroxypterocarpan dimethyl-allyl transferase from elicitor-induced cotyledons. *FEBS Lett.*, **101**, 90–92.

6 Metabolism of phytoalexins

HANS D. VANETTEN, DAVID E. MATTHEWS
AND DAVID A. SMITH

Introduction

Phytoalexins are natural compounds. It is therefore reasonable to assume that they can be metabolized by microorganisms. In one of several pioneering studies, Müller (1958) alluded to the possibility of phytoalexin inactivation as a result of the metabolic activity of fungi. It was subsequently suggested that the ability to metabolize phytoalexins might be related to pathogenic potential (Nüesch, 1963; Uehara, 1964). In recent years considerable progress has been made in exploring these possibilities. Many pathogens have been tested for the ability to degrade their hosts' phytoalexins. Some of the early metabolites formed have been identified, permitting these reactions to be categorized into a few basic mechanisms. A few cell-free preparations catalysing the transformations have even been obtained. In addition, the significance of phytoalexin metabolism for the outcome of host–parasite interactions has been examined. These developments will be the main focus of this chapter.

The degradation of phytoalexins by the plants that produce them has also been studied to some extent. In some cases the rate of phytoalexin turnover by plants has been proposed to play an important role in regulating the accumulation of these compounds in infected tissue (Stoessl et al., 1977; Yoshikawa et al., 1979; see also chapter 9). In addition, many of the known transformations of phytoalexins by plants involve the conversion of one phytoalexin to another and thus pertain to the biogenesis of these compounds, which is discussed in chapter 5. However, some examples of plant-mediated phytoalexin metabolism will be described here, especially where they offer interesting comparisons with microbial reactions. A few conversions that have been observed only in infected tissue will also be mentioned, even though it may not be certain whether the plant or the microorganism is responsible.

General observations

It is difficult to derive any comprehensive generalizations from the existing literature on phytoalexin metabolism *per se*. However, some overall comments on these reactions can be made in the context of the wealth of information available about the metabolism of other natural, and man-made, xenobiotics (e.g. Testa and Jenner, 1976; Hill and Wright, 1978; Millburn, 1978; Scheline, 1978).

Perhaps the most surprising observation, in view of the diversity of compounds known to be degraded by bacteria, is that no prokaryote has yet been shown to metabolize a phytoalexin. However, the number of published reports in which this possibility has been tested is small (Bruin *et al.*, 1977; Platero Sanz and Fuchs, 1978). There is therefore no reason to assume that bacteria are incapable of phytoalexin metabolism.

The degradation of a foreign compound by different microorganisms is often found to proceed by different reaction pathways. Such diversity of metabolic routes has also been observed in the metabolism of some phytoalexins by fungi. For example, the initial attack on phaseollin can involve at least four sites on the molecule and two reaction types, depending on the fungal species used (Figure 6.1). Similar results have been obtained for medicarpin and maackiain; one isolate of *Nectria haematococca* even performs two different initial reactions simultaneously (Figure 6.2). On the other hand, the first step in the degradation of pisatin is the demethylation of the aryl-*O*-methyl ether in every case reported to date (Figure 6.3). Likewise, the only fungal metabolite of capsidiol yet detected is capsenone (Figure 6.8). It would be surprising if only one initial reaction proved to be possible for the microbial metabolism of each of these compounds, and further studies are likely to reveal other pathways. It is already known that *Capsicum frutescens* can metabolize capsidiol by a different route (Figure 6.14).

Most phytoalexins are lipophilic compounds. The metabolism of lipophilic xenobiotic compounds usually involves conversion to more polar products. In higher organisms the ultimate products frequently are highly polar or ionic compounds, due to conjugation reactions such as glycosylation. These reactions are assumed to facilitate excretion by animals (Testa and Jenner, 1976; Millburn, 1978) and storage in vacuoles by plants (Harborne, 1977). The formation of polar conjugates by microorganisms has also been reported (Hopkinson and Pridham, 1967; Cerniglia and Gibson, 1979) but apparently is not common (Towers, 1964; Alexander, 1981). Usually, the increase in water solubility is more gradual, for example, by the introduction of hydroxyl groups in oxidative catabolism. As will be described below, most of the known conversions of phytoalexins by fungi involve the creation of new hydroxyl groups, by oxygenation, hydration, carbonyl reduction or ether cleavage. Fungal conversion of phytoalexins to highly water-soluble conjugates has not been reported, but in most studies only metabolites in organic-solvent extracts have been examined. There is indirect evidence for a fungal conjugation reaction leading to a less polar metabolite: a methylated pterocarpan derivative was found in tissue infected with *Botrytis cinerea* but not with other fungi (Ingham, 1976).

The metabolism of phytoalexins usually, though not always, yields products that are less toxic, but the diminished toxicity does not appear to be due solely to increased polarity. Attempts to correlate polarity with the toxicity of various structural analogues of phytoalexins have been inconclusive (Ishizaka *et al.*, 1969; Ward *et al.*, 1975; Carter *et al.*, 1978; also see chapter 7). In some cases, the metabolic alteration of a phytoalexin not only increases its polarity but also causes significant changes in its molecular shape. For example, based on NMR spectral data (Van den Heuvel *et al.*, 1974), the conformation of ring B in 1a hydroxyphaseollone is quite different from that in phaseollin (Figure 6.1). In view of the structural requirements of most biological processes it is not surprising that a change in biological activity follows a change in molecular shape. However, as with polarity, attempts to determine specific structural requirements for the antifungal activity of phytoalexins or their metabolites have been inconclusive (Perrin and Cruickshank, 1969; VanEtten, 1976; see chapter 7).

The ability to metabolize a plant's phytoalexins is not restricted to pathogens of that plant; nor do all pathogens appear to have this ability. Lack of a simple correlation between pathogenicity and the ability to degrade phytoalexins has sometimes been regarded as evidence against the importance of phyto- alexin metabolism by pathogens, and even of phytoalexins themselves, in plant-parasite interactions. Such "evidence" requires some dubious further assumptions, namely that (1) phytoalexins are for some reason uniquely refractory to the versatile metabolic capabilities of microorganisms in general, and (2) phytoalexin metabolism is required for pathogenicity either in all cases or in no cases. The significant question is whether any diseases do exist in which the ability of the pathogen to metabolize the host's phytoalexins plays an important role. This appears to be the case, as will be described in a later section (p. 202).

Enzymic mechanisms of phytoalexin metabolism by fungi

The known metabolic conversions of phytoalexins by fungi are classified here according to the reaction types involved. For the most part, the reaction mechanisms have only been inferred from the structures of the metabolites produced, by analogy with previously described enzymic or chemical reactions of other compounds. In only a few cases have these inferences been tested by isolating cell-free preparations in which the predicted cofactor requirements were examined. Indeed, even the number of enzymic steps assumed to be involved in a particular conversion must remain tentative until the reaction has been performed with purified enzyme(s), a goal which has been approached in only one case (Kuhn and Smith, 1979).

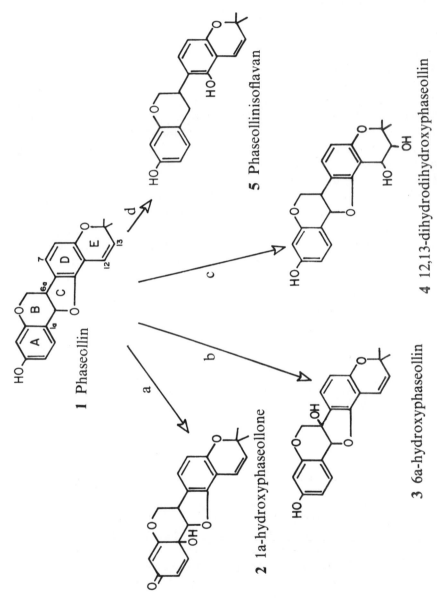

Figure 6.1 The initial metabolites of phaseollin produced by *Fusarium solani* f. sp. *phaseoli*[p] and *Cladosporium herbarum* (reaction *a*; van den Heuvel *et al.*, 1974; van den Heuvel and Glazener, 1975), *Colletotrichum lindemuthianum*[p] and *Botrytis cinerea*[p] (reaction *b*; Burden *et al.*, 1974; van den Heuvel and Glazener, 1975), *Septoria nodorum* (reaction *c*; Bailey *et al.*, 1977), and *Stemphylium botryosum* (reaction *d*; Higgins *et al.*, 1974). Fungi pathogenic on plants that produce the phytoalexin are identified in the figure legends with a superscript p.

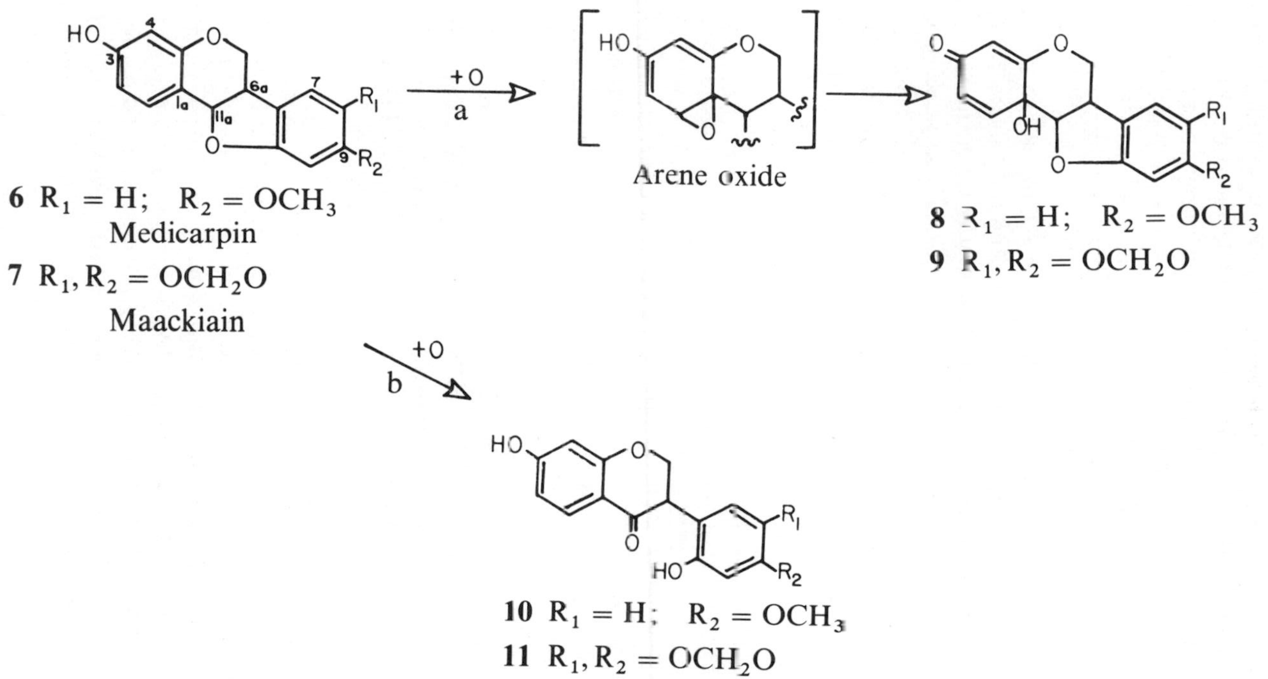

Figure 6.2 The metabolism of medicarpin and maackiain by *Nectria haematococca*[b] (Denny and VanEtten, 1981*a*).

Monooxygenation

Monooxygenases are enzymes which catalyse the incorporation of one atom of molecular oxygen directly into a substrate, while the other oxygen atom is reduced to water (Hayaishi, 1974). The most commonly observed conversions of phytoalexins by fungi involve the addition of one atom of oxygen to the compound. Such an alteration is circumstantial evidence for the operation of a monooxygenase. However, a two-step reaction involving dehydrogenation and hydration is also possible.

Recently direct evidence has been obtained for the participation of a monooxygenase in the metabolism of phaseollin by *Fusarium solani* f. sp. *phaseoli* (Figure 6.1, reaction *a*). Using media enriched with $^{18}O_2$ or $H_2{}^{18}O$, it was shown that only oxygen from O_2 was incorporated into the product, 1a-hydroxyphaseollone (Kistler and VanEtten, 1981). The same metabolite was produced from phaseollin by *Cladosporium herbarum* (van den Heuvel and Glazener, 1975), and analogous conversions of medicarpin and maackiain (Figure 6.2, reaction *a*) have been observed with *N. haematococca* (Denny and VanEtten, 1981*a*). In all these cases the likely intermediate is an arene oxide (Figure 6.2) (Testa and Jenner, 1976; Kistler and VanEtten, 1981).

Demethylation of aryl-*O*-methyl ethers to produce the corresponding phenols is a commonly observed metabolic alteration of natural products. A number of fungi are known to demethylate pisatin in this way (Figure 6.3). Medicarpin (**6**) and its methylated derivative 3,9-dimethoxypterocarpan are also demethylated (Ingham, 1976; Weltring and Barz, 1980). *Fusarium proliferatum* removes both methyl groups from the latter compound, the one at position 9 being attacked first (Weltring and Barz, 1980).

Demethylation of aromatic ethers is normally catalysed by monooxygenases, both in microbes (Hill and Wright, 1978) and in higher organisms (Testa and Jenner, 1976). The probable intermediate in these reactions is an unstable phenolic hemiacetal, from which the methyl group is released as formaldehyde (Figure 6.3). The formaldehyde is usually converted via formic acid to CO_2. The 3-*O*-methyl group of pisatin is known to be evolved as CO_2 by *N. haematococca* (VanEtten *et al.*, 1980). A cell-free preparation capable of demethylating pisatin has been obtained from *N. haematococca* (Matthews and VanEtten, 1981). The reaction required a reduced pyridine nucleotide cofactor (NADPH preferred), and exhibited an oxygen uptake of approximately 1 mole O_2 per mole of pisatin demethylated. The pisatin demethylase activity was localized primarily in the microsomal fraction. These properties are characteristic of most monooxygenases from higher eukaryotes that catalyse demethylation reactions (Hayaishi, 1974). Although fungal demethylases have not been well characterized, some other fungal monooxygenases share these properties (Murphy *et al.*, 1974; Davies *et al.*, 1976; Ferris *et al.*, 1976; Cerniglia and Gibson, 1978).

No pisatin demethylase activity was found in culture filtrates from *N. haematococca* (VanEtten and Matthews, unpublished), a result consistent with

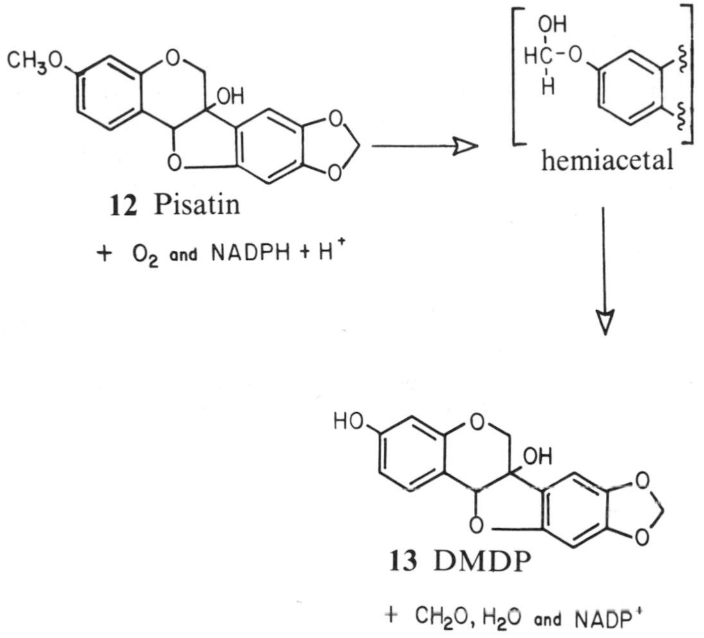

12 Pisatin

+ O₂ and NADPH + H⁺

hemiacetal

13 DMDP

+ CH₂O, H₂O and NADP⁺

Figure 6.3 The demethylation of pisatin to 3,6a-dihydroxy-8,9-methylenedioxypterocarpan (DMDP) by a cell-free preparation from *Nectria haematococca*[P] (Matthews and VanEtten, 1981). The metabolism of pisatin to DMDP is also accomplished by *Aphanomyces euteiches*[P], *A. laevis*, *Ascochyta pisi*[P], *Fusarium oxysporum* f. sp. *lycopersici* (Platero Sanz and Fuchs, 1978), *F. anguioides*, *F. avenaceum* (Lappe and Barz, 1978), *F. oxysporum* f. sp. *pisi*[P] (Fuchs *et al.*, 1980a) and *Stemphylium botryosum* (Heath and Higgins, 1973; VanEtten *et al.*, 1975).

the NADPH-dependence of the enzyme. When this fungus is cultured on an agar medium containing enough pisatin to make the medium turbid, a zone of clearing can be observed in advance of the growing mycelium (Nonaka, 1967; VanEtten, unpublished). This phenomenon has been reported for other fungi and other phytoalexins (Uehara, 1964; Bailey *et al.*, 1976) and the involvement of an extracellular phytoalexin-degrading enzyme has sometimes been proposed (Harborne and Ingham, 1978). The results with *N. haematococca* suggest that an intracellular enzyme, in combination with simple diffusion, might solubilize a precipitated compound at a distance.

The conversions of medicarpin and maackiain to their respective iso-flavanones (Figure 6.2, reaction *b*) also involve the dealkylation of an aromatic ether linkage. The formation of these metabolites can be envisioned to occur via oxygenation at carbon 11a to form an unstable hemiketal analogous to the intermediate depicted in Figure 6.3. Another possibility, hydrolysis of the C_{11a}—O bond followed by dehydrogenation to the ketone, seems less likely; ^{18}O labelling experiments should discriminate between these two alternatives. This transformation of medicarpin and maackiain was performed by an isolate

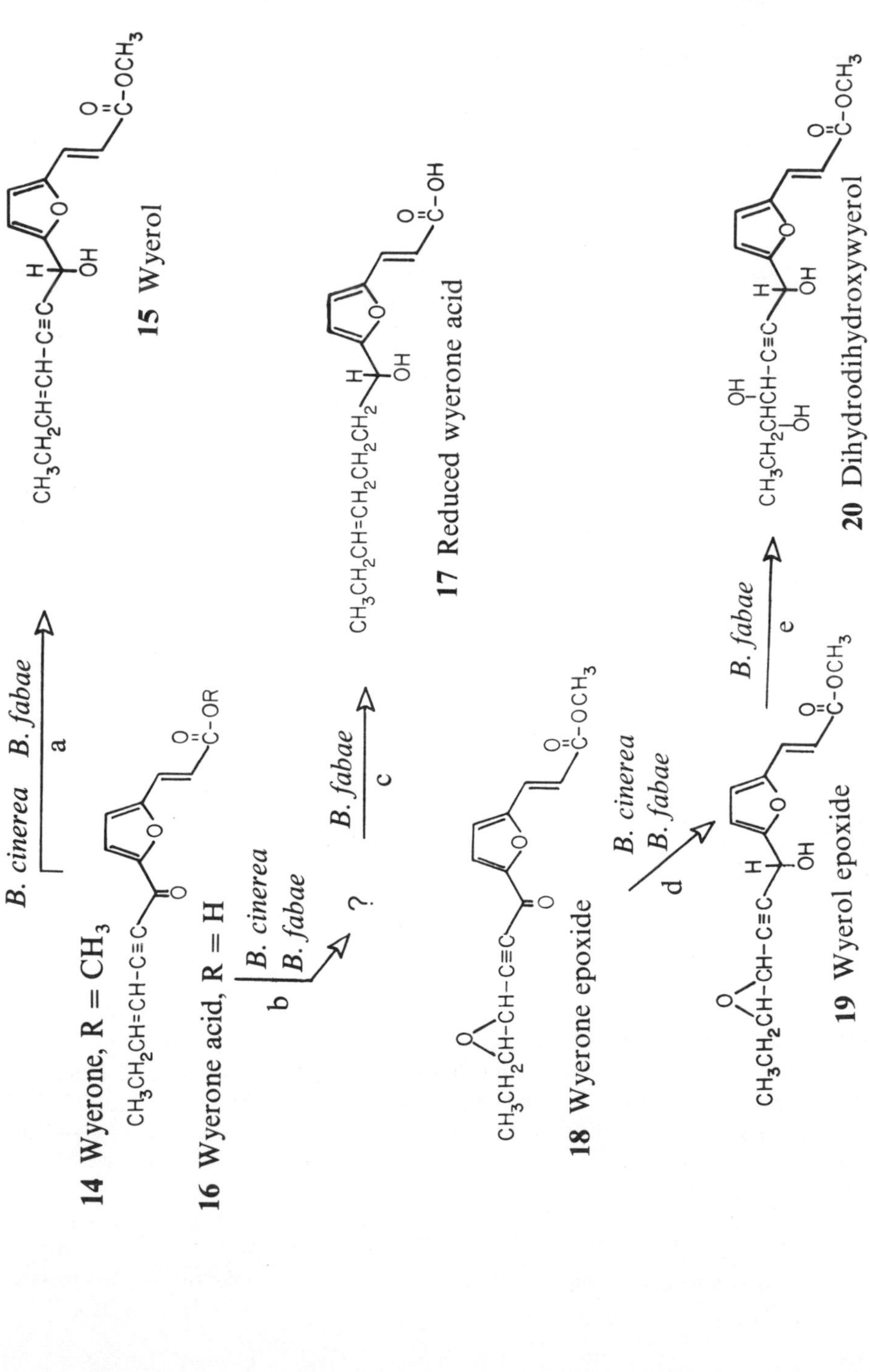

$CH_3CH_2CH=CH-C≡C$

14 Wyerone, R = CH₃

$CH_3CH_2CH=CH-C≡C$

16 Wyerone acid, R = H

B. cinerea B. fabae

a

B. cinerea
b | *B. fabae*

?

$C-OR$

15 Wyerol

$CH_3CH_2CH=CH-C≡C$

B. fabae

c

$C-OH$

17 Reduced wyerone acid

$CH_3CH_2CH=CH_2CH_2CH_2$

18 Wyerone epoxide

$CH_3CH_2CH-CH-C≡C$

B. cinerea
B. fabae

d

19 Wyerol epoxide

$CH_3CH_2CH-CH-C≡C$

B. fabae

e

20 Dihydrodihydroxywyerol

$CH_3CH_2CHCH-C≡C$

of *N. haematococca* that simultaneously produced the respective 1a-hydroxy metabolites (Figure 6.2, reaction *a*) (Denny and VanEtten, 1981*a*). Thus, it is possible that these two kinds of metabolites are produced by a single enzyme capable of hydroxylating either the 1a or 11a carbon atoms.

Another commonly observed alteration of pterocarpans that is suggestive of the action of a monooxygenase is the hydroxylation of carbon 6a. Burden *et al.* (1974) first demonstrated this transformation using *Colletotrichum linde-muthianum* and phaseollin (Figure 6.1, reaction *b*). *B. cinerea* also performs this transformation of phaseollin (van den Heuvel and Glazener, 1975; van den Heuvel, 1976), and metabolizes medicarpin (**6**) and maackiain (**7**) via the same hydroxylation (Ingham, 1976; Macfoy and Smith, 1979). *Sclerotinia trifoliorum* (Debnam and Smith, 1976; Bilton *et al.*, 1976) and *N. haematococca* (Denny and VanEtten, 1981*a*) likewise performed the latter two reactions. A cell-free preparation from *S. trifoliorum* was reported to degrade maackiain (Macfoy and Smith, 1978), but no details have been published.

The 6a-hydroxy derivatives of these three phytoalexins can be subsequently hydroxylated at the aromatic carbon 7 (Burden *et al.*, 1974; Macfoy and Smith, 1979), a reaction consistent with the action of a monooxygenase. The aromatic carbon 4 of medicarpin also appears to be subject to hydroxylation, based on a metabolite found in *B. cinerea*-infected tissue (Ingham, 1976).

One other example of the possible participation of a monooxygenase in the metabolism of a phytoalexin is the conversion of phaseollin to the dihydrodiol shown in Figure 6.1, reaction *c*. At least three mechanisms can be proposed to account for the incorporation of two atoms of oxygen and two atoms of hydrogen in this reaction:

1. hydration of the double bond and subsequent monooxygenation at carbon 12 or 13;
2. formation of an epoxide analogous to the arene oxide in Figure 6.2 followed by the action of an epoxide hydratase, or
3. reductive dioxygenation.

None of these possibilities readily explains the fact that both the *cis* and *trans* dihydrodiols were obtained as products (Bailey *et al.*, 1977). The reductive dioxygenases that have been characterized to date produce only the *cis* isomers (Subramanian *et al.*, 1979). Mammalian epoxide hydratases yield only *trans* products (Oesch, 1973) but there is a report of a fungal enzyme that hydrates epoxides exclusively in the *cis* manner (Kolattukudy and Brown, 1975).

In summary, fungal monooxygenases appear to be able to degrade ptero-carpanoid phytoalexins by attack at carbons 1a, 4, 6a, 7, 11a, and the 3- and 9-*O*-methyl groups as well as at carbon 12 or 13 of phaseollin. In view of the

---

Figure 6.4 The metabolism of wyerone derivatives by *Botrytis cinerea* and *Botrytis fabae*[p]. Redrawn from Mansfield (1980). Placement of the hydroxy groups on **20** is not intended to indicate a preference for *trans* vs. *cis* hydration.

number of these reactions and the variety of fungal species involved, it is remarkable that there is apparently only one report of fungal oxygenation of any other chemical class of phytoalexins: Lyr (1962) reported the degradation of pinosylvin monomethyl ether by phenol oxidases isolated from *Trametes versicolor* and *Fistulina hepatica*. It would not be surprising if other phenolic phytoalexins could also be metabolized by this widely occurring type of monooxygenase.

Reduction

The transformation of phytoalexins by reductive reactions has been studied most extensively in the metabolism of wyerone derivatives by *B. cinerea* and *B. fabae* (Figure 6.4). Both of these fungi reduced the ketones wyerone (**14**) and wyerone epoxide (**18**) to the respective alcohols wyerol and wyerol epoxide (Hargreaves *et al.*, 1976a,b). The latter product was further metabolized by *B. fabae* via hydration of the epoxide (reaction *e*) (Hargreaves *et al.*, 1976b). The conversion of wyerone acid (**16**) by *B. fabae* involved not only reduction of the carbonyl function but also the complete saturation of the acetylenic bond (Mansfield and Widdowson, 1973). The latter reduction is probably a two-step process: there are other reports of the reduction of alkynes to alkenes (Tsorbatzoudi *et al.*, 1976; Tillmanns *et al.*, 1978), and alkene reduction is well known. An intermediate in the metabolism of wyerone acid to reduced wyerone acid (**17**) has been isolated. It is a hydroxy derivative of wyerone acid, and is also produced by *B. cinerea* (Mansfield, 1980). A cell-free extract from germ tubes of *B. fabae* converts wyerone to wyerol, and wyerone epoxide to dihydro-dihydroxywyerol, NADPH being required. The preparation did not affect wyerone acid (Mansfield, 1980).

The reduction of the sesquiterpenoid phytoalexin lubimin (**21**) has been demonstrated with three fungi (Ward and Stoessl, 1977). In this process an aldehyde function was converted to a primary alcohol (Figure 6.5).

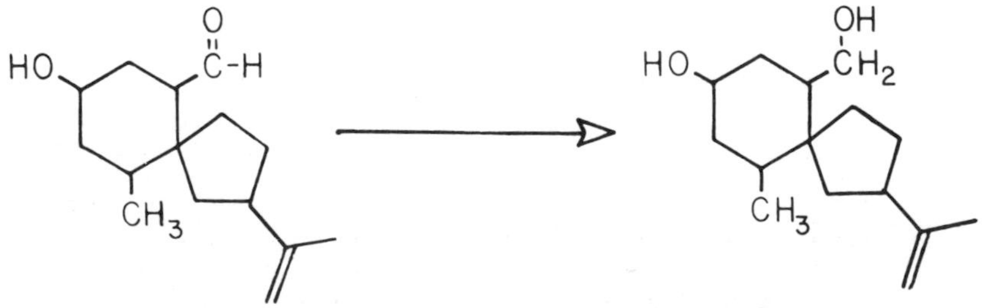

21 Lubimin   **22** 15-dihydrolubimin

Figure 6.5   Metabolism of lubimin by *Phytophthora capsici*, *Glomerella cingulata* and *Fusarium sulphureum*[p] (Ward and Stoessl, 1977).

An interesting example of phytoalexin reduction reactions is the reductive cleavage of various pterocarpans to the corresponding isoflavans. *Stemphylium botryosum* performs this conversion of phaseollin (Figure 6.1, reaction *d*) as well as of medicarpin and maackiain (Figure 6.6). These transformations of phaseollin and medicarpin have also been observed in uninfected plant tissue (see p. 208). Reactions *d* and *e* (Figure 6.6) are the second steps in the metabolism of

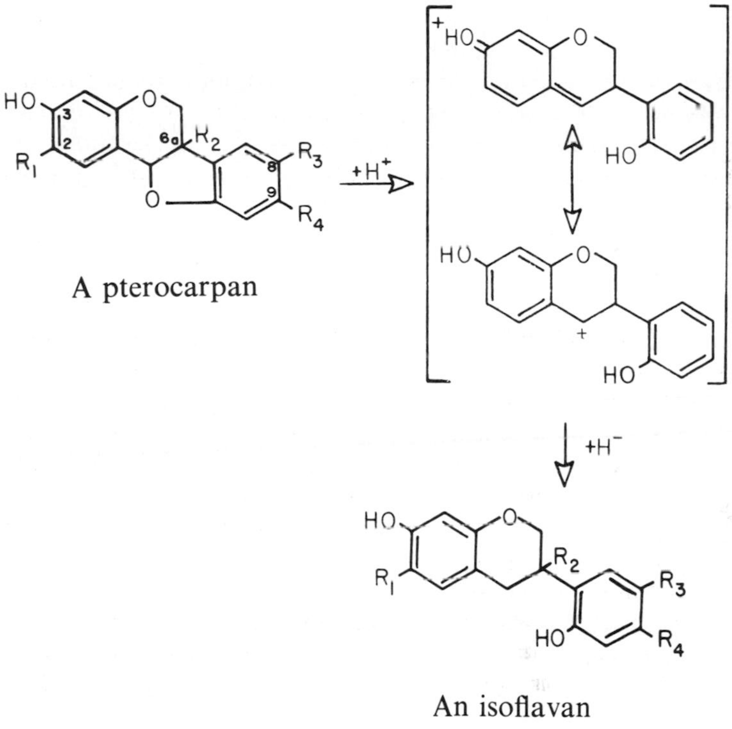

A pterocarpan

An isoflavan

Reaction	Substrate	Fungus	Reference
a	**6** $R_1, R_2, R_3 = H; R_4 = OCH_3$: medicarpin	*Stemphylium botryosum*[P]	Higgins, 1975
b	**7** $R_1, R_2 = H; R_3R_4 = OCH_2O$: maackiain	*S. botryosum*	Duczek and Higgins, 1976
c	**23** $R_2, R_3 = H; R_1, R_4 = OCH_3$: 3-hydroxy-2,9-dimethoxypterocarpan	*Nectria haematococca*[P]	Pueppke and VanEtten, 1976a
d	**24** $R_1, R_2, R_3 = H; R_4 = OH$: 3,9-dihydroxypterocarpan	*Fusarium proliferatum*	Weltring and Barz, 1980
e	**13** $R_1 = H; R_2 = OH; R_3R_4 = OCH_2O$: 3,6a-dihydroxy-8,9-methylene-dioxypterocarpan	*Fusarium oxysporum* f. sp. *pisi*[P]	Fuchs *et al.*, 1980b

Figure 6.6 Conversion of pterocarpans to isoflavans by fungi.

medicarpin and pisatin, respectively. Biological reductive cleavage of simple alkylphenyl ethers has been suggested to occur (Macrae and Alexander, 1963), but no enzymes catalysing such reactions have been demonstrated. Pterocarpans, however, are benzylphenyl ethers, and may be thus more susceptible to reductive cleavage. The involvement of a carbonium ion↔protonated quinone methide intermediate (Figure 6.6) has been proposed for the interconversions of pterocarpans and isoflavans in plants (Dewick and Martin, 1979; Martin and Dewick, 1980). Such an intermediate may also be involved in the fungal metabolism of pterocarpans to isoflavans. The possible importance of the quinone methide form in stabilizing this intermediate is indicated by the observation that a free hydroxyl substituent at position 3 appears to be required for the conversion (Figure 6.6). Furthermore, the 3-O-methylated analogues of **24** and **13** (3-methoxy-9-hydroxypterocarpan and pisatin) were not converted directly into isoflavans; in both cases this reaction was preceded by demethylation at this position (Weltring and Barz, 1980; Fuchs *et al.*, 1980a).

Hydration

The possible involvement of hydration reactions in the cleavage of epoxide bonds during the metabolism of phaseollin (Figure 6.1, reaction *d*) and wyerone epoxide (Figure 6.4, reaction *e*) has been mentioned above. The other two known examples are the hydration of the isopentenyl side chains of kievitone (Kuhn *et al.*, 1977) and phaseollidin (Smith *et al.*, 1980) by *F. solani* f. sp. *phaseoli* (Figure 6.7). Only the tertiary alcohols were obtained as products, indicating reactions with strict Markovnikov-type specificity.

Partial purification and characterization of kievitone hydratase, which mediates the hydration of kievitone, has been achieved from cell-free culture filtrates of *F. solani* f. sp. *phaseoli* (Cleveland and Smith, 1981; Cleveland and Smith, unpublished results; Kuhn and Smith, 1979). Kievitone hydratase was localized extracellularly, although some activity was detected in mycelial homogenates. The enzyme was isolated from culture filtrates and purified by ion exchange and affinity chromatography, gel filtration and isoelectric focusing. Characterization studies revealed that kievitone hydratase (MW = 173 000 ± 15 000) was an acidic glycoprotein with an isoelectric point of 5.1. It had a temperature optimum of 55°C, but denatured rapidly at temperatures exceeding 65°C. The pH optimum was 5.5 at 27°C with an apparent K_m of 1.7×10^{-5} M under these conditions.

The similarity of the transformation of kievitone to kievitone hydrate (Kuhn *et al.*, 1977) and of phaseollidin to phaseollidin hydrate (Smith *et al.*, 1980) suggested that the same enzyme system might catalyse both reactions. Initial preparations of kievitone hydratase seemed able to accomplish both transformations (Kuhn and Smith, 1979, 1980), but hydration of phaseollidin by cell-free preparations has not been obtained consistently (Smith *et al.*, 1981). Highly-purified kievitone hydratase will detoxify kievitone but not phaseollidin

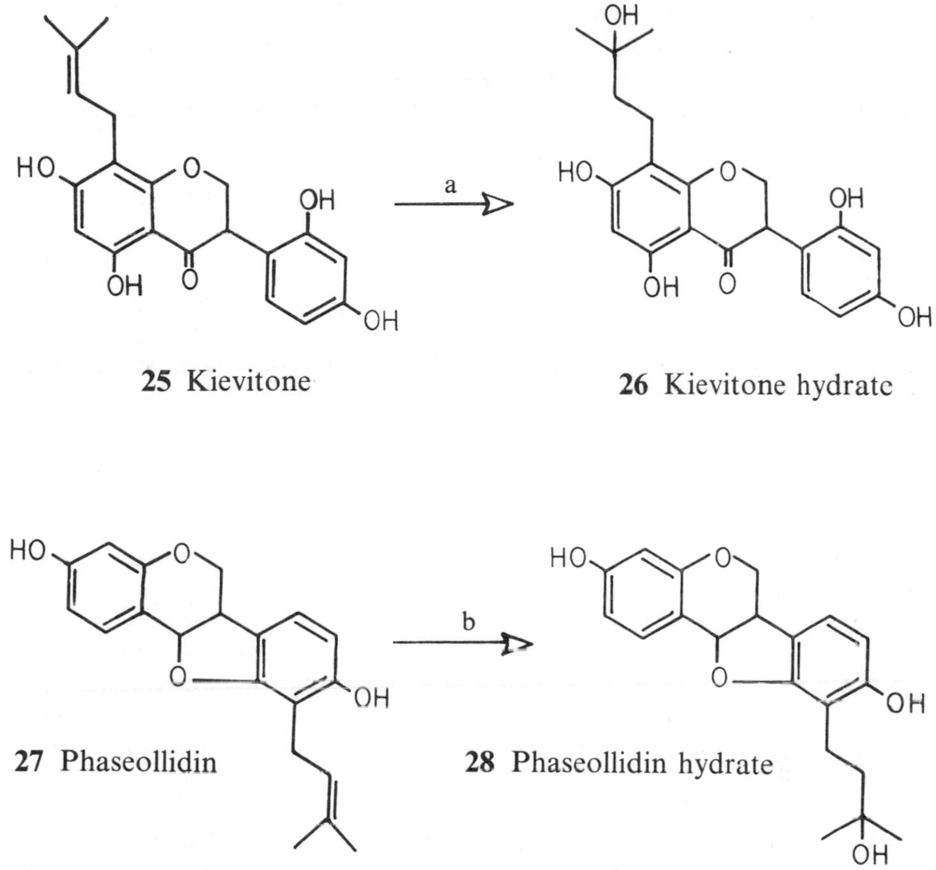

25 Kievitone   **26** Kievitone hydrate

27 Phaseollidin   **28** Phaseollidin hydrate

Figure 6.7 Metabolism of kievitone and phaseollidin by *Fusarium solani* f. sp. *phaseoli*[P] (Kuhn *et al.*, 1977; Smith *et al.*, 1980).

(Cleveland and Smith, unpublished results). Isolation and purification of kievitone hydratase is straightforward; this is apparently not the case for phaseollidin hydratase.

Oxidation

One example of dehydrogenation of a phytoalexin is known, the conversion of capsidiol to capsenone (Figure 6.8). *B. cinerea* and *Fusarium oxysporum* f. sp. *vasinfectum* both effected this alcohol dehydrogenase type of reaction in pure culture. In addition, capsenone was detected in diffusates from pepper fruit infected with three other *Fusarium* species, but not with a number of other fungi (Ward and Stoessl, 1972; Stoessl *et al.*, 1973). Capsenone was further degraded

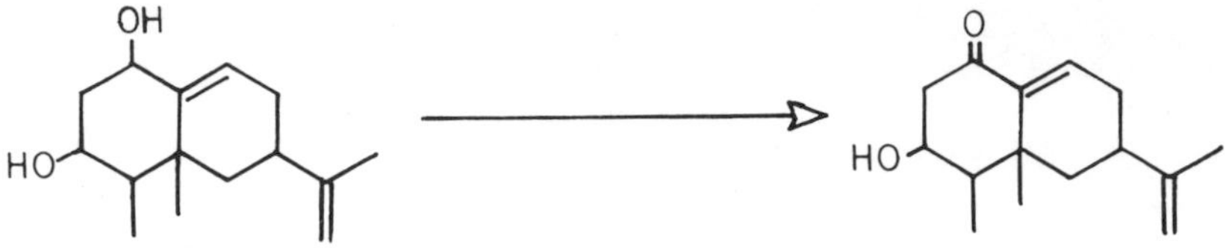

29 Capsidiol   **30** Capsenone

Figure 6.8   Metabolism of capsidiol by *Botrytis cinerea*[p] and *Fusarium oxysporum* f. sp. *vasinfectum*[p] (Stoessl *et al.*, 1973).

by *B. cinerea*, and at a slower rate by *F. oxysporum* f. sp. *vasinfectum*, but the products were not identified (Stoessl *et al.*, 1973).

Retroaldol cleavage

4-Hydroxymyoporone (**31**) is a compound produced by the metabolism of the phytoalexin ipomeamarone by sweet potato tissue (see p. 211). Although it is unclear whether 4-hydroxymyoporone is a phytoalexin, because of a lack of data on its microbial toxicity, the metabolism of this compound by fungi is of particular interest because the products (Figure 6.9, **32** to **35**) are highly toxic to mammals, primarily affecting lung tissues. The conversion of 4-hydroxy-myoporone to ipomeanine (Figure 6.9, reaction *a*) can be performed chemically by a retroaldol reaction and this mechanism, analogous to the aldolase reaction of glycolysis, was proposed for this transformation by *F. oxysporum* and *F. solani*. The production of the other predicted metabolite, 4-methyl-2-pentanone, was not reported, however (Burka *et al.*, 1977). *Ceratocystis fimbriata* did not perform this reaction, but all three of these fungi were able to reduce the carbonyl groups of ipomeanine to produce the other lung toxins (Figure 6.9, reaction *b*) (Burka *et al.*, 1977).

Further conversions of phytoalexin metabolites

A complete sequence of metabolism of a phytoalexin to CO_2 and H_2O is not known. Indeed there are some reports in which no further degradation of the initial metabolite was observed, for example, after the demethylation of pisatin by *Ascochyta pisi* (de Wit-Elshove and Fuchs, 1971; Platero Sanz and Fuchs, 1978) and the hydration of kievitone and phaseollidin by *F. solani* f. sp. *phaseoli* (Kuhn and Smith, 1978; Smith *et al.*, 1980). However, in most of the studies described in the previous section, the products were further metabolized, and it seems likely that complete degradation occurs at least in some cases.

It seems possible that different, structurally related phytoalexins are eventually converted to a common metabolic intermediate which is then degraded by a single pathway. Such a point of convergence has not been discovered among the

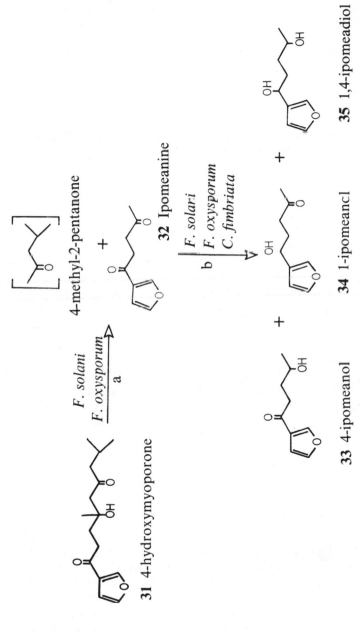

Figure 6.9 Metabolism of 4-hydroxymyoporone and ipomeanine by *Fusarium solani*[p], *F. oxysporum*[p], and *Ceratocystis fimbriata*[p] (Burka *et al.*, 1977).

early metabolites that have been characterized to date. Indeed, as mentioned previously, the initial conversion of a single phytoalexin by different fungi can be quite different (e.g. Figure 6.1). Even where a common intermediate is known to exist, the subsequent metabolic steps can differ. The metabolite 3,6a-dihydroxy-8,9-methylenedioxypterocarpan (DMDP) is an intermediate in the degradation of both pisatin (Figure 6.10, reaction *a*) and maackiain (Figure 6.10, reaction *c*). However, DMDP is subsequently metabolized by at least two different routes (Figure 6.10, reactions *b* and *d*) (Fuchs *et al.*, 1980*a*; Macfoy and Smith, 1979). It should be noted that DMDP derived from pisatin is probably enantiomeric with DMDP derived from maackiain, because these two phyto-alexins have opposite absolute configurations (Wong, 1970). As pointed out by Fuchs *et al.* (1980*a*), this difference in stereochemistry may be a significant factor in the pathways by which fungi metabolize DMDP.

Regulation of phytoalexin metabolism in fungi

The study of the regulation of phytoalexin metabolism may provide information about the normal physiological role of the enzymes involved. If a phytoalexin-degrading enzyme is always absent from a fungus unless it has been specifically induced to form by treatment of the fungus with the phytoalexin, it can be argued that the metabolism of that phytoalexin is the enzyme's most important function. Lack of induction by the substrate, or non-specific induction by a variety of stimuli, would suggest that phytoalexin metabolism could be a gratuitous activity of a relatively non-specific enzyme having some other primary role in the biology of the fungus.

Evidence for the inducibility of phytoalexin-degrading enzymes has been obtained in several cases. A lag of approximately an hour between the addition of a phytoalexin and the most rapid rate of its disappearance has been observed for the oxidation of capsidiol (Stoessl *et al.*, 1973), the 6a-hydroxylation of medicarpin (Ingham, 1976), the reduction of maackiain (Higgins, 1975), the 1a-hydroxylation of phaseollin (van den Heuvel and VanEtten, 1973), and the demethylation of pisatin (VanEtten and Barz, 1981). For the last three reactions it was found that a second dose of the phytoalexin was metabolized without a lag and even more rapidly. This stimulation could be prevented by adding cycloheximide at the time of the first treatment (Higgins, 1975; Kistler and VanEtten, 1981; VanEtten and Matthews, unpublished). Although *F. solani* f. sp. *phaseoli* exhibits a relatively high constitutive rate of kievitone hydration, the rate is nonetheless increased by prior exposure to the substrate (Kuhn and Smith, 1979). Cell-free preparations from pre-treated cultures also show in-creased kievitone hydratase activity (Kuhn and Smith, 1979); therefore, the observed stimulation represents an increase in the concentration of active enzyme, rather than some other physiological adaptation to the presence of the phytoalexin. A similar result was obtained for pisatin demethylase (Matthews and VanEtten, unpublished).

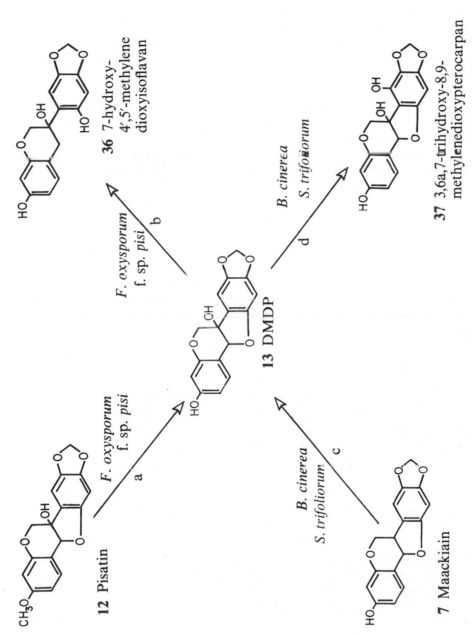

Figure 6.10 The sequential steps of metabolism of pisatin and maackiain by *Fusarium oxysporum* f. sp. *pisi*[P], *Botrytis cinerea* and *Sclerotinia trifoliorum*[P].

12 Pisatin

36 7-hydroxy-4',5'-methylene dioxyisoflavan

13 DMDP

37 3,6a,7-trihydroxy-8,9-methylenedioxypterocarpan

7 Maackiain

F. oxysporum f. sp. *pisi*

F. oxysporum f. sp. *pisi*

B. cinerea *S. trifoliorum*

B. cinerea *S. trifoliorum*

In other cases the absence of a detectable lag in the rate of metabolism suggests that phytoalexin-metabolizing activity may be constitutive. Even better evidence for constitutive activity would be the demonstration that cell-free preparations from cultures that have not been exposed to the phytoalexin can perform the reaction. Such evidence has been reported for the transformation of wyerone and wyerone epoxide by extracts from *B. fabae* (Mansfield, 1980), and the metabolism of medicarpin by culture filtrates from *Leptosphaerulina briosiana* (Higgins, 1972). (The latter reaction was very rapid, but unfortunately no metabolites were identified.) It is not known whether the constitutive rates observed in these cases could, as for kievitone hydratase, be further increased by pre-treatment with the substrates.

The reduction of maackiain (**7**) to the corresponding isoflavan by *S. botryosum* was induced not only by maackiain, but also by phaseollin (**1**) and medicarpin (**6**). *S. botryosum* was able to convert the latter compounds to the respective isoflavans, and these reactions were likewise induced by maackiain. This cross-specificity of induction suggested that one enzyme might be responsible for all three transformations (Higgins, 1975). This interesting possibility will doubtless be explored when a cell-free preparation catalysing these reactions is obtained.

It is not uncommon to find that an enzyme can be induced by some compounds that are not substrates, especially if they are structurally similar to the substrate. An example from the phytoalexin literature is the stimulation of 4-hydroxymyoporone cleavage (Figure 6.9) by pretreatment of *F. solani* with ipomeamarone (**42**). Ipomeamarone itself is not susceptible to cleavage by the proposed reaction mechanism, and it was not found to be converted to 4-hydroxymyoporone by this fungus (Burka *et al.*, 1977).

The specificity of induction of pisatin demethylation has been examined in one isolate of *N. haematococca* (VanEtten and Barz, 1981; VanEtten *et al.*, unpublished). Of over 50 compounds tested, 11 stimulated the activity significantly, though none to a greater degree than pisatin itself. No compound with a structure grossly different from that of pisatin was effective as an inducer, and a number of seemingly minor differences also negated inducing ability. On the other hand, not all of the effective inducers contained a methoxy group, or a substituent at the position corresponding to carbon 3 of pisatin. Overall, the structural requirements for induction of pisatin demethylase appeared to be fairly stringent. A more limited study of the stimulation of phaseollin 1a-hydroxylation in *F. solani* f. sp. *phaseoli* showed a wider variety of compounds to be effective inducers (Kistler and VanEtten, 1981).

The nutrient status of the culture medium can also be an important factor in the regulation of phytoalexin metabolism. This effect was first demonstrated by de Wit-Elshove and Fuchs (1971). It was found that a high concentration of glucose or sucrose in the medium prevented pisatin degradation by *Fusarium solani* f. sp. *pisi* for up to a week, and it was suggested that pisatin metabolism was subject to catabolite repression. This fungus was later shown to demethylate pisatin very rapidly (0.1 mM in 2 h) when the mycelium was suspended in phosphate buffer, and only a little more slowly in a succinate-containing growth medium; again there was no metabolism in the presence of glucose (VanEtten and Barz, 1981).

This phenomenon was also observed in *Fusarium oxysporum* f. sp. *pisi*, but not in *A. pisi* (de Wit-Elshove and Fuchs, 1971). The metabolism of a number of phytoalexins (e.g. Figure 6.1, reactions *a*, *b*; Figure 6.7, reaction *a*; Figure 6.8) has been demonstrated with fairly young cultures. in which the concentration of carbohydrate was probably still high (Stoessl *et al.*, 1973; van den Heuvel and VanEtten, 1973; van den Heuvel and Glazener, 1975; van den Heuvel and Vollard, 1975; Ingham, 1976; Macfoy and Smith, 1979; Kuhn and Smith, 1979); therefore catabolite repression of these reactions is not complete, if it exists at all. It is thus not yet clear how common carbohydrate repression may be and its relevance to metabolism in diseased tissues requires further investigation.

The possibility that phytoalexin metabolism is subject to regulation is also relevant to experiments in which fungi are surveyed simply for the ability to metabolize phytoalexins. A number of isolates of *N. haematococca* that are capable of demethylating pisatin (**12**) fail to do so under some cultural conditions, even though other isolates readily perform this reaction under the same conditions (VanEtten *et al.*, 1980). Procedural differences may also be responsible for the disagreements about the ability of *A. pisi*, *Aphanomyces euteiches*, and *Fusarium oxysporum* f. sp. *lycopersici* to degrade pisatin (Cruickshank and Perrin, 1965; de Wit-Elshove and Fuchs, 1971; Pueppke and VanEtten, 1976; Platero Sanz and Fuchs, 1978; Fuchs *et al.*, 1980a).

The relationship between phytoalexin metabolism and phytoalexin tolerance

Although phytoalexins are toxic to a broad range of organisms, the level of phytoalexin sensitivity varies significantly among fungal species (chapter 7). A possible explanation for the relative phytoalexin tolerance found in some fungi is that they are capable of metabolically detoxifying these compounds. However, the ability of a fungus to metabolize a phytoalexin may not necessarily be an important factor in its tolerance to the phytoalexin. Some fungi are not especially tolerant of the phytoalexins they metabolize. The explanation for this observation may be simple in the case of reactions that yield products with as much general microbial toxicity as the phytoalexins themselves. The antimicrobial activity of the metabolite phaseollinisoflavan (Figure 6.1, reaction *d*) was clearly shown to be similar to that of phaseollin (Skipp and Bailey, 1977) and the toxicity of vestitol, the isoflavan produced from medicarpin (**6**), seems to be comparable to that of medicarpin (Bonde *et al.*, 1973; Ingham and Millar, 1973). The reduction of lubimin to 15-dihydrolubimin (Figure 6.5) likewise does not appear to represent a detoxification (Ward and Stoessl, 1977). However, most of the reactions described above do produce metabolites which are less toxic (Stoessl *et al.*, 1973; VanEtten and Smith, 1975; van den Heuvel and

Table 6.1 Inhibition of radial mycelial growth of fungi on solid medium containing 3×10^{-4} M pisatin (**12**) or DMDP (**13**)[a]

Fungus	Ability to demethylate pisatin	Percentage inhibition	
		3×10^{-4} M pisatin *(94 μg/ml)*	3×10^{-4} M DMDP *(90 μg/ml)*
Fusarium solani f. sp. *pisi*[b]	+	7	8
Mycosphaerella pinodes[b]	+[c]	8	6
Ascochyta pinodella[b]	?	16	12
Aphanomyces euteiches[b]	+	96	98
Stemphylium botryosum	+	66	3
F. solani f. sp. *phaseoli*	−	67	6
F. solani f. sp. *cucurbitae*[d]	−	100	12
Helminthosporium turcicum	?	74	35
Neurospora crassa	?	78	28
Penicillium expansum	?	36	6
Rhizopus stolonifer	?	74	6

[a] Data from VanEtten and Pueppke (1976), VanEtten *et al.* (1975), VanEtten and Stein (1978) and Platero Sanz and Fuchs (1978).

[b] Pathogens of *Pisum sativum* L., a species which produces pisatin as a phytoalexin.

[c] Ability to demethylate pisatin suggested by the presence of DMDP in *M. pinodes*-infected pea tissue (Platero Sanz and Fuchs, 1980).

[d] Concentration of test compounds for this fungus was 100 μg/ml.

Glazener, 1975; Ingham, 1976; VanEtten and Pueppke, 1976; Bailey *et al.*, 1977; Macfoy and Smith, 1979; Mansfield, 1980; Kuhn and Smith, 1978; Smith *et al.*, 1980; Denny and VanEtten, 1981*b*).

Table 6.1 illustrates the decrease in toxicity associated with a phytoalexin metabolite (DMDP) compared to the parent phytoalexin (pisatin). (For additional examples see Harborne and Ingham, 1978). DMDP is relatively non-toxic to most fungi, and pisatin is equally non-toxic to two of the fungi that can perform this conversion. The sensitivity of *A. euteiches* to pisatin may be related to its unusual sensitivity to DMDP and its possible inability to metabolize DMDP to further products (see e.g. Platero Sanz and Fuchs, 1978). However, some other factor must be involved in the pisatin-sensitivity of *S. botryosum*. This kind of result could be due to a low rate of pisatin demethylation, or to differences between the experimental conditions used to measure sensitivity and metabolism.

Another situation in which phytoalexin metabolism might be of little significance with regard to phytoalexin tolerance would be the case of a fungus that possesses other kinds of tolerance mechanisms—see chapter 7. Such alternative mechanisms might provide adequate protection even in the absence of metabolic detoxification. The ability to degrade phytoalexins for carbon and energy could still confer a selective advantage in these cases.

It has also been proposed that other tolerance mechanisms might be required for a fungus to express its ability to metabolize a phytoalexin (Rossall *et al.*, 1977; Rossall and Mansfield, 1978). This hypothesis was based on the inter-action of wyerone acid with *B. cinerea* and *B. fabae*. Although both fungi can metabolize wyerone acid, *B. cinerea* does so much more slowly and is about twice as sensitive to this compound. Several observations led the authors to suggest that the lower rate of metabolism by *B. cinerea* was the result, rather than the cause, of its greater sensitivity. Perhaps the best evidence was the finding that *B. cinerea* metabolized wyerone acid more rapidly when the initial concentration was only $3.5\,\mu g/ml$ (the ED_{50} for germ tube elongation in this fungus) than when it was $10\,\mu g/ml$. However, at similar levels of growth inhibition, the absolute rate of metabolism was still higher in *B. fabae* than in *B. cinerea* (Rossall and Mansfield, 1978).

The existence of two different mechanisms governing tolerance to pisatin, only one of which involves degradation, was demonstrated in a survey of naturally occurring isolates of *N. haematococca* (VanEtten *et al.*, 1980). Even isolates of *N. haematococca* mating population (MP) VI that could not de-methylate pisatin were significantly more tolerant of this compound than isolates of *N. haematococca* MP I[1]. For example, the growth of MP VI isolates that could not demethylate pisatin was inhibited less than 10% on agar medium containing $100\,\mu g$ of pisatin/ml, whereas almost all MP I isolates were com-pletely inhibited. Furthermore the tolerance found in these MP VI isolates seems to involve an active response to the phytoalexin, because it is expressed only after a period of adaptation. The high degree of inhibition exhibited initially upon exposure to toxic levels of pisatin disappeared after a few hours (Figure 6.11), and was eliminated entirely by prior treatment with a lower concentration ($32\,\mu g/ml$) of pisatin (Denny and VanEtten, unpublished). Similar examples of adaptive tolerance in the absence of phytoalexin metabolism have been reported previously for other fungi and other phytoalexins (Skipp and Bailey, 1976; Smith, 1976; Skipp and Bailey, 1977; Higgins, 1978; VanEtten and Stein, 1978; Denny and VanEtten, 1981*b*).

All naturally occurring isolates of *N. haematococca* MP VI tested to date possess a significant degree of tolerance to pisatin, but those isolates that can demethylate pisatin are more tolerant than those that cannot (VanEtten *et al.*, 1980). The role of pisatin demethylating ability in pisatin tolerance has been examined further by genetic analysis (Tegtmeier and VanEtten, 1981). In some crosses (Figure 6.12*a*) the progeny showed an absolute correlation between the ability to perform this reaction and a higher level of tolerance. These results are consistent with the existence of one tolerance mechanism, common to both of the parents, interacting in a simple manner with the tolerance due to pisatin demethylating ability. Crosses between some other pairs of isolates showed a

[1] The heterothallic fungi identified morphologically as *Nectria haematococca* comprise seven distinct mating populations (MP), which are intrafertile but not interfertile. All known isolates with an imperfect stage of *Fusarium solani* f. sp. *pisi* are in MP VI. *Fusarium solani* f. sp. *cucurbitae* race I isolates are in MP I.

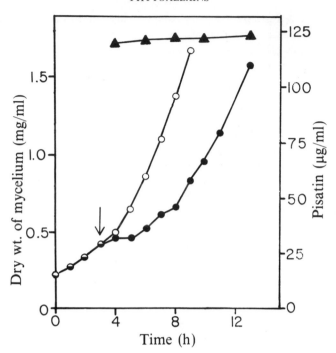

Figure 6.11   The effect of pisatin on the growth of an isolate of *Nectria haematococca* MP VI that does not metabolize pisatin. The fungus was grown on a liquid glucose-asparagine medium at 27°C and 125 rpm. At the time (3 h) indicated by the arrow, pisatin was added to a final concentration of 125 μg/ml. ○, control culture; ●, pisatin treated culture; ▲, pisatin concentration in pisatin treated culture. Denny and VanEtten, unpublished results.

more complex segregation of tolerance (Figure 6.12*b*), and a few of the pisatin-demethylating progeny were as sensitive to pisatin as some non-demethylating progeny. In these cases it may be that the parents differ with regard to another mechanism of tolerance besides the ability to demethylate pisatin. However, even in the more complex genetic backgrounds evident in crosses of this type, pisatin demethylating ability still seems to be required for a high level of tolerance.

Phytoalexin metabolism and pathogenicity

Of prime concern for phytopathologists is whether the ability of a pathogen to metabolize a host's phytoalexins relates to its pathogenic potential on that plant. It is easy to assume that this is the case in those diseases in which large concentrations of phytoalexins accumulate in the infected tissue and in which the pathogen is known to metabolize the phytoalexin to less toxic products. However, conclusive evidence for this assumption is difficult to obtain.

The need for further verification of the importance of phytoalexin metabolism

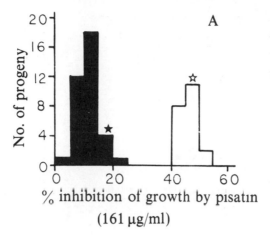

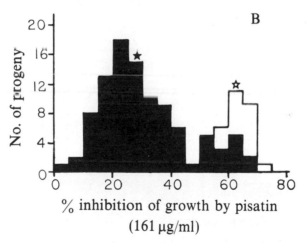

Figure 6.12 The relationship between pisatin demethylating ability and pisatin sensitivity in the progeny of two crosses between isolates of *Nectria haematococca* MP VI differing in these characteristics. Growth rates were measured and pisatin demethylating ability determined on an agar medium. The phenotypes of the parents are indicated by stars. Filled symbols represent isolates which had the ability to demethylate pisatin; open symbols, isolates which lacked the ability to demethylate pisatin (Tegtmeier and VanEtten, 1981; Tegtmeier and VanEtten, unpublished).

in pathogenicity is emphasized by the existence of a variety of apparent "exceptions", interactions in which the ability to degrade phytoalexins does not seem to play a decisive role. Among the pathogens that have been tested for the ability to metabolize their hosts' phytoalexins (Table 6.2) there are a number of cases in which no metabolism was detected. This observation is not surprising: it has not been proposed that phytoalexin metabolism is the only means of circumventing a phytoalexin-based defence mechanism. Indeed there is evidence

Table 6.2 The ability of fungal pathogens to metabolize their hosts' phytoalexins[a]

Phytoalexin	Host	Pathogen	Metabolite	Reference
capsidiol (**29**)	pepper (*Capsicum frutescens*)	*Botrytis cinerea*	capsenone (**30**)	Ward and Stoessl (1972)
		Fusarium oxysporum f. sp. *vasinfectum*	capsenone (**30**)	Ward and Stoessl (1972)
		Phytophthora capsici	no metabolism	Ward and Stoessl (1972)
		Cladosporium herbarum	no metabolism	Ward and Stoessl (1972)
rishitin (**39**)	tomato (*Lycopersicon esculentum*)	*Botrytis cinerea*	unidentified	Lyon (1976)
lubimin (**21**)	potato (*Solanum tuberosum*)	*Fusarium sulphureum*	15-dihydrolubimin (**22**)	Ward and Stoessl (1977)
		Phytophthora infestans	no metabolism	Ward and Stoessl (1977)
rishitin (**39**)		*Phytophthora infestans*	no metabolism	Vasyukova *et al.* (1977)
ipomeamarone (**42**)	sweet potato (*Ipomoea batatas*)	*Corticium rolfsii*	unidentified	Uehara and Kiku (1969)
		Botryodiplodia theobromae	unidentified	Arinze and Smith (1980)
		Ceratocystis fimbriata	no metabolism	Takeuchi *et al.* (1978)
ipomeamaronol		*Botryodiplodia theobromae*	unidentified	Arinze and Smith (1980)
wyerone acid (**16**)	broad bean (*Vicia faba*)	*Botrytis fabae*	reduced wyerone acid (**17**)	Mansfield and Widdowson (1973)
wyerone (**14**)		*Botrytis fabae*	wyerol (**15**)	Hargreaves *et al.* (1976a)
wyerone epoxide (**18**)		*Botrytis fabae*	wyerol epoxide (**19**) dihydrodihydroxywyerol (**20**)	Hargreaves *et al.* (1976b)
phaseollin (**1**)	French bean (*Phaseolus vulgaris*)	*Botrytis cinerea*	6a-hydroxyphaseollin (**3**)	van den Heuvel and Glazener (1975)
		Fusarium oxysporum f. sp. *phaseoli*	no metabolism	van den Heuvel and Glazener (1975)
		Thielaviopsis basicola	no metabolism	van den Heuvel and Glazener (1975)
		Rhizoctonia solani	unidentified	van den Heuvel and Glazener (1975) Pierre and Bateman (1967)
		Sclerotinia sclerotiorum	unidentified	van den Heuvel and Glazener (1975)
		Colletotrichum lindemuthianum	6a-hydroxyphaseollin (**3**) and 6a,7-dihydroxyphaseollin	Burden *et al.* (1974)
		Fusarium solani f. sp. *phaseoli*	1a-hydroxyphaseollone (**2**)	van den Heuvel *et al.* (1974)

Phytoalexin (plant)	Fungus	Metabolite	Reference
phaseollidin (27)	*Fusarium solani* f. sp. *phaseoli*	phaseollidin hydrate (28)	Smith *et al.* (1980)
	Colletotrichum lindemuthianum	unidentified	Bailey (1974)
phaseollinisoflavan (5)	*Fusarium solani* f. sp. *phaseoli*	unidentified	VanEtten and Smith (1975)
	Colletotrichum lindemuthianum	unidentified	Bailey (1974)
kievitone (25)	*Fusarium solani* f. sp. *phaseoli*	kievitone hydrate (26)	Kuhn *et al.* (1977)
	Colletotrichum lindemuthianum	unidentified	Bailey (1974)
medicarpin (6) alfalfa (*Medicago sativa*)	*Stemphylium botryosum*	vestitol	Steiner and Millar (1974)
	Phoma herbarum var. *medicaginis*	unidentified	Higgins (1975)
	Leptosphaerulina briosiana	unidentified	Higgins (1972)
	Stemphylium loti	unidentified	Higgins (1972)
sativan	*Stemphylium botryosum*	unidentified	Higgins and Millar (1970); Steiner and Millar (1974)
pisatin (12) pea (*Pisum sativum*)	*Fusarium oxysporum* f. sp. *pisi*	metabolites 13 and 36	Fuchs *et al.* (1980b)
	Nectria haematococca	metabolite 13	VanEtten *et al.* (1975)
	Ascochyta pisi	metabolite 13	van't Land *et al.* (1975)
	Aphanomyces euteiches	metabolite 13	Platero Sanz and Fuchs (1978)
	Mycosphaerella pinodes	unidentified	de Wit-Elshove (1969)
3-hydroxy-2,9-dimethoxy pterocarpan (23)	*Nectria haematococca*	7,2'-dihydroxy-6,4'-dimethoxyisoflavan (Fig. 6.6, c)	Pueppke and VanEtten (1976a)
maackiain (7) red clover (*Trifolium pratense*)	*Sclerotinia trifoliorum*	metabolite 13 and 37	Bilton *et al.* (1976)
	Stemphylium sarcinaeforme	unidentified	Macfoy and Smith (1979); Duczek and Higgins (1976)
medicarpin (6)	*Sclerotinia trifoliorum*	6a-hydroxymedicarpin and 6a,7-dihydroxymedicarpin pin	Bilton *et al.* (1976)
	Stemphylium sarcinaeforme	unidentified	Macfoy and Smith (1979); Duczek and Higgins (1976)
glyceollin soybean (*Glycine max*)	*Phytophthora megasperma* var. *sojae*	no metabolism	Yoshikawa *et al.* (1979)
maackiain (7) chickpea (*Cicer arietinum*)	*Nectria haematococca*	metabolites 9, 11 and 13	Denny and VanEtten (1981a)
medicarpin (6)	*Nectria haematococca*	metabolites 8, 10 and 6a-hydroxymedicarpin	Denny and VanEtten (1981a)

[a] The metabolism studies listed are ones in which phytoalexin metabolism was demonstrated in pure culture. There is circumstantial evidence from studies of metabolites in infected tissue that a number of other pathogens are able to carry out the same reactions.

that some pathogens, especially *Phytophthora* species, can interfere with the host's ability to accumulate phytoalexins (Ward and Stoessl, 1972; Yoshikawa *et al.*, 1978; Doke *et al.*, 1979; chapters 8 and 9).

In most of the instances in which pathogens have been shown to metabolize a phytoalexin, other fungi not pathogenic on that plant were found to be capable of performing the same reaction. This observation is of little significance with regard to the relationship between phytoalexin metabolism and pathogenicity, since it would be naïve to assume that the ability to metabolize phytoalexins is the only trait required for pathogenicity in any fungus.

A different kind of exception has been reported in the interaction between garden pea and *A. euteiches* (Pueppke and VanEtten, 1976*b*). This aggressive pea pathogen was unable to metabolize pisatin and was highly sensitive to it under all bioassay conditions tested. Pisatin accumulated in infected tissue to concentrations much higher than those required for complete inhibition of growth *in vitro* but growth in the tissues was apparently unaffected. Several lines of evidence argued against a physical separation between the fungus and pisatin in the infected tissue. Taken at face value, these results imply that *A. euteiches* is able to grow vigorously under conditions that completely inhibit its growth. Since this conclusion is self-contradictory one or more of the observations must be somehow misleading. The more recent finding that *A. euteiches* can degrade pisatin under some conditions (Platero Sanz and Fuchs, 1978) may be important in this regard.

Although phytoalexin metabolism is not proposed to play a significant role in all diseases of phytoalexin-producing plants, this ability probably is important for pathogenicity in at least some cases. Potentially toxic levels of phytoalexins are produced in some interactions in which the pathogen is able to detoxify these compounds. Furthermore, in a few cases the fungal metabolite of a phytoalexin has been found in tissue infected with that pathogen (Mansfield, 1980; Stoessl *et al.*, 1973; VanEtten and Smith, 1975; Van den Heuvel and Grootveld, 1980), whereas the plant itself apparently does not produce the same metabolite (Ward *et al.*, 1977; Hargreaves and Selby, 1978). These observations provide more direct evidence that these pathogens do contact and metabolize the phytoalexins *in situ*. Both the fungal enzyme kievitone hydratase and its product have been isolated from bean tissue infected with *F. solani* f. sp. *phaseoli* (Cleveland and Smith, unpublished), indicating that the enzyme interacts with its substrate in the infected tissue.

A different type of experimental approach has recently been employed to evaluate the importance of phytoalexin metabolism for pathogenicity. This approach utilizes genetic analysis to determine the effect on virulence of specifically altering the ability of an organism to detoxify a phytoalexin. When naturally occurring isolates of *N. haematococca* MP VI were surveyed, it was found that all those isolates that could not demethylate pisatin were more sensitive to this phytoalexin and were non-pathogenic, or of low virulence, on pea (VanEtten *et al.*, 1980). As a further test of the relationship of pisatin

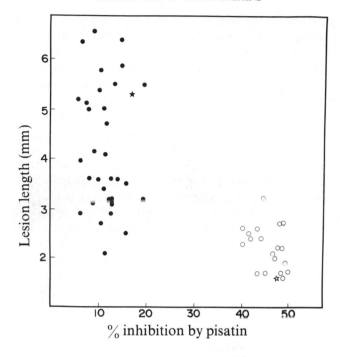

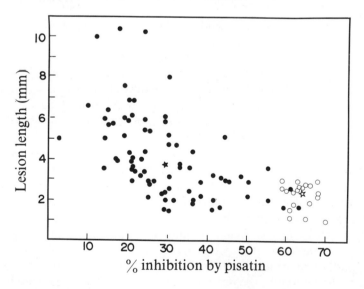

Figure 6.13 Relationship between sensitivity to pisatin, the ability to demethylate pisatin and virulence on pea among progeny of two crosses between *Nectria haematococca* MP VI isolates differing in these traits. The phenotypes of the parents are indicated by stars, and those of the progeny by circles. Filled symbols represent isolates able to demethylate pisatin, and open symbols represent isolates unable to do so. From Tegtmeier and VanEtten, 1981.

demethylating ability to the level of virulence on pea, isolates of MP VI differing in these properties were crossed (Tegtmeier and VanEtten, 1981). Two types
of inheritance patterns were observed (Figure 6.13). In all cases, ascospore
progeny which were not able to demethylate pisatin were low in virulence and
relatively sensitive to pisatin. All isolates that were highly or moderately virulent
on pea could demethylate pisatin and were tolerant of pisatin.

A third class of progeny was also obtained in all crosses; these progeny were
low in virulence even though they were tolerant and able to demethylate pisatin.
The existence of this class of progeny is significant in two respects. First, it
illustrates that pisatin demethylating ability alone is not sufficient for a high
level of virulence on pea. Similar evidence that other kinds of virulence traits
must also be required was obtained in the previous survey of field isolates of this
fungus (VanEtten et al., 1980). Second, it indicates that one or more of these
other virulence genes are segregating independently of pisatin demethylating
ability in these crosses. Thus the progeny possess a variety of genetic backgrounds, with regard to other virulence factors. No combination of virulence
genes has been obtained that can replace the requirement of pisatin demethylating ability for a high level of virulence. These results represent the strongest
evidence to date that phytoalexin metabolism can be an essential trait for
virulence.

Metabolism of phytoalexins by plants

Phytoalexins are not necessarily end-products of plant metabolism. Although
they accumulate in plant tissue under some conditions, in other situations their
concentration may decrease due to transformations apparently catalysed by the
plant itself. The metabolites produced by these reactions have been clearly
identified only in experiments in which exogenous phytoalexins were added to
plant tissue, and a hazard always associated with such studies is that the
observed conversions may not be typical of those that occur in situ (Barz and
Hösel, 1975; Moesta and Grisebach, 1980).

The conversion of medicarpin to the isoflavan vestitol (Figure 6.6, reaction a),
and the reverse reaction, have been demonstrated in alfalfa (Martin and Dewick,
1980). The latter reaction was also performed by red clover (Dewick and Martin,
1979). Phaseolus vulgaris transformed phaseollin to phaseollinisoflavan (Figure
6.1, reaction d) (Hargreaves and Selby, 1978). Thus the ability to convert
pterocarpans to isoflavans occurs in plants as well as in fungi. No other products of the metabolism of pterocarpanoid phytoalexins by plants have been
identified, although a number of structurally related metabolites are found in
plant tissue. For example, the isoflavanones produced by fungal metabolism of
medicarpin and maackiain (Figure 6.2, reaction b) are also apparently synthesized by plants (Suginome, 1959; Ingham, 1978); but isoflavanones are thought
to be precursors rather than products of pterocarpans in plant metabolic
pathways (Martin and Dewick, 1980).

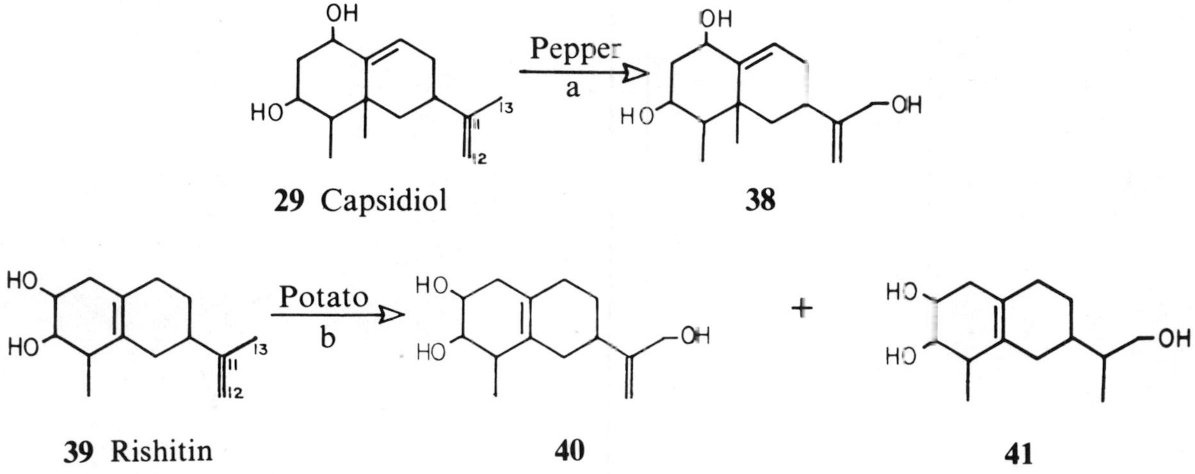

Figure 6.14 Metabolism of capsidiol by *Capsicum frutescens* (Ward *et al.*, 1977) and rishitin by *Solanum tuberosum* (Murai *et al.*, 1977; Ward *et al.*, 1977).

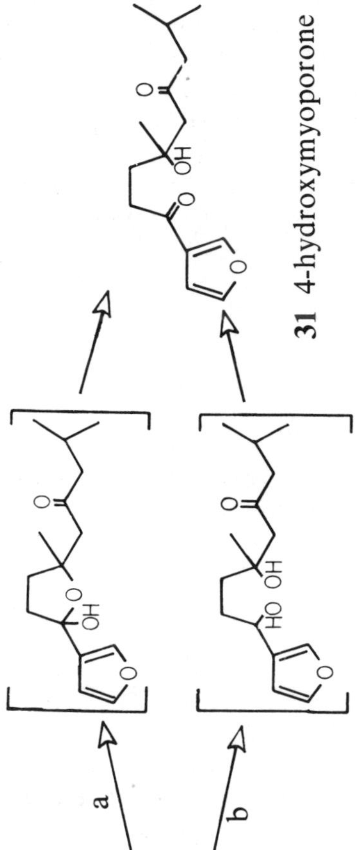

42 Ipomeamarone

31 4-hydroxymyoporone

Figure 6.15 Metabolism of ipomeamarone by sweet potato (*Ipomoea batatas*) (Burka and Kuhnert, 1977).

The metabolism of capsidiol by *Capsicum frutescens* is different from the known fungal transformation (Figure 6.8). The addition of an oxygen atom to the isopropenyl side chain (Figure 6.14, reaction *a*) could occur via one of several mechanisms. Direct hydroxylation of carbon 13 is the most obvious possibility. However, the oxygen atom might also be incorporated at carbon 12, either by formation of the 11,12-epoxide and subsequent rearrangement, or by 11,12 hydration followed by dehydrogenation. This uncertainty was resolved by the elegant study of Stothers *et al.* (1978). When pepper fruits were incubated with capsidiol enriched with ^{13}C at carbon 13 but not at carbon 12, the ^{13}C-NMR spectrum of the product (**38**) showed that oxygen had been incorporated only at carbon 13 of capsidiol.

Potato tissue can metabolize rishitin to two products (Figure 6.14, reaction *b*). Metabolite **40** is analogous to the capsidiol metabolite **38** and may be formed by the same mechanism. Murai *et al.* (1977) suggested that **41** was produced from rishitin by a hydration reaction, but the possibility that it was formed by reduction of **40** has not been excluded experimentally. The ability to metabolize rishitin to these compounds, like the ability to synthesize rishitin, seems to be induced in potato tissue by wounding (Ishiguri *et al.*, 1978*b*).

Rishitin metabolites **40** and **41** were further converted by potato tissue to unidentified water-soluble products (Ishiguri *et al.*, 1978*b*). This observation is consistent with the suggestion that the hydroxy derivatives of sesquiterpenoid phytoalexins might be further detoxified by conjugation reactions such as glucosidation (Ward *et al.*, 1977). The importance of this latter kind of transformation in the detoxification of these compounds is still unclear. Metabolites **40** and **41** are themselves less toxic than rishitin, not only to potato tissue but also to *Phytophthora infestans* (Ishiguri *et al.*, 1978*a*). The ability of *P. infestans* to metabolize **40** and **41** has not been reported.

Addition of ipomeamarone to sweet potato tissue results in the conversion of some of this compound to 4-hydroxymyoporone (Burka and Kuhnert, 1977). The authors pointed out that the reaction could proceed either by an oxygenation mechanism with a hemiketal intermediate (Figure 6.15, reaction *a*) or via hydrolysis of the tetrahydrofuran ring followed by an oxidation step (Figure 6.15, reaction *b*). These alternatives are analogous to the possible mechanisms for the fungal conversion of pterocarpans to isoflavanones (Figure 6.2, reaction *b*).

Concluding remarks

The normal biological functions of the microbial enzymes catalysing phytoalexin metabolism are still unknown. Undoubtedly the primary function of some of these enzymes is unrelated to the breakdown of phytoalexins. However, evidence is emerging that for some pathogens these enzymes serve a specific function in pathogenesis, namely to remove or reduce a potential chemical barrier produced by the plant in response to the invading microorganism.

Studies on the importance of phytoalexin metabolism, and on the enzymic mechanisms involved, may be useful in designing new methods of disease control. As suggested previously (Sisler, 1977), specific inhibitors of fungal phytoalexin-degrading enzymes could act as synergists of the host phytoalexin response and thereby convert some susceptible reactions to resistant reactions. The most attractive targets for inhibition are the monooxygenases. This class of enzyme seems to be quite important in the metabolism of phytoalexins by fungi. Moreover, a considerable amount of relevant research has already been performed on the analogous problem of developing inhibitors of insect monooxygenases to prevent detoxification of insecticides (Wilkinson, 1976). This research on insecticide synergists has demonstrated, among other things, that a large number of detoxification reactions can be catalysed by a single monooxygenase. If this is true for plant pathogens, the fungal metabolism of a variety of phytoalexins, and perhaps of some fungicides as well, may be subject to inhibition by one or a few compounds. Further studies on the substrate specificity and inhibitor sensitivity of fungal monooxygenases are clearly needed. A particular advantage of the use of a phytoalexin synergist is that its effectiveness would not depend on the simultaneous application of a pesticide; the directly toxic compounds (phytoalexins) would be supplied by the plant.

Acknowledgements

We thank Al Collmer, Tim Denny and Corby Kistler for their critiques of the manuscript, Patty Matthews for the drawings, and Barbara Mosher for typing the manuscript.

REFERENCES

Alexander, M. (1981) Biodegradation of chemicals of environmental concern. *Publ. Am. Assoc. Adv. Sci.*, **211**, 132–138.

Arinze, A. E. and Smith, I. M. (1980) Antifungal furanoterpenoids of sweet potato in relation to pathogenic and non-pathogenic fungi. *Physiol. Plant Pathol.*, **17**, 145–155.

Bailey, J. A. (1974) The relationship between symptom expression and phytoalexin concentration in hypocotyls of *Phaseolus vulgaris* infected with *Colletotrichum lindemuthianum*. *Physiol. Plant Pathol.*, **4**, 477–488.

Bailey, J. A., Carter, G. A. and Skipp, R. A. (1976) The use and interpretation of bioassays for fungitoxicity of phytoalexins in agar media. *Physiol. Plant Pathol.*, **8**, 189–194.

Bailey, J. A., Burden, R. S., Mynett, A. and Brown, C. (1977) Metabolism of phaseollin by *Septoria nodorum* and other non-pathogens of *Phaseolus vulgaris*. *Phytochemistry*, **16**, 1541–1544.

Barz, W. and Hosel, W. (1975) "Metabolism of flavonoids", in *The Flavonoids*, Part 2, eds. J. B. Harborne, T. J. Mabry and Helga Mabry, Academic Press, New York, 916–969.

Bilton, J. N., Debnam, J. R. and Smith, I. M. (1976) 6a-Hydroxypterocarpans from red clover. *Phytochemistry*, **15**, 1411–1412.

Bonde, M. R., Millar, R. L. and Ingham, J. L. (1973) Induction and identification of sativan and vestitol as two phytoalexins from *Lotus corniculatus*. *Phytochemistry*, **12**, 2957–2959.

Bruin, G. C. A., Gieskes, S. A. and Fuchs, A. (1977) Induction of the synthesis of pisatin, and its breakdown by bacteria. *Acta Bot. Neerl.*, **26**, 269–270.

Burden, R. S., Bailey, J. A. and Vincent, G. G. (1974) Metabolism of phaseollin by *Colletotrichum lindemuthianum*. *Phytochemistry*, **13**, 1789–1791.

Burka, L. T. and Kuhnert, L. (1977) Biosynthesis of furanosesquiterpenoid stress metabolites in sweet potato (*Ipomoea batatas*). Oxidation of ipomeamarone to 4-hydroxymyoporone. *Phytochemistry*, **16**, 2022–2023.

Burka, L. T., Kuhnert, L., Wilson, B. J. and Harris, T. M. (1977) Biogenesis of lung-toxic furans produced during microbial infection of sweet potatoes (*Ipomoea batatas*). *J. Am. Chem. Soc.*, **99**, 7.

Carter, G. A., Chamberlain, K. and Wain, R. L. (1978) Investigations on fungicides. XX. The fungitoxicity of analogues of the phytoalexin 2-(2'-methoxy-4'-hydroxyphenyl)-6-methoxybenzofuran (vignafuran). *Ann. appl. Biol.*, **88**, 57–64.

Cerniglia, C. E. and Gibson, D. T. (1978) Metabolism of naphthalene by cell extracts of *Cunninghamella elegans*. *Arch. Biochem. Biophys.*, **186**, 121–127.

Cerniglia, C. E. and Gibson, D. T. (1979) Oxidation of benzo[a]pyrene by the filamentous fungus *Cunninghamella elegans*. *J. Biol. Chem.*, **254**, 12174–12180.

Cleveland, T. E. and Smith, D. A. (1981) Characterization of kievitone hydratase from *Fusarium solani* f. sp. *phaseoli* culture filtrates. *Phytopathology*, Abstr. (in press).

Cruickshank, I. A. M. and Perrin, D. R. (1965) Studies on phytoalexins. VIII. The effect of some further factors on the formation, stability and localization of pisatin *in vivo*. *Aust. J. Biol. Sci.*, **18**, 817–828.

Davies, J. S., Wellman, A. M. and Zajic, J. E. (1976) Oxidation of ethane by an *Acremonium* species. *Appl. Environ. Microbiol.*, **32**, 14–20.

Dobnam, J. R. and Smith, I. M. (1976) Changes in the isoflavones and pterocarpans of red clover on infection with *Sclerotinia trifoliorum* and *Botrytis cinerea*. *Physiol. Plant Pathol.*, **9**, 9–23.

Denny, T. P. and VanEtten, H. D. (1981a) Metabolism of the phytoalexins medicarpin and maackiain by *Fusarium solani*. *Phytochemistry*, in press.

Denny, T. P. and VanEtten, H. D. (1981b) Tolerance by *Nectria haematococca* MP VI of the chickpea (*Cicer arietinum*) phytoalexins medicarpin and maackiain. *Physiol. Plant Pathol.*, in press.

Dewick, P. M. and Martin, M. (1979) Biosynthesis of pterocarpan and isoflavan phytoalexins in *Medicago sativa*; The biochemical interconversion of pterocarpans and 2'-hydroxyisoflavans. *Phytochemistry*, **18**, 591–596.

de Wit-Elshove, A. (1969) The role of pisatin in the resistance of pea plants—some further experiments on the breakdown of pisatin. *Neth. J. Plant Pathol.*, **75**, 164–168.

de Wit-Elshove, A. and Fuchs, A. (1971) The influence of the carbohydrate source on pisatin breakdown by fungi pathogenic to pea (*Pisum sativum*). *Physiol. Plant Pathol.*, **1**, 17–24.

Doke, N., Garas, N. A. and Kuć, J. (1979) Partial characterization and aspects of the mode of action of a hypersensitivity-inhibiting factor (HIF) isolated from *Phytophthora infestans*. *Physiol. Plant Pathol.*, **15**, 127–140.

Duczek, L. J. and Higgins, V. J. (1976) The role of medicarpin and maackiain in the response of red clover leaves to *Helminthosporium carbonum*, *Stemphylium botryosum*, and *S. sarcinaeforme*. *Can. J. Bot.*, **54**, 2609–2619.

Ferris, J. P., MacDonald, L. H., Patrie, M. A. and Martin, M. A. (1976) Aryl hydrocarbon hydroxylase activity in the fungus *Cunninghamella bainieri*: Evidence for the presence of cytochrome P-450. *Arch. Biochem. Biophys.*, **175**, 443–452.

Fuchs, A., de Vries, F. W. and Platero Sanz, M. (1980a) The mechanism of pisatin degradation by *Fusarium oxysporum* f. sp. *pisi*. *Physiol. Plant Pathol.*, **16**, 119–133.

Fuchs, A., de Vries, F. W., Landheer, C. A. and van Veldhuizen, A. (1980b) 3-Hydroxymaackiain-isoflavan, a pisatin metabolite produced by *Fusarium oxysporum* f. sp. *pisi*. *Phytochemistry*, **19**, 917–919.

Harborne, J. B. (1977) *Introduction to Ecological Biochemistry*. Academic Press, New York.

Harborne, J. B. and Ingham, J. L. (1978) "Biochemical aspects of the coevolution of higher plants with their fungal parasites", in *Biochemical Aspects of Plant and Animal Coevolution*, ed. J. B. Harborne, Academic Press, New York, 343–405.

Hargreaves, J. A. and Selby, C. (1978) Phytoalexin formation in cell suspensions of *Phaseolus vulgaris* in response to an extract of bean hypocotyls. *Phytochemistry*, **17**, 1099–1102.

Hargreaves, J. A., Mansfield, J. W. and Coxon, D. T. (1976a) Conversion of wyerone to wyerol by *Botrytis cinerea* and *B. fabae in vitro*. *Phytochemistry*, **15**, 651–653.

Hargreaves, J. A., Mansfield, J. W., Coxon, D. T. and Price, K. R. (1976b) Wyerone epoxide as a phytoalexin in *Vicia faba* and its metabolism by *Botrytis cinerea* and *B. fabae in vitro*. *Phytochemistry*, **15**, 1119–1121.

Hayaishi, O. (1974) "General properties and biological functions of oxygenases", in *Molecular Mechanisms of Oxygen Activation*, ed. O. Hayaishi, Academic Press, New York, 1–28.

Heath, M. C. and Higgins, V. J. (1973) *In vitro* and *in vivo* conversion of phaseollin and pisatin by an alfalfa pathogen *Stemphylium botryosum*. *Physiol. Plant Pathol.*, **3**, 107–120.

Higgins, V. J. (1972) Role of the phytoalexin medicarpin in three leaf spot diseases of alfalfa. *Physiol. Plant Pathol.*, **2**. 289–300.

Higgins, V. J. (1975) Induced conversion of the phytoalexin maackiain to dihydromaackiain by the alfalfa pathogen *Stemphylium botryosum*. *Physiol. Plant Pathol.*, **6**, 5–18.

Higgins, V. J. (1978) The effect of some pterocarpanoid phytoalexins on germ tube elongation of *Stemphylium botryosum*. *Phytopathology*, **68**, 339–345.

Higgins, V. J. and Millar, R. L. (1970) Degradation of alfalfa phytoalexin by *Stemphylium loti* and *Colletotrichum phomoides*. *Phytopathology*, **60**, 269–271.

Higgins, V. J., Stoessl, A. and Heath, M. C. (1974) Conversion of phaseollin to phaseollinisoflavan by *Stemphylium botryosum*. *Phytopathology*, **64**, 105–107.

Hill, I. R. and Wright, S. J. L. (eds.) (1978) *Pesticide Microbiology—Microbiological Aspects of Pesticide Behaviour in the Environment*, Academic Press, New York.

Hopkinson, S. M. and Pridham, J. B. (1967) Enzymic glucosylation of phenols. *Biochem. J.*, **105**, 655–662.

Ingham, J. L. (1976) Fungal modification of pterocarpan phytoalexins from *Melilotus alba* and *Trifolium pratense*. *Phytochemistry*, **15**, 1489–1495.

Ingham, J. L. (1978) Flavonoid and isoflavonoid compounds from leaves of sainfoin (*Onobrychis viciifolia*). *Z. Naturforsch.*, **33**, 146–148.

Ingham, J. L. and Millar, R. L. (1973) Sativin: an induced isoflavan from the leaves of *Medicago sativa*. *Nature*, **242**, 125–126.

Ishiguri, Y., Tomiyama, K., Murai, A., Katsui, N. and Masamune, T. (1978*a*) Toxicity of rishitin, rishitin-M-1 and rishitin-M-2 to *Phytophthora infestans* and potato tissue. *Ann. Phytopathol. Soc. Jpn.*, **44**, 52–56.

Ishiguri, Y., Tomiyama, K., Doke, N., Murai, A., Katsui, N., Yagihashi, F. and Masamune, T. (1978*b*) Induction of rishitin-metabolizing activity in potato tuber disks by wounding and identification of rishitin metabolites. *Phytopathology*, **68**, 720–725.

Ishizaka, N., Tomiyama, K., Katsui, N., Murai, A. and Masamune, T. (1969) Biological activities of rishitin, an antifungal compound isolated from diseased potato tubers, and its derivatives. *Plant Cell Physiol.*, **10**, 183–192.

Kistler, H. C. and VanEtten, H. D. (1981) Phaseollin metabolism and tolerance by *Fusarium solani* f. sp. *phaseoli*. *Physiol. Plant Pathol.*, in press.

Kolattukudy, P. E. and Brown, L. (1975) Fate of naturally occurring epoxy acids: A soluble epoxide hydrase, which catalyzes cis hydration, from *Fusarium solani pisi*. *Arch. Biochem. Biophys.*, **166**, 599–607.

Kuhn, P. J. and Smith, D. A. (1978) Detoxification of the phytoalexin, kievitone, by *Fusarium solani* f. sp. *phaseoli*. *Ann. appl. Biol.*, **89**, 362–366.

Kuhn, P. J. and Smith, D. A. (1979) Isolation from *Fusarium solani* f. sp. *phaseoli* of an enzymic system responsible for kievitone and phaseollidin detoxification. *Physiol. Plant Pathol.*, **14**, 179–190.

Kuhn, P. J. and Smith, D. A. (1980) Enzymic modification of kievitone and phaseollidin by *Fusarium solani* f. sp. *phaseoli*. *Ann. Phytopath.*, in press.

Kuhn, P. J., Smith, D. A. and Ewing, D. F. (1977) 5,7,2′,4′-Tetrahydroxy-8-(3″-hydroxy-3″-methyl-butyl) isoflavanone, a metabolite of kievitone produced by *Fusarium solani* f. sp. *phaseoli*. *Phytochemistry*, **16**, 296–297.

Lappe, U. and Barz, W. (1978) Degradation of pisatin by fungi of the genus *Fusarium*. *Z. Naturforsch.* (Sect. C), **33**, 301–302.

Lyon, G. D. (1976) Metabolism of the phytoalexin rishitin by *Botrytis* spp. *J. Gen. Microbiol.*, **96**, 225–226.

Lyr, H. (1962) Enzymatische Detoxifikation der Kernholztoxine. *Flora* (Jena), **152**, 570–579.

Macfoy, C. A. and Smith, I. M. (1978) Metabolism of clover phytoalexins by fungi. *3rd Inter. Congr. Plant Pathol.*, p. 247. Commission Agency Paul Parey, Berlin and Hamburg.

Macfoy, C. A. and Smith, I. M. (1979) Phytoalexin production and degradation in relation to resistance of clover leaves to *Sclerotinia* and *Botrytis* spp. *Physiol. Plant Pathol.*, **14**, 99–111.

Macrae, I. C. and Alexander, M. (1963) Metabolism of phenoxyalkyl carboxylic acids by a *Flavobacterium* species. *J. Bacteriol.*, **86**, 1231–1235.

Mansfield, J. W. (1980) "Mechanisms of resistance to *Botrytis*", in *The Biology of Botrytis*, eds. J. R. Coley-Smith and W. R. Jarvis, Academic Press, New York, 181–218.

Mansfield, J. W. and Widdowson, D. A. (1973) The metabolism of wyerone acid (a phytoalexin from *Vicia faba* L.) by *Botrytis fabae* and *B. cinerea*. *Physiol. Plant Pathol.*, **3**, 393–404.

Martin, M. and Dewick, P. M. (1980) Biosynthesis of pterocarpan, isoflavan and coumestan metabolites of *Medicago sativa*: The role of an isoflav-3-ene. *Phytochemistry*, **19**, 2341–2346.

Matthews, D. E. and VanEtten, H. D. (1981) Demethylation of pisatin by a cell-free preparation from *Nectria haematococca*. *Abstracts of the Annual Meeting of the American Society for Microbiology*, 1981, p. 163.

Millburn, P. (1978) "Biotransformation of xenobiotics by animals", in *Biochemical Aspects of Plant and Animal Coevolution*, ed. J. B. Harborne, Academic Press, New York, 35–73.

Moesta, P. and Grisebach, H. (1980) Effects of biotic and abiotic elicitors on phytoalexin metabolism in soybean. *Nature*, **286**, 710–711.

Müller, K. O. (1958) Studies on phytoalexins. I. The formation and immunological significance of phytoalexin production by *Phaseolus vulgaris* in response to infections with *Sclerotinia fructicola* and *Phytophthora infestans*. *Aust. J. Biol. Sci.*, **11**, 275–300.

Murai, A., Katsui, N., Yagihashi, F., Masamune, T., Ishiguri, Y. and Tomiyama, K. (1977) Structure of rishitin M-1 and M-2, metabolites of rishitin in healthy potato tuber tissues. *J. Chem. Soc. Chem. Commun.*, 670–671.

Murphy, G., Vogel, G., Krippahl, G. and Lynen, F. (1974) Patulin biosynthesis: The role of mixed-function oxidases in the hydroxylation of *m*-cresol. *Eur. J. Biochem.*, **49**, 443–455.

Nonaka, F. (1967) Inactivation of pisatin by pathogenic fungi. *Agric. Bull. Saga Univ.*, **24**, 109–121.

Nuesch, J. (1963) "Defence reactions in orchid bulbs", in *Symbiotic Associations*, eds. P. S. Nutman and B. Mosse, Cambridge University Press, Cambridge, 335–343.

Oesch, F. (1973) Mammalian epoxide hydrasse: Inducible enzymes catalysing the inactivation of carcinogenic and cytotoxic metabolites derived from aromatic and olefinic compounds. *Xenobiotica*, **3**, 305–340.

Perrin, D. R. and Cruickshank, I. A. M. (1969) The antifungal activity of pterocarpans towards *Monilinia fructicola*. *Phytochemistry*, **8**, 971–978.

Pierre, R. E. and Bateman, D. F. (1967) Induction and distribution of phytoalexins in *Rhizoctonia*-infected bean hypocotyls. *Phytopathology*, **57**, 1154–1160.

Platero Sanz, M. and Fuchs, A. (1978) Degradation of pisatin, an antimicrobial compound produced by *Pisum sativum* L. *Phytopathology mediterr.*, **17**, 14–17.

Platero Sanz, M. and Fuchs, A. (1980) Short-lived protection of pea plants against *Mycosphaerella pinodes* by prior inoculation with *Pseudomonas phaseolicola*. *Neth. J. Plant Pathol.*, **86**, 181–190.

Pueppke, S. G. and VanEtten, H. D. (1976a) Accumulation of pisatin and three additional antifungal pterocarpans in *Fusarium solani*-infected tissues of *Pisum sativum*. *Physiol. Plant Pathol.*, **8**, 51–61.

Pueppke, S. G. and VanEtten, H. D. (1976b) The relation between pisatin and the development of *Aphanomyces euteiches* in diseased *Pisum sativum*. *Phytopathology*, **66**, 1174–1185.

Rossall, S. and Mansfield, J. W. (1978) The activity of wyerone acid against *Botrytis*. *Ann. appl. Biol.*, **89**, 359–362.

Rossall, S., Mansfield, J. W. and Price, N. C. (1977) The effect of reduced wyerone acid on the antifungal activity of the phytoalexin wyerone acid against *Botrytis fabae*. *J. Gen. Microbiol.*, **102**, 203–205.

Scheline, R. R. (1978) *Mammalian Metabolism of Plant Xenobiotics*. Academic Press, New York.

Sisler, H. D. (1977) "Fungicides: problems and prospects", in *Antifungal Compounds*, Vol. 1, *Discovery, Development and Uses*, eds. M. R. Siegel and H. D. Sisler, Marcel Dekker, Inc., New York, 531–547.

Skipp, R. A. and Bailey, J. A. (1976) The effect of phaseollin on the growth of *Colletotrichum lindemuthianum* in bioassays designed to measure fungitoxicity. *Physiol. Plant Pathol.*, **9**, 253–263.

Skipp, R. A. and Bailey, J. A. (1977) The fungitoxicity of isoflavanoid phytoalexins measured using different types of bioassay. *Physiol. Plant Pathol.*, **11**, 101–112.

Smith, D. A. (1976) Some effects of the phytoalexin, kievitone, on the vegetative growth of *Aphanomyces euteiches, Rhizoctonia solani*, and *Fusarium solani* f. sp. *phaseoli*. *Physiol. Plant Pathol.*, **9**, 45–55.

Smith, D. A., Harrer, J. M. and Cleveland, T. E. (1981) Simultaneous detoxification of phytoalexins by *Fusarium solani* f. sp. *phaseoli*. *Phytopathology*, **71**, in press.

Smith, D. A., Kuhn, P. J., Bailey, J. A. and Burden, R. S. (1980) Detoxification of phaseollidin by *Fusarium solani* f. sp. *phaseoli*. *Phytochemistry*, 19, 1673–1675.

Steiner, P. W. and Millar, R. L. (1974) Degradation of medicarpin and sativan by *Stemphylium botryosum*. *Phytopathology*, 64, 586 (Abstr.).

Stoessl, A., Unwin, C. H. and Ward, E. W. B. (1973) Postinfectional fungus inhibitors from plants: Fungal oxidation of capsidiol in pepper fruit. *Phytopathology*, 63, 1225–1231.

Stoessl, A., Robinson, J. R., Rock, G. L. and Ward, E. W. B. (1977) Metabolism of capsidiol by sweet pepper tissue: some possible implications for phytoalexin studies. *Phytopathology*, 67, 64–66.

Stothers, J. B., Stoessl, A. and Ward, E. W. B. (1978) A ^{13}C-NMR study of the biological oxidation of capsidiol. *Z. Naturforsch.*, 33, 149–150.

Subramanian, V., Liu, T., Yeh, W. K. and Gibson, D. T. (1979) Toluene dioxygenase: Purification of an iron-sulfur protein by affinity chromatography. *Biochem. Biophys. Res. Commun.*, 91, 1131–1139.

Suginome, H. (1959) Oxygen heterocycles. A new isoflavanone from *Sophora japonica*. L. *J. Org. Chem.*, 24, 1655–1662.

Takeuchi, A., Oguni, I., Oba, K., Kojima, M. and Uritani, I. (1978) Interactions between diseased sweet potato terpenoids and *Ceratocystis fimbriata*. *Agric. Biol. Chem.*, 42, 935–939.

Tegtmeier, K. J. and VanEtten, H. D. (1981) The rôle of pisatin tolerance and degradation in the virulence of *Nectria haematococca*. A genetic analysis. *Phytopathology*, 71, in press.

Testa, B. and Jenner, P. (1976) *Drug Metabolism—Chemical and Biochemical Aspects*, Marcell Dekker inc., New York.

Tillmanns, G. M., Wallnöfer, P. R., Engelhardt, G., Olie, K. and Hutzinger, O. (1978) Oxidative dealkylation of five phenylurea herbicides by the fungus *Cunninghamella echinulata thaxter*. *Chemosphere*, 7, 59–64.

Towers, G. H. N. (1964) "Metabolism of phenolics in higher plants and micro-organisms", in *Biochemistry of Phenolic Compounds*, ed. J. B. Harborne, Academic Press, New York, 249–294.

Tsorbatzoudi, E., Vockel, D. and Korte, F. (1976) Metabolismus von Buturon-^{14}C in Algen. *Chemosphere*, 5, 49–52.

Uehara, K. (1964) Relationship between host specificity of pathogen and phytoalexin. *Ann. Phytopathol. Soc. Jpn.*, 29, 103–110.

Uehara, K. and Kiku, T. (1969) Inactivation of ipomeamarone by *Corticium rolfsii* (Sacc.) Curzi. *Bull. Fac. Agric. Kagoshima Univ.*, 19, 73–80.

Van den Heuvel, J. (1976) Sensitivity to, and metabolism of, phaseollin in relation to the pathogenicity of different isolates of *Botrytis cinerea* to bean (*Phaseolus vulgaris*). *Neth. J. Planet Pathol.*, 82, 153–160.

Van den Heuvel, J. and Glazener, A. (1975) Comparative abilities of fungi pathogenic and nonpathogenic to bean (*Phaseolus vulgaris*) to metabolize phaseollin. *Neth. J. Plant Pathol.*, 81, 125–137.

Van den Heuvel, J. and Grootveld, D. (1980) Formation of phytoalexins within and outside lesions of *Botrytis cinerea* in French bean leaves. *Neth. J. Plant Pathol.*, 86, 27–35.

Van den Heuvel, J. and VanEtten, H. D. (1973) Detoxification of phaseollin by *Fusarium solani* f. sp. *phaseoli*. *Physiol. Plant Pathol.*, 3, 327–339.

Van den Heuvel, J. and Vollaard, P. J. (1975) Metabolism of phaseollin by different races of *Colletotrichum lindemuthianum*. *Neth. J. Plant Pathol.*, 82, 103–108.

Van den Heuvel, J., VanEtten, H. D., Serum, J. W., Coffen, D. L. and Williams, T. H. (1974) Identification of 1a-hydroxy phaseollone, a phaseollin metabolite produced by *Fusarium solani*. *Phytochemistry*, 13, 1129–1131.

VanEtten, H. D. (1976) Antifungal activity of pterocarpans and other selected isoflavonoids. *Phytochemistry*, 15, 655–659.

VanEtten, H. D. and Barz, W. (1981) Expression of pisatin demethylating ability in *Nectria haematococca*. *Arch. Microbiol.*, 129, 56–60.

VanEtten, H. D. and Pueppke, S. G. (1976) "Isoflavonoid phytoalexins", in *Biochemical Aspects of Plant-Parasitic Relationships*, eds. J. Friend and D. R. Threlfall, Academic Press, New York, 239–289.

VanEtten, H. D. and Smith, D. A. (1975) Accumulation of antifungal isoflavonoids and 1a-hydroxy-phaseollone, a phaseollin metabolite, in bean tissue infected with *Fusarium solani* f. sp. *phaseoli*. *Physiol. Plant Pathol.*, 5, 225–237.

VanEtten, H. D. and Stein, J. I. (1978) Differential response of *Fusarium solani* isolates to pisatin and phaseollin. *Phytopathology*, **68**, 1276–1283.

VanEtten, H. D., Pueppke, S. G. and Kelsey, T. C. (1975) 3,6a-Dihydroxy-8,9-methylenedioxy-pterocarpan as a metabolite of pisatin produced by *Fusarium solani* f. sp. *pisi. Phytochemistry*, **14**, 1103–1105.

VanEtten, H. D., Matthews, P. S., Tegtmeier, K. J., Dietert, M. F. and Stein, J. I. (1980) The association of pisatin tolerance and demethylation with virulence on pea in *Nectria haematococca*. *Physiol. Plant Pathol.*, **16**, 257–268.

Van 'T Land, B. G., Wiersma-van Duin, E. D. and Fuchs, A. (1975) *In vitro* and *in vivo* conversion of pisatin by *Ascochita pisi. Acta Bot. Neerl.*, **24**, 251.

Vasyukova, N. I., Davydova, M. A., Shcherbakova, L. A. and Ozeretskovskaya, O. L. (1977) Phytosterols as a factor protecting the potato phytophthorosis pathogen from the action of phytoalexins. *Dokl. Akad. Nauk. SSSR*, **235**, 216–220.

Ward, E. W. B. and Stoessl, A. (1972) Postinfectional inhibitors from plants. III. Detoxification of capsidiol, an antifungal compound from peppers. *Phytopathology*, **62**, 1186–1187.

Ward, E. W. B. and Stoessl, A. (1977) Phytoalexins from potatoes: evidence for the conversion of lubimin to 15-dihydrolubimin by fungi. *Phytopathology*, **67**, 468–471.

Ward, E. W. B., Unwin, C. H. and Stoessl, A. (1975) Postinfectional inhibitors from plants. XV. Antifungal activity of the phytoalexin orchinol and related phenanthrenes and stilbenes. *Can. J. Bot.*, **53**, 964–971.

Ward, E. W. B., Stoessl, A. and Stothers, J. B. (1977) Metabolism of the sesquiterpenoid phytoalexins capsidiol and rishitin to their 13-hydroxy derivatives by plant cells. *Phytochemistry*, **16**, 2024–2025.

Weltring, K.-M. and Barz, W. (1980) Degradation of 3,9-dimethoxypterocarpan and medicarpin by *Fusarium proliferatum* Z. *Naturforsch.*, **35**, 399–405.

Wilkinson, C. F. (1976) "Insecticide synergism", in *The Future for Insecticides: Needs and Prospects*, eds. R. L. Metcalf and J. J. McKelvey, Jr., John Wiley & Sons, Inc., New York, 195–222.

Wong, E. (1970) "Structural and biogenetic relationships of isoflavonoids", in *Prog. Chem. Org. Nat. Prod.*, eds. W. Hera, H. Grisebach and A. I. Scott, Springer-Verlag, New York, **28**, 1–73.

Yoshikawa, M., Yamauchi, K. and Masago, H. (1978) Glyceollin: its role in restricting fungal growth in resistant soybean hypocotyls infected with *Phytophthora megasperma* var. *sojae. Physiol. Plant Pathol.*, **12**, 73–82.

Yoshikawa, M., Yamauchi, K. and Masago, H. (1979) Biosynthesis and biodegradation of glyceollin by soybean hypocotyls infected with *Phytophthora megasperma* var. *sojae. Physiol. Plant Pathol.*, **14**, 157–169.

7 Toxicity of phytoalexins

DAVID A. SMITH

Introduction

The physiological functions assigned to Müller and Börger's conceptual phytoalexin were "... 'paralysis' or the premature death of the fungus" (Müller and Börger, 1940; see also Paxton, 1980). It is now clear, however, that in addition to antifungal activity, phytoalexins exhibit toxicity across much of the biological spectrum, prokaryotic and eukaryotic, and that activity is by no means confined to microorganisms. Phytoalexins may also have indirect inhibitory effects, for example, by inactivating extracellular enzymes (Bull and Smith, 1980). Nevertheless, most of the research which has followed Müller's classic studies has concerned interactions between phytoalexins and fungi, and direct fungitoxicity will therefore be the main concern of this chapter. Compounds recognized as phytoalexins come from a wide variety of chemical classes. It might be expected that they would possess different modes of action and also exhibit differential toxicity. As discussed more fully later, modes of action have not been determined precisely and, although differences in toxicity do occur between and within chemical families, effective doses of phytoalexins generally fall within one order of magnitude (10^{-5} to 10^{-4} M). This review will therefore consider toxicity according to the type of organisms affected rather than to the chemical class of a particular phytoalexin. Data dealing with isoflavonoids and terpenoids predominate, reflecting the extensive literature dealing with these compounds.

In studying the toxic nature of any chemical, an important consideration is the purity of the compound. An impure preparation will necessarily yield suspect information, since the apparent toxicity of the principal compound might be enhanced or moderated by contaminants. Obtaining sufficient amounts of pure phytoalexins has presented a considerable limitation to meaningful toxicological studies. Few phytoalexins have been synthesized via the standard protocols of organic chemistry, and certainly not with the ease needed to obtain the substantial amounts necessary for significant mode of action investigations. Diseased plants continue to serve as sources, providing usually milligram yields after purification. Such small amounts often prohibit certain experimentation, particularly studies of mammalian toxicity. Furthermore, lengthy extraction of a phytoalexin from a natural source may, "subtly modify it so that certain properties, such as solubility, are irrevocably and undetectably changed" (J. M. Daly in Smith and Ingham, 1980). This problem is difficult to resolve, but it should not be forgotten.

Fungitoxicity

Methodology

Numerous types of bioassay have been utilized to assess the fungitoxic activities of phytoalexins. These have encompassed effects on germination of spores (Cruickshank and Perrin, 1961; McCance and Drysdale, 1975; Tomiyama et al., 1968), growth of germ-tubes (Duczek and Higgins, 1976; Harris and Dennis, 1976; Higgins, 1978; Müller, 1958; Rossall and Mansfield, 1978), growth of mycelium across an agar surface (Cruickshank, 1962; Skipp and Bailey, 1976, 1977; Smith et al., 1975; Ward et al., 1974) and growth of mycelium in liquid media (Bailey and Deverall, 1971; Smith, 1976; VanEtten and Bateman, 1971). These citations serve only as examples from a considerable literature; several of the authors utilized more than one type of bioassay. Before reflecting upon data generated by these means, some important qualifications must be introduced; these apply to all assays irrespective of the particular organism under investigation.

Where the purpose of bioassays in vitro is to provide information which would help clarify any role phytoalexins might play in plant disease resistance, the bioassays should be designed to yield accurate measures of the effective concentrations of the phytoalexins in plant tissues. Unfortunately this ideal has not been achieved as no bioassay medium can be presumed to reflect exactly the activity of a given phytoalexin in vivo. Incorporating components of the relevant plant tissues into media (Deverall, 1969; Pueppke and VanEtten, 1976; Smith et al., 1975) may provide only approximations of the natural inhibitory nature of the particular compound. It is important to recognize also that the composition and pH of media may have marked effects on fungitoxicity. For example, incorporation of rose bengal into agar modified the response of fungi to pisatin (VanEtten, 1973) and unknown components of pollen grains profoundly reduced the fungitoxicity of wyerone acid (Deverall and Rogers, 1972). By contrast, bioassays in malt agar revealed enhanced fungitoxicity for pinosylvin and its monomethylether (Loman, 1970). Alkaline conditions reduced the antifungal activity of wyerone acid (Deverall and Rogers, 1972) and also kievitone (Bull, 1981). The decreased activity of kievitone may reflect dissociation of phenolic hydroxyls, as suggested by changes in the ultraviolet absorbance spectrum first evident about pH 6.5 (Smith and Harrer, unpublished).

An additional consideration pertinent to media composition is the common inclusion of an organic solvent as a carrier for the frequently lipophilic phytoalexins. This probably produces conditions unlike the generally aqueous environments of cells and intercellular spaces where phytoalexins must act *in vivo*, and the use of organic solvents has consequently been a matter of some concern (Daly, 1972). This may however be unfounded—pisatin exhibited comparable activity in the presence or absence of organic solvent (Pueppke and VanEtten, 1976) and similar observations have been made with kievitone (Smith, 1978). Low water solubility itself need not necessarily preclude phytoalexins as toxic agents, for many conventional fungicides are only sparingly soluble in water. Wyerone retained activity when deposited on filter paper (Hargreaves *et al.*, 1977), suggesting that some phytoalexins may inhibit fungal invasion *in vivo*, even if they are not in solution, but this finding cannot be assumed to be representative of all phytoalexins because when solid wood or sawdust were impregnated with either pinosylvin or its monomethylether, inhibition was reduced approximately 50- to 100-fold (Hart and Shrimpton, 1979). Nonetheless, the well-established practice of thin layer chromatogram bioassays (Figure 7.1) underlines the fungitoxicity that phytoalexins can exert in the absence of organic solvents since these are not required in this bioassay procedure.

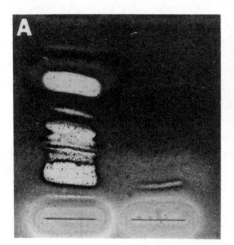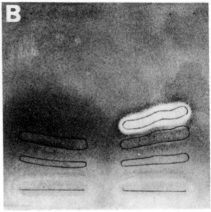

Figure 7.1   Phytoalexin bioassays on thin-layer chromatograms. (A) Thin layer chromatogram bioassay of extracts from *Helminthosporium carbonum*-inoculated (left) and water-treated (right) stems of pigeon pea (*Cajanus cajan*). The developed chromatogram was air-dried to remove traces of solvent and then sprayed with a dense spore suspension of *Cladosporium herbarum* in a glucose–mineral salts medium. The plate was then incubated in the dark at 30°C for five days. Several phytoalexins are apparent in the diseased tissue extract, indicated by the white areas where fungal growth has been inhibited; only slight antifungal activity is associated with the control extract.

(B) A similar bioassay of extracts from *H. carbonum*-inoculated (right) and water-treated (left) stems of chickpea (*Cicer arietinum*). The antifungal zone represents a mixture of maackiain and medicarpin. The non-inhibitory bands indicated in control and diseased tissue extracts are the constitutive isoflavones biochanin A (R_f 0.27) and formononetin (R_f 0.15).

Courtesy of Dr. J. L. Ingham, Department of Botany, University of Reading, U.K. See Ingham (1976*a*,*b*).

Selection of an appropriate growth stage of the assay organism is alo an important decision. Fungal spores, sporelings and established mycelium have all been employed. In attempts to throw light on the natural relevance of phyto-alexins, a judicious choice of experimental material must be made, one appro-priate to the particular host–pathogen interaction. The use of spores seems least satisfactory, since phytoalexin accumulation would presumably not be substantial until some penetration of host tissues by the fungus has occurred (chapter 8). Medicarpin proved fungitoxic to spores and to germ-tubes suggest-ing its involvement in disease resistance in alfalfa (Higgins, 1969; Higgins and Millar, 1968); however, the insensitivity of mycelium rendered this role more tenuous (Higgins, 1972). Sporelings are a legitimate stage of fungal development to investigate in certain situations. During the formation of limited lesions in *Vicia faba*, colonization by *Botrytis* species was initiated by short infection hyphae. Consequently, phytoalexin bioassays of short germ-tubes, produced from conidia in a nutrient solution, may approximate the presumed interaction *in vivo* (Mansfield, 1980). On other occasions, more substantial mycelium would be the most relevant growth stage. For example, *Rhizoctonia solani* established extensive mycelium within *Phaseolus vulgaris* tissues before fungal growth was restricted and lesions became limited (Christou, 1962; VanEtten *et al.*, 1967). These observations probably justify the use of mycelial discs or pads (Smith, 1976, Smith *et al.*, 1973, 1975; VanEtten and Bateman, 1971).

Timing of growth measurements may also be critical in determinations of apparent phytoalexin toxicity. On agar impregnated with phaseollin, mycelium of *Colletotrichum lindemuthianum* was completely inhibited for an initial period, but then grew at a rate similar to that of the controls (Bailey *et al.*, 1976a). In such circumstances, measurement of growth after only one time period would provide misleading data. Although this has not always proved to be the case, monitoring changes which occur during bioassays should potentially allow fuller evaluation of the results (Skipp and Bailey, 1977; Bailey and Skipp, 1978; Smith, 1978).

Clearly, any bioassay represents a model system providing only limited information. It is therefore helpful to utilize a variety of systems in assessing phytoalexin toxicity, and this has been carried out in some cases. Kievitone, for example, was toxic to vegetative and reproductive fungal structures in three different agar media (one incorporating a host plant extract), in two liquid culture systems, and on thin layer chromatograms (Bull, 1981; Smith, 1976, 1978; Smith *et al.*, 1973, 1975). It therefore seems most probable that kievitone is also active in intact plant tissues.

Cytological effects

Inhibition of radial mycelial growth of *R. solani* by kievitone (40 µg/ml, 1.1×10^{-4} M) is illustrated in Figure 7.2. Similar concentrations (10^{-5} to 10^{-4} M) of chemically very different phytoalexins have been shown to have

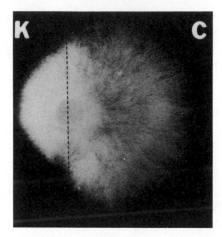

Figure 7.2 Macroscopic inhibition of fungal extension growth by phytoalexins. Radial mycelial growth of *Rhizoctonia solani* is inhibited by kievitone. A 4 mm diameter mycelial disc was placed on agar so that one half lay on medium incorporating the phytoalexin (40 μg/ml in 2 % ethanol) while the other half was simultaneously exposed to control medium containing 2 % ethanol. The plate was incubated at 25°C until at least 10 mm net growth had occurred on the control (C) side. The dashed line indicates the border between the control and kievitone-containing (K) agar. Kievitone inhibited radial mycelial spread, producing a colony with greater hyphal density. From Smith (1976). (× 1.75).

comparable effects on a wide range of fungi (Cruickshank and Perrin, 1961; Cruickshank, 1962; Rossall and Mansfield, 1978; VanEtten and Pueppke, 1976; Ward *et al.*, 1974). The fungitoxicity of phytoalexins is easily demonstrated by the inhibition of germ-tube elongation, radial mycelial growth and/or mycelial dry weight accumulation. In order to develop a deeper understanding of the causes of these macroscopic effects it is essential to examine the response of individual fungal cells by both light and electron microscopy.

Oomycete zoospores have proved to be very convenient cells for studies on the effects of both isoflavonoid and terpenoid phytoalexins. Harris and Dennis (1977) reported that treatment of motile zoospores of three *Phytophthora* species (*P. infestans*, *P. porri* and *P. cactorum*) with four terpenoids (rishitin, phytuberin, anhydro-β-rotunol and solavetivone) resulted in a standard series of consequences; loss of motility, rounding-up of the cell (with some swelling), cytoplasmic granulation, agitation of cell contents, bursting of the cell membrane and dispersal of cell contents through the medium. Figure 7.3 features some of these effects.

Comparable changes, i.e. rapid cessation of cytoplasmic streaming, granulation of the cytoplasm, disorganization of cell contents and breakdown of the cell membrane, have been observed in individual cells within fungal colonies treated with several isoflavonoid phytoalexins (Bull, 1981; Grisebach and Ebel, 1978; Skipp and Bailey, 1976; Slayman and VanEtten, 1974; Smith, 1976; VanEtten and Bateman, 1971; VanEtten and Pueppke, 1976). Examples of these symptoms of toxicity are illustrated in Figure 7.4–7.7.

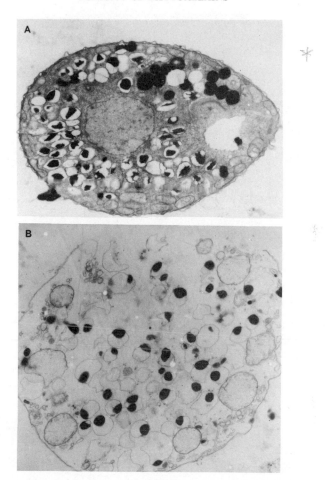

Figure 7.3 Ultrastructural observations of phytoalexin fungitoxicity. (A) Electron micrograph of a cross-section of a motile zoospore of *Phytophthora infestans.* Note the many crystal-containing vesicles and intact plasmalemma. (× 6550).

(B) Electron micrograph of a cross-section of a motile zoospore of *P. infestans* two minutes after exposure to phytuberin (100 µg/ml). Note the disorganized internal structure and discontinuous plasmalemma. (× 4740).

Courtesy of Dr. C. Dennis, A.R.C. Food Research Institute, Norwich, U.K. See Harris and Dennis (1977).

The damage outlined above may often prove fatal to individual fungal cells. In this respect, some terpenoid and isoflavanoid phytoalexins may rightly be considered fungicidal. Similarly, the furanoacetylenic phytoalexins wyerone, wyerone acid and wyerone epoxide were lethal to sporelings of *Botrytis cinerea* and *B. fabae* (Rossall *et al.*, 1980)—see Figure 7.8 . The phenanthrenes orchinol and dehydroorchinol have also been reported to kill apical cells of *Monilinia fructicola* by causing the rupture of hyphal tips (Ward *et al.*, 1975).

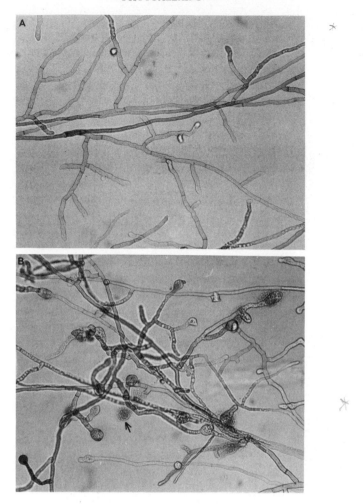

Figure 7.4 Observation of phytoalexin fungitoxicity by light microscopy. (A) Typical hyphal growth of *Rhizoctonia solani* on agar incorporating 0.5 % ethanol, 24 h after inoculation.

(B) Typical hyphal growth of *R. solani* on agar incorporating 25 μg kievitone/ml in 0.5 % ethanol, 24 h after inoculation. The hyphae were stained in 0.05 % Evan's blue for two minutes prior to photographing (× 175). Note in (B) the disorganized pattern of growth and frequency of swollen hyphal tips, some of which have lysed (arrowed).

Courtesy of C. A. Bull, Plant Biology Department, University of Hull, U.K. See Bull (1981); Smith and Bull (1978).

Cell or colony death is not however an inevitable consequence of phytoalexin treatment. Many zoospores of *Pythium middletonii*, despite a rapid loss of motility upon exposure to phytuberin, were not killed (Harris and Dennis, 1977). *Botrytis* sporelings treated with wyerone acid contained cells which survived and continued growth (Rossall *et al.*, 1980; see Figure 7.8, B and C).

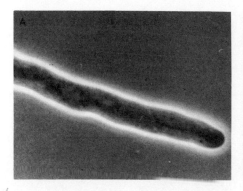
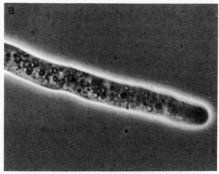

Figure 7.5 Observation of phytoalexin fungitoxicity by phase-contrast microscopy. (A) A hyphal tip of *Rhizoctonia solani* growing on a thin layer of agar on a microscope slide; a drop of water was applied to the hypha before taking the photograph. Little fine detail is evident within the hypha because active cytoplasmic streaming was occurring. Phase contrast, × 1120.

(B) The same hyphal tip of *R. solani*. A drop of water containing kievitone (75 μg/ml; no ethanol) was applied ten minutes before taking the photograph. Cessation of cytoplasmic streaming, as well as apparent granulation and clumping of the cytoplasm, occurred. Phase contrast, × 1120.

Courtesy of C. A. Bull, Plant Biology Department, University of Hull, U.K. See Bull (1981).

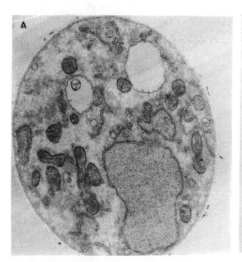
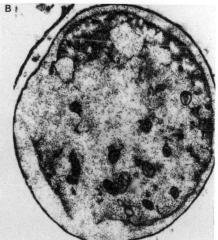

Figure 7.6 Ultrastructural observation of phytoalexin fungitoxicity. (A) Electron micrograph of a transverse section of a *Rhizoctonia solani* hypha from the hyphal tip region of a mycelial pad grown in liquid culture, ninety minutes after the addition of ethanol (to 0.5 %). Note the healthy appearance of the cell with an intact plasmalemma tightly appressed to the wall, two vacuoles, a single nucleus and numerous mitochondria. (× 14 000).

(B) Electron micrograph of a transverse section of a *R. solani* hypha from the hyphal tip region of a mycelial pad grown in liquid culture, ninety minutes after exposure to kievitone (75 μg/ml in 0.5 % ethanol). Note the disorganized internal structure, the increased electron density of membranes, as well as the discontinuities in the plasmalemma and its separation from the cell wall. (× 13 000).

Courtesy of C. A. Bull, Plant Biology Department, University of Hull, U.K. See Bull (1981).

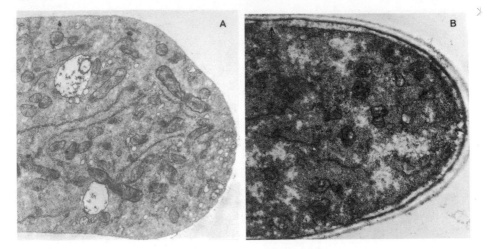

Figure 7.7 Ultrastructural observation of phytoalexin fungitoxicity. (A) Electron micrograph from the hyphal tip region of a mycelial pad of *Rhizoctonia solani* grown in liquid culture, ninety minutes after the addition of ethanol (to 0.5 %). Note the healthy appearance of the cell with an intact plasmalemma tightly appressed to the wall, vacuoles and numerous mitochondria. (× 10 500).

(B) Electron micrograph from the hyphal tip region of a mycelial pad of *R. solani* grown in liquid culture, ninety minutes after treatment with kievitone (75 μg/ml in 0.5 % ethanol). The plasmalemma has withdrawn from the cell wall and appears to have lysed in some places (arrowed). (× 11 800).

Courtesy of C. A. Bull, Plant Biology Department, University of Hull, U. K. See Bull (1981).

Skipp and Bailey (1976) noted that phaseollin-treated sporelings of *C. lindemuthianum* often regrew from apparently unaffected hyphae, particularly from interstitial cells. Maackiain, medicarpin, or mixtures of these phytoalexins, caused only temporary cessation of germ-tube elongation in *Stemphylium botryosum* (Higgins, 1978).

Observations of regrowth from surviving cells suggest that some sort of "escape" from, or tolerance of, phytoalexin damage can occur. A well-established basis for tolerance is the ability of fungi to detoxify phytoalexins (see chapter 6) but detoxification may not explain the differential sensitivity of individual cells within septate hyphae (Skipp and Carter, 1978). Cytological studies indicate that blockage of phytoalexin entry may also contribute to tolerance. Apical cells of hyphae are particularly vulnerable to phytoalexins (Figure 7.8C). These cells are growing actively and possess weak cell walls which may facilitate the entry of phytoalexins (Bartnicki-Garcia and Lippman, 1972; Skipp and Bailey, 1976; Higgins, 1978; Rossall *et al.*, 1980). Direct entry into sub-apical and interstitial cells may in contrast be restricted by their mature cell walls, and entry via the apical cell may be prevented by the formation of septal plugs separating severely damaged from less affected cells, (Figure 7.9, A and B). New hyphae can develop from the protected cells (Figure 7.9, C and D; Higgins, 1978; Bull, 1981). An inability to "block-off", and so protect, individual cells might also

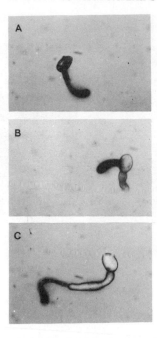

Figure 7.8 Observation of phytoulexin fungitoxicity by light microscopy. Sporelings of *Botrytis fabae* exposed to solutions of wyerone acid (20 μg/ml) for 24 h and stained with trypan blue (1 % w/v); dead cells accumulate the stain. All × 310.

(A) and (B). Six-hour-old-sporelings. In (A), note complete cell death. In (B), note the production of a secondary germ tube from the surviving conidium.

(C) Nine-hour-old sporeling. Note the death of the apical cell.

Courtesy of J. W. Mansfield, Wye College, University of London, U.K. See Rossall *et al.* (1980).

explain the extreme vulnerability of the coenocytic hyphae of *A. euteiches* to isoflavonoids (Pueppke and VanEtten, 1976; Smith, 1976).

The available cytological evidence is insufficient to define a precise mode of action for any phytoalexin. It is, however, consistent with rapid and substantial damage to the plasmalemma. Lysis of terpenoid-treated motile zoospores of *P. infestans* suggested an increased osmotic fragility within the spores causing rupture of the plasmalemma (Harris and Dennis, 1976, 1977). A putative membranolytic action was proposed for kievitone (Smith and Bull, 1978), evidenced by swelling and bursting of hyphal tips of *R. solani* (Figure 7.4B). Several of the cytological effects incited by kievitone were comparable to those caused by the established membranolytic agents, chloroform and Triton X-100 (Bull, 1981; Smith and Bull, 1978). Likewise, Harris and Dennis (1977) reported that Triton X-100 and the surfactants Deciquam and Manoxol, brought about cytological changes, including swelling and bursting of the plasmalemma of *Phytophthora* spp., which were comparable to those observed after treatment with terpenoid phytoalexins.

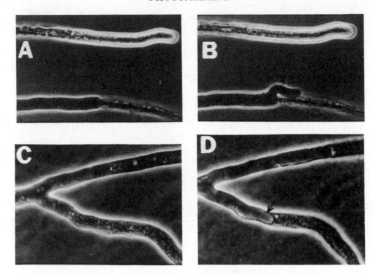

Figure 7.9 Cytological indications of recovery of fungi after phytoalexin treatment. (A) Portions of two hyphae of *Fusarium solani* f. sp. *phaseoli*, thirty minutes after treatment with kievitone (75 µg/ml). Note the increased refractivity and cytoplasmic granular clumping in the upper hypha as well as the cell on the right-hand side of the lower hypha. A septal plug appears to cut off an apparently unaffected cell on the left-hand side of the lower hypha. Phase contrast, × 770.

(B) The same hyphae, thirty minutes later, showing a new brnach that has arisen from the apparently unaffected cell in the lower hypha; the branch has formed just behind the septal plug. Phase contrast, × 770).

(C) Hyphae of *Rhizoctonia solani*, thirty minutes after treatment with kievitone (75 µg/ml). Note the cytoplasmic granulation in the cells on the right-hand side of the photograph, but not on the left-hand side. Septal plugs again seem to be providing a protective function. Phase contrast, × 560.

(D) The same hyphal segment, sixty minutes later, showing intrahyphal hyphae developing through the granulated cells. This is most evident in the lower branch (arrowed). Phase contrast, × 560.

Courtesy of C. A. Bull, Plant Biology Department, University of Hull, U.K. See Bull (1981).

The cell wall may afford protection against cell rupture in vegetative hyphae. Whereas apical cells of *R. solani*, lacking properly constituted walls, often swelled and burst after kievitone treatment, mature hyphal cells did not. These latter cells, nonetheless, often exhibited breaks in the plasmalemma. Phaseollin caused marked degenerative changes in the fine structure of non-encysted and encysted *A. euteiches* zoospores (Grisebach and Ebel, 1978), but in only the former case did cell rupture occur. Furthermore, Harris and Dennis (1977) cited the protective effect of the cell wall for the inability of potato metabolites and surfactants to lyse motile zoospores of *P. middletonii* and *Saprolegnia* species.

Physiological effects

The extensive membrane damage which occurs soon after fungi are exposed to phytoalexins is, not surprisingly, reflected in substantial leakage of electrolytes

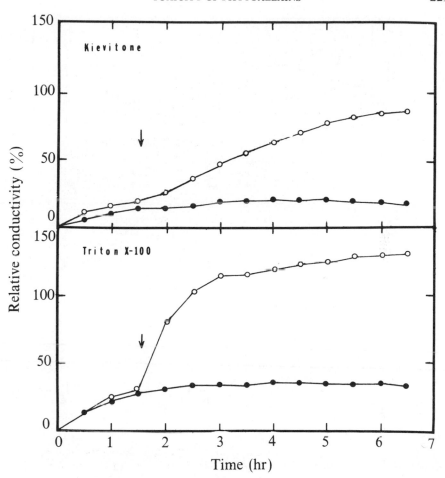

Figure 7.10 The effects of kievitone and Triton X-100 on the relative conductivities of 0.2 M sucrose solution bathing mycelial pads of *Rhizoctonia solani*. The pads (1/ml) were agitated in 10 ml sucrose solution before adding kievitone (75 µg/ml in 0.5 % ethanol) or 0.05 % Triton X-100 (arrow). Agitation was continued and the conductivities of the solutions were monitored for the next 5 h; incubation was at room temperature, *ca.* 22°C. Each point is the mean of two replicates. Treatments (O————O); controls (●————●).

and metabolites. Figure 7.10 contrasts the effects of kievitone and of Triton X-100 on *R. solani*. The loss of ^{14}C-labelled metabolites, induced by phaseollin (VanEtten and Bateman, 1971), maackiain (Higgins, 1978) and kievitone (Bull, 1981), represents another means of monitoring this response. Using this procedure, Bull (1981) detected increased leakage ten minutes after *R. solani* mycelium was exposed to the minimum inhibitory concentration of kievitone, *ca.* 50 µg/ml. The inevitable consequence of uncontrolled leakage of metabolites is loss of mycelial dry weight (Bailey and Deverall, 1971; Smith, 1976; VanEtten

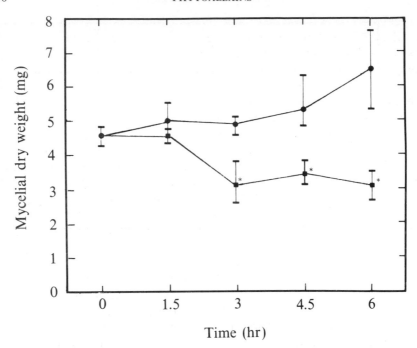

Figure 7.11 The effect of kievitone on the mycelial dry weight of *Rhizoctonia solani* in liquid culture. The fungus was exposed to 50 µg kievitone/ml in 0.5% ethanol (■———■); the control received only 0.5% ethanol (●———●). These additions were made at time zero. Each value is the mean of three replicates. Those values marked with an asterisk represent a statistically significant (5% level) loss of mycelial dry weight from that occurring at time zero.
Adapted and redrawn from Smith (1976).

and Bateman, 1971). Data presented in Figure 7.11 illustrate that kievitone (50 µg/ml) caused a significant loss of dry weight from *R. solani* after three hours' exposure. Ion leakage and dry weight loss are strong evidence that disruption of plasmalemma integrity is an important effect of isoflavonoid phytoalexins.

Influx as well as efflux is affected. Phaseollin inhibited the removal of ^{14}C-U-glucose from liquid culture medium by *R. solani* (VanEtten and Bateman, 1971) and kievitone had a comparable effect (Bull, 1981). Although no marked change in the rate of glucose uptake by germinated spores of *S. botryosum* was noted in the first hour after maackiain treatment, uptake was inhibited for several hours thereafter (Higgins, 1978).

Movement of molecules across the plasmalemma is not the only physiological process adversely affected by exposure to phytoalexins. Exogenous and endogenous respiration in *R. solani* were differentially affected by phaseollin (VanEtten and Bateman, 1971); uptake of oxygen by actively growing mycelium was inhibited, whilst the endogenous respiration of starved mycelium was stimulated. Lack of respiratory inhibition by phaseollin-treated mitochondria, isolated from *Neurospora crassa* (Slayman and VanEtten, 1974), suggested indirect effects. However, subsequent research (Bowman et al., 1978) revealed that phaseollin had no appreciable effect on membrane ATPases, as first speculated (Slayman and VanEtten, 1974).

Suppression of exogenous respiration in R. solani by phaseollin may reflect insufficient uptake of substrate due to membrane damage. Stimulation of oxygen uptake in starved mycelium by this phytoalexin might also be mediated via membrane dysfunction, either because it allowed decompartmentalization of potential respiratory substrates within the mycelia, or because it uncoupled oxidative phosphorylation. Either of these roles is also consistent with the observation that low kievitone concentrations stimulated endogenous respiration in R. solani (Bull, personal communication).

Physiological consequences of phytoalexin treatment represent either direct or indirect manifestations of the toxicities of these chemicals. Early in this chapter, however, allusion was made to the possibility of more subtle fungal inhibition. Such indirect activity might occur, for example, by inactivation of cell wall-degrading enzymes; these are important to certain plant pathogens, particularly "soft rot" organisms (Bateman and Basham, 1976). Preliminary investigations revealed that kievitone inhibited commercial preparations of pectinase and cellulase purified from Aspergillus niger (Bull, 1981). Further work indicated that cellulase and polygalacturonate- and polymethylgalacturonate-degrading enzymes, produced by R. solani in culture filtrates were likewise inhibited (Table 7.1; Bull, 1981, Bull and Smith, 1980) and that enzymes

Table 7.1 The effect of kievitone on polygalacturonate-degrading enzyme activity in Rhizoctonia solani culture filtrates. Adapted from Bull (1981).

Reaction mixture[a]	Enzyme activity (RA/ml)[b]	
	1 h	18 h
Control (no ethanol)	352[c]	394
Control (2% ethanol)	358	400
Kievitone (50 μg/ml)	312	341
(100 μg/ml)	280	211
(250 μg/ml)	261	94
(500 μg/ml)	85	5
(1000 μg/ml)	14	0

[a] Reaction mixtures contained 100 μl pectin culture filtrate and kievitone (in 2 μl ethanol). Controls received either the same amount of ethanol without kievitone, or no ethanol. The mixtures were incubated for either 1 h or 18 h at 25°C before assay.

[b] Viscometric assay. Each reaction mixture was made up to 1 ml with citrate buffer (pH 4.8) and mixed with 9 ml 1.2% sodium polypectate. Relative activity units per ml (RA/ml) were determined from the reciprocal of the time in minutes for a 50% loss in viscosity, multiplied by 1000.

[c] Each value is the mean of two replicates.

extracted from *R. solani* lesions on bean hypocotyls were also inhibited (Bull, 1981). Inhibition was most apparent with phytoalexin concentrations substantially greater than those required to inhibit fungal growth. A tetrahydroxyphenolic molecule such as kievitone might bind to proteins in a non-specific manner, leading to enzyme inactivation (Hunter, 1974; Loomis, 1974; Van Sumere *et al.*, 1975). Other workers have shown inhibition of pectic enzymes by phytoalexins (Ravisé and Kirkiacharian, 1976; Ravisé *et al.*, 1980). Enzyme inhibition is not restricted to isoflavonoids; capsidiol exhibited a similar effect (Ravisé and Kirkiacharian, 1976).

Differential sensitivity

Differences in sensitivity amongst fungi to phytoalexins have been apparent since the earliest investigations (Cruickshank and Perrin, 1961) and at one time were reputed to parallel pathogenicity; pathogenic fungi were thought to be insensitive to their hosts' phytoalexins, whereas non-pathogens were sensitive. While such instances have been recorded (Cruickshank, 1962; Cruickshank and Perrin, 1971; Higgins, 1972; Kojima *et al.*, 1979; Mansfield, 1980; Perrin *et al.*, 1974; Pueppke and VanEtten, 1974; VanEtten, 1973), several exceptions also exist (Bailey and Burden, 1973; Keen *et al.*, 1971; Klarman and Sanford, 1968; Pueppke and VanEtten, 1974; Smith *et al.*, 1975). Therefore, while a generalization in this regard seems unwarranted, some interesting individual cases merit consideration.

Tolerance of pisatin seems a useful, possibly essential, characteristic for the pea pathogen *Fusarium solani* f. sp. *pisi* (*Nectria haematococca*) (Tegtmeier and VanEtten, 1979; VanEtten and Barz, 1980; VanEtten *et al.*, 1980; see chapter 6), a result consistent with Cruickshank's early proposal (Cruickshank, 1962). Interactions between *Botrytis* species and the broad bean, *Vicia faba*, suggest that the less sensitive nature of *B. fabae* hyphae to the phytoalexins produced, particularly wyerone acid, contributes to the more aggressive behaviour of this fungus, compared to the more sensitive *B. cinerea* (Hutson and Mansfield, 1980; Mansfield, 1980). On the other hand, there is the anomalous example of *A. euteiches* (Pueppke and VanEtten, 1974, 1976). Growth of this destructive root- and epicotyl-rotting pea pathogen is severely restricted *in vitro* by pisatin, and yet this fungus develops extensively in pea tissues, despite the presence of considerable pisatin concentrations.

Although its relationship to pathogenicity is ambiguous, differential sensitivity to phytoalexins amongst fungi is an established fact. Several mechanistic explanations can be suggested, though few as yet have substantial experimental support. Parallels may be drawn, however, from work with conventional fungicides (Dekker, 1977*a,b*). Fungi which lack sensitivity to particular phytoalexins may contain a plasmalemma or cell wall less permeable to these molecules than that of sensitive organisms. Effective penetration by a phytoalexin into what may be a potentially vulnerable cell would therefore be prevented. A second possibility is that the sites of action in the insensitive fungal cell possess weak affinity for the phytoalexin. An otherwise comparable site, but one with a high affinity for the toxic moiety, would be possessed by a sensitive individual. There is also the possibility of circumvention of the site of action. In this instance, the insensitive individual possesses a metabolic bypass for the process inhibited by the phytoalexin, resulting in effective compensation. Little experimental evidence pertaining to these three separate possibilities exists for any phytoalexin. Considerable data are available, however, concerning a fourth possibility—detoxification of a phytoalexin before or after entry into the fungal cell. This issue has been discussed at length in chapter 6. It is possible too that metabolic alteration to a *more* toxic chemical may occur, but this appears to have been little investigated. In this case, differentially sensitive individuals would be those having the biosynthetic ability for such a reaction, while insensitive ones would lack it.

Although phytopathogenic fungi, and some saprophytes, have dominated phytoalexin bioassays, zoopathogenic fungi are also inhibited by phytoalexins (Gordon *et al.*, 1980). Whether phytoalexins or their artificial analogues will ever prove useful in the treatment of human mycoses is doubtful, however.

Structure-activity relationships

Little can be stated definitely concerning the molecular features of phytoalexins which confer fungitoxic capabilities. Lipophilic nature seems to be the common denominator, and possibly this permits effective penetration of fungal membranes (Harborne and Ingham, 1978). A lipophilic side chain is vital for the antifungal activities of wighteone (Ingham *et al.*, 1977) and kievitone (Smith, 1978); their non-prenylated analogues, genistein and dalbergiodin, respectively, possess little fungitoxicity (Ingham *et al.*, 1977; Kuhn, 1979; see also chapter 2). Indirect evidence for this characteristic comes from the increased polarity of several phytoalexin detoxification products (see chapter 6). Since the plasmalemma is probably the principal barrier between an external toxic chemical and its site(s) of action (Lukens, 1971), partitioning coefficients between aqueous and lipophilic phases are recognized to be important parameters governing the toxicity of conventional fungicides.

The characteristic antifungal activity of pterocarpan phytoalexins was postulated by Perrin and Cruickshank (1969) to be dependent on particular steric and compositional requirements; the two aromatic rings should be almost perpendicular to one another and there should be small oxygen-containing substituents. However, the conclusions of this early survey were not supported by subsequent research (VanEtten, 1976). The structure–antifungal activity relationships of pterocarpans, and indeed isoflavonoid phytoalexins in general, require clarification (Johnson *et al.*, 1976; VanEtten and Pueppke, 1976).

Lack of any obvious relationship between structure and activity is also evident from work with terpenoid phytoalexins. Thus, Ward *et al.* (1974) found that while slight molecular alterations to capsidiol were found to modify fungitoxicity, no essential structural or steric factors were established. These authors also examined a range of phenanthrenes and stilbenes structurally related to the phytoalexin orchinol (Ward *et al.*, 1975). As before, molecular variants exhibited different activities, but apart from purely structural considerations, this work also highlighted the problem of differential solubility within a series of related compounds. Insolubility in assay media could mask potent fungitoxic activity, which might nevertheless be expressed *in vivo* (VanEtten, 1976). Smith (1978) utilized additional solubilizing agents in an attempt to circumvent the problem. As already mentioned, a further complicating factor is that many fungicides are extremely effective, although largely insoluble in water.

An additional difficulty in this type of approach was discussed by Carter *et al.* (1978) in their investigations of structure–activity patterns within vignafuran analogues. The variability of estimates of fungitoxicity, which differed depending upon whether spore germination or inhibition of germ tube growth was used as the parameter of toxicity, prevented the establishment of clear correlations between physicochemical properties and fungitoxicity. Despite this they demonstrated that the presence of one (but not more) phenolic hydroxyl in the molecule was critical for antifungal activity. Fully methylated derivatives were inactive. No correlation was evident, however, between fungitoxicity and either Hansch π values or the nuclear magnetic resonance chemical shifts of the phenolic protons.

Modes of action

The mode of action of any toxic agent is usually derived by determining the earliest physiological effect on the target organism. One qualifying factor applies, namely that the lowest concentration of toxicant needed to effect inhibition should be used. Employing high concentrations will increase the likelihood of secondary effects incidental to the primary mode of action, which may complicate interpretation of the results. A distinction should also be drawn between toxic chemicals which are multi-site, affecting many biochemical reactions and physiological processes, and those which are site-selective, where a particular reaction or process will be primarily affected (Siegel, 1975).

Although very many reports of the fungitoxicity of phytoalexins exist, there have been few attempts to determine mode(s) of action, (Bull, 1981; Harris and Dennis, 1976, 1977; Skipp and Bailey, 1976, 1977; Smith, 1976, 1978; VanEtten and Bateman, 1971). Consequently, although numerous possibilities have been raised, the precise mode of action of any phytoalexin has yet to be defined. The available evidence suggests two principal features of activity: that phytoalexins probably represent multi-site toxicants and that dysfunction of membrane systems, particularly the plasmalemma, is instrumental in their toxicity.

Examples of evidence supporting multi-site activity are: (a) many phytoalexins are broad-spectrum toxicants (as will become apparent later in this chapter); (b) a variety of biochemical and physiological repercussions are apparent soon after fungi are treated, such that many "secondary" effects apparently exist; (c) comparatively high concentrations $(10^{-4} M)$ are often required to achieve inhibition; (d) indiscriminate binding probably occurs; and (e) where tolerance is found, detoxification is often involved (Bull, 1981; Fuchs *et al.*, 1980; VanEtten and Bateman, 1971). Such multi-site activity may be of selective advantage to the plant. Site-selective compounds are probably more easily countered by fungi as alterations in only one fungal gene may be sufficient to induce a change at the site of action and so confer resistance (Delp, 1980).

The preceding considerations of common cytological and physiological effects of phytoalexins underline the likelihood that membrane dysfunction is an important part of the mode of action. Substantial loss of dry weight, leakage, swelling and bursting of affected hyphae, all testify to membrane damage, particularly damage to the plasmalemma. The apparent importance of a lipophilic nature for many of the compounds would also be consistent with such an activity. Multi-site toxicants, of course, would be capable of affecting other reactions, resulting in interference with respiration (VanEtten and Bateman, 1971), inhibition of enzymes (Bull and Smith, 1980; Ravisé and Kirkiacharian, 1976), and possible prevention of cell wall biosynthesis, an effect consistent with the vulnerability of hyphal tip cells. Preliminary findings (Sprang and VanEtten, 1978, and unpublished results) suggest that one site of action of pisatin on *Fusarium solani* f. sp. *cucurbitae* is at the level of directional cell wall synthesis. Disruption of the synthesis or assembly of membrane components would also be dramatically expressed in the leading cells.

For many phytoalexins, there may be no one site of action, but a range of targets. Selective inhibition of one particular intracellular process, such as protein synthesis, respiration, or nucleic acid transcription seems unlikely, since a considerable time would probably have to elapse before secondary effects would be apparent, and ultrastructural damage would not be evident after only a few minutes' treatment, as is the case for several terpenoids and isoflavonoids (Bull, 1981; Harris and Dennis, 1977; Smith and Bull, 1978). A disruption of preformed components necessary for cellular maintenance (e.g. of the cytoskeleton) would also be consistent with some of the immediate effects observed.

The possibility of a site-specific action for phytoalexins cannot be dismissed, however, although adequate data are absent. Parallels exist, for example, between some of the effects of polyene antibiotics and certain phytoalexins. Polyenes cause leakage of intracellular constituents from sensitive cells and severe metabolic disturbances (Lyr, 1977; Misato and Kakiki, 1977). The plasmalemma is the primary point of attack for these site-specific fungicides, where interactions occur with membrane sterols. However, whereas the incorporation of sterols into membranes renders such membranes sensitive to polyenes (Misato and Kakiki, 1977), Vasyukova *et al.* (1977) claim the reverse for rishitin. Exogenous cholesterol, apparently incorporated into the membranes of *P. infestans*, made the fungus less vulnerable to rishitin. Depriving this pathogen of sterols seems to make it more susceptible to the influence of phytoalexins. Other disparities with the polyenes are evidenced by the failure of sterol, incorporated in bioassay media, to protect *R. solani* against phaseollin (VanEtten and Bateman, 1971) or kievitone (Bull, 1981). The addition of phospholipids to media did, however, reduce kievitone's toxic effects towards *R. solani* (Bull, 1981). If phytoalexins have an important action against membranes, sterols do not seem likely sites of interaction, though phosopholipids may be (Lyon, 1980).

Many advances have been made in clarifying the fungitoxic nature of Müller and Börger's "inhibiting principle". More information is however still needed about the toxic natures of these compounds. Mechanistic explanations of differential sensitivity and thorough proof of individual modes of action are needed. A great dearth of data also exists concerning accurate measures of toxicity *in vivo*. Clarification of this aspect is particularly pertinent in light of observations that viable hyphae may survive in plant tissues thought to contain high concentrations of phytoalexins (Bailey and Rowell, 1980; Bull, 1981; Kenning and Hanchey, 1980; Pueppke and VanEtten, 1976).

Antibacterial activity

Like plant pathology itself, the phytoalexin concept evolved from a mycological perspective. However, it soon became apparent that the same compounds accumulated in plants in response to bacterial infections. As with fungal pathogens, assigning phytoalexins a role in resistance to bacteria is dependent upon establishing their antibiotic activity in the infected plant.

A variety of assays *in vitro* indicates appreciable toxicity towards bacteria, but constraints comparable to those discussed in connection with fungal assays again impinge when attempting to draw parallels to natural infections. Several types of assay procedure have been employed, including dispersal of phytoalexin in agar medium which was subsequently streaked with bacteria (Cruickshank, 1962; Keen and Kennedy, 1974); application of phytoalexin in organic solvent to bacteria-seeded agar (Wyman and VanEtten, 1978); impregnation of phytoalexin in antibiotic assay discs, before placing the discs on seeded agar (Gnanamanickam and Smith, 1980; Lyon and Wood, 1975; Wyman and VanEtten, 1978); spraying of thin layer chromatograms with bacterial cells (Lyon and Bayliss, 1975; Lyon and Wood, 1975) as a parallel to the well-established antifungal assay (Figure 7.1); and introduction of phytoalexin to liquid media, subsequently monitoring the effects on bacterial growth by absorbance (Albersheim and Valent, 1978; Wyman and VanEtten, 1978) or dilution counts (Lyon and Bayliss, 1975). These various assays leave no doubt that phytoalexins exert antibacterial activity. The effects may be either bacteriostatic (Lyon and Wood, 1975) or bactericidal. Rishitin ($360\,\mu g/ml$) caused a decrease in the viable count of *Erwinia atroseptica* from 10^9 cells/ml to <3 cells/ml in 4 hours at 25°C when incorporated in 0.1 % peptone water, a

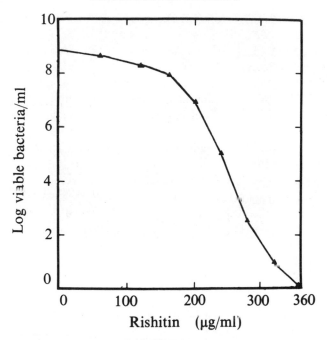

Figure 7.12 The effect of rishitin on the viability of *Erwinia atroseptica*. Cells were incubated in 0.1% peptone water containing rishitin for 4 h at 25°C. The log viable bacteria/ml in the initial inoculum was 8.32.
Adapted and redrawn from Lyon and Bayliss (1975).

medium widely used to maintain bacterial viability without allowing a significant increase in viable counts (Lyon and Bayliss, 1975; Figure 7.12). The effectiveness of related compounds, however, varies considerably. Whereas rishitin was bactericidal, phytuberin and spirovetiva-1 (10),11-dien-2-one were not, even when much higher concentrations were employed (up to 800 μg/ml).

It is difficult to generalize from results obtained with individual phytoalexins in various assays against different species or isolates of bacteria. Even the effect of a single phytoalexin on the same bacterium is greatly affected by the test conditions (Lyon and Bayliss, 1975). Consequently, substantial variations in estimates of phytoalexin activity by different workers arise. For instance, in contrast to previously published reports (Gnanamanickam and Patil, 1977; Keen and Kennedy, 1974; Lyon and Wood, 1975), Wyman and VanEtten (1978) found that coumestrol lacked significant antibacterial activity in semi-solid and liquid media.

As with the fungi, estimates *in vitro* of phytoalexin toxicity towards bacteria are affected by many procedural variables, including the type of assay employed, the composition of the medium, inoculum density and incubation period. Notwithstanding such difficulties, Wyman and VanEtten (1978) believe that real

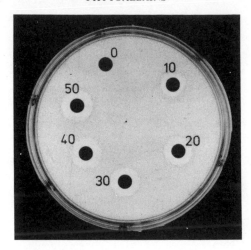

Figure 7.13 Antibacterial activity of a phytoalexin. Inhibition of growth of *Corynebacterium fascians* by kievitone, 10 to 50 μg/disc. The antibiotic assay disc (0.6 cm diameter) was loaded with prescribed amounts of the phytoalexin in ethanol; the control disc received only ethanol. Sufficient time was allowed for the ethanol to evaporate before transferring discs to the bacteria-seeded agar surface. The plate was incubated at 30°C; inhibition zones were apparent the next day.
    See Gnanamanickam and Smith (1980).

Table 7.2 Selective toxicity of phytoalexins to Gram-positive bacteria. Adapted from Gnanamanickam and Mansfield (1981).

	Area of inhibition (mm²)[a]			
Gram reaction and bacterium	Kievitone	Phaseollin	7-Hydroxyflavan	Wyerone
Gram-negative				
Erwinia carotovora atroseptica	0[b]	0	0	0
E. carotovora carotovora	0	0	0	0
Pseudomonas phaseolicola	0	0	0	0
P. syringae	0	0	0	0
Vibrio anguillarum	0	0	0	0
Xanthomonas phaseoli	0	0	0	0
Gram-positive				
Bacillus megaterium	137	50	29	139
Corynebacterium betae	302	126	270	126
C. fascians	270	89	610	172
Micrococcus lysodeikticus	149	46	22	110
Mycobacterium phlei	105	25	32	90
Streptomyces scabies	105	67	35	67

[a] Antibiotic assay discs were impregnated with 50 μg of the appropriate phytoalexin and placed on the surface of soft agar plates seeded with the test bacterium. Inhibition zones (area of total inhibition − area of disc) were measured after incubation at 28°C for 24 h. Control discs caused no inhibition.
    [b] Each value is the integer mean of two experiments.

differences in the antibacterial activities of individual compounds and the intrinsic sensitivities of bacterial isolates can be assigned.

Differential sensitivity to phytoalexins amongst bacteria had previously been recorded (Cruickshank, 1962; Cruickshank and Perrin, 1971) and a suggested association with virulence had also been made (Gnanamanickam and Patil, 1977). Overall, however, little biological pattern seemed evident from the results, until a recent suggestion that isoflavonoid, flavonoid and furanoacetylenic phytoalexins are selectively toxic to Gram-positive bacteria (Gnanamanickam and Mansfield, 1981; Gnanamanickam and Smith, 1980; Table 7.2; Figure 7.13). Rishitin exhibited similar selective action, although its activity was weak (Gnanamanickam and Mansfield, 1981). These proposals were based on preliminary findings only and require additional research for confirmation. However, Gram-negative bacteria are known to be less sensitive to antibiotics in general than their Gram-positive counterparts (Glasby, 1976), placing the results with phytoalexins in broad agreement with other bacteriological data.

Vegetative cells of *Bacillus cereus* were more sensitive to rishitin than were the spores (Lyon and Bayliss, 1975). This observation has little direct bearing on plant disease, since no typical phytopathogenic bacteria produce spores, but is consistent with the well-established resistant nature of bacterial spores.

Just as bacteria may be differentially sensitive to a single phytoalexin, structurally-related phytoalexins exhibit differential antibacterial activity. Pankhurst and Biggs (1980) tested eleven isoflavonoids in both soft agar and liquid media for inhibitory activity against eight *Rhizobium* strains. Medicarpin and kievitone were most toxic, being bactericidal towards *R. japonicum* and *R. lupini* at 100 μg/ml. Phaseollin and maackiain were moderately inhibitory whereas pisatin, coumestrol, biochanin A, formononetin, genistein, rotenone and vestitol had little activity. These findings were in general agreement with a separate survey of isoflavonoid antibacterial activity conducted by Wyman and VanEtten (1978).

Gnanamanickam and Patil (1977) noted that *Pseudomonas phaseolicola* isolate HB-36 formed atypical colonies on tetrazolium chloride agar when incubated with kievitone (42 to 66 μg/ml) and suggested that the phytoalexin was inducing a mutation in the bacterium. However, attempts to confirm this solitary indication of mutagenic activity in a phytoalexin were unsuccessful (Gnanamanickam and Smith, unpublished results).

Insufficient evidence is available to define the mode(s) of action of antibacterial phytoalexins. Lyon (1978), however, believes that rishitin may act directly on the cell membrane. This phytoalexin decreased oxygen uptake by *Erwinia carotovora* var. *atroseptica* and caused an increase in conductivity of the suspending medium. Both effects are consistent with damage to the plasmalemma, since the sites of bacterial respiration are thought to occur at invaginations of this membrane. In general, however, much more research is needed; ultrastructural studies, as well as more thorough physiological and biochemical investigations, should in time yield the essential information.

Phytotoxicity

The focus of toxicological studies with phytoalexins has centred on their antimicrobial activities. The rationale was that these inhibitory chemicals might help to protect the plant by curbing the development of invading micro-organisms. More recently, many phytoalexins, including some furanoacetylenes, isoflavonoids and terpenoids, have been shown to be phytotoxic.

Rishitin is phytotoxic—germination of pollen from three *Solanum* species was inhibited by rishitin concentrations occurring in potato tissues infected with incompatible isolates of *P. infestans* (Hodgkin and Lyon, 1979). One hundred μg rishitin/ml completely inhibited germination; 50 μg/ml inhibited pollen tube development. Pollen tubes which did develop were often thicker-walled than those of controls and were frequently contorted, with constrictions and swollen tips which eventually burst. Cell death and darkening of potato tuber slices to which rishitin was applied (Ishiguri *et al.*, 1978) and rishitin-incited lysis of potato and tomato protoplasts (Lyon and Mayo, 1978) have also been noted.

Lyon (1980) investigated the effect of rishitin on potato leaf tissues and concluded that its phytotoxicity reflects a primary site of action on membranes. This is in agreement with earlier work on fungi (Harris and Dennis, 1976) and bacteria (Lyon, 1978), as well as plant tissues (Ishiguri *et al.*, 1978; Lyon and Mayo, 1978). Rishitin treatment of potato leaf epidermal strips allowed rapid accumulation of Evan's blue, staining being particularly strong in nuclei. These results indicate a loss of cell viability (Lyon, 1980). Substantial lysis of isolated chloroplasts was also occasioned by exposure to this phytoalexin. Furthermore, rishitin increased the permeability of liposomes to a range of low molecular weight non-electrolytes, affecting equally negatively- and positively-charged liposomes. Since rishitin acted on liposomes prepared without cholesterol, it was probably interacting with phospholipid, a possibility raised earlier for kievitone. Lyon (1980) postulates that rishitin may increase membrane fluidity and so contribute to cell dysfunction. While the results are interesting, one drawback is the high concentration of this phytoalexin (300 μg/ml) used in most of the bioassays.

Apart from the possibility of lytic consequences, phytotoxic effects of phyto-alexins are evidenced in several other ways. Phaseollin caused rapid inhibition of respiration, a reduced growth of suspension cultures, and cell death (Skipp *et al.*, 1977). Affected cells of both bean and tobacco appeared granular. After continued incubation, phaseollin-treated bean cell suspension cultures, however, eventually achieved substantial growth. Two factors may be involved in this long-term recovery; escape of individual cells from the phytotoxic effects of phaseollin, and conversion of the phytoalexin to non-toxic products. Cell death and initial reduced growth of suspension cultures were noted by Glazener and VanEtten (1978) after phaseollin treatment of *Phaseolus aureus* and *P. vulgaris*. Although both species seemed able to metabolize the phytoalexin, enhanced tolerance of *P. vulgaris* was not evident upon treatment with low doses prior to higher concentrations; this is unlike the situation with some fungi (Van den Heuvel and VanEtten, 1973; see chapter 6).

These separate, but comparable, studies of phaseollin phytotoxicity (Glazener and VanEtten, 1978; Skipp *et al.*, 1977) confirm that cell suspension cultures of *P. vulgaris*, a legume which produces abundant phaseollin, are sensitive to this phytoalexin—indeed, as sensitive as are several fungi. Sensitivity to exogenously applied phaseollin may not, however, accurately reflect vulnerability to the compound when it is produced *in vivo*, where compartmentalization or other factors may be influential (Glazener and VanEtten, 1978). This point seems especially relevant in light of the production of phaseollin in "healthy" suspension cultures of *P. vulgaris* (Dixon and Fuller, 1977).

Pisatin was also inhibitory to the growth of pea callus cultures (Bailey, 1970) and retarded primary root growth in wheat (Cruickshank and Perrin, 1961). Trifolirhizin repressed root and hypocotyl growth, as well as seed germination in several clover species (Chang *et al.*, 1969). Glyceollin had an inhibitory effect on mitochondria isolated from soybean and beet (Kaplan *et al.*, 1980b). This soybean phytoalexin did not function as an uncoupler of oxidative phosphorylation but rather inhibited the electron transport system at some point after the succinate dehydrogenase site.

Leakage of betacyanin and/or electrolytes has been reported after exposure of beetroot discs to kievitone (Bull, 1981) and phaseollin, but not pisatin or medicarpin (Hargreaves, 1980). Obvious lysis of the plasmalemma did not seem responsible for the efflux of the red pigment, although lysis of pea protoplasts by pisatin has been reported (Shiraishi *et al.*, 1975). This differential lytic effect of individual isoflavonoids remains to be explained.

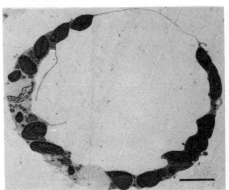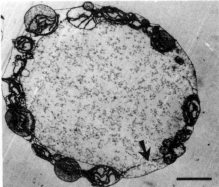

Figure 7.14 Ultrastructural observation of phytoalexin phytotoxicity. (A) Electron micrograph of an untreated protoplast prepared from beet leaves. The bar represents 5 μm.

(B) An equivalent beet leaf protoplast ten minutes after exposure to phaseollin (32 μg/ml). Note the electron dense material in the vacuole, rupture of the tonoplast (arrowed) as well as the swollen and distorted chloroplasts. The bar represents 5 μm.

Courtesy of Dr. J. A. Hargreaves, Long Ashton Research Station, Bristol, U.K. See Hargreaves (1980).

Notwithstanding the inevitable variability introduced because of procedural differences employed, phytotoxic effects of phytoalexins seem to reflect the actions of general toxicants. Substantial uptake of phaseollin by bean cells (Skipp *et al.*, 1977) is indicative of indiscriminate binding, though not necessarily of a multi-site inhibitor. Several different effects are noted; lysis, impaired oxygen uptake, leakage and growth inhibition. Membranes are again likely sites of action. Hargreaves (1980) suggested, however, that the tonoplast may be primarily affected by phaseollin and not the plasmalemma (Figure 7.14). The increased electron density apparent in the treated cell (Figure 7.14B) is comparable to that noted in *R. solani* following kievitone treatment (Figures 7.6B, 7.7B), just as the chloroplast disruption broadly parallels mitochondrial disorganization in the fungus.

More information is needed concerning the toxic effects of phytoalexins to plant tissues, particularly in assessing how such activity might influence plant-pathogen interactions. Phytoalexin accumulation within infected plants is often associated with a limited lesion or hypersensitive response, the invading organism being restricted within the necrotic tissue. It has been suggested that the death of plant cells during this type of resistant reaction may be caused by the production and accumulation of phytoalexins (Mansfield *et al.*, 1974). However, where detailed time course studies have been carried out, as for example with *P. infestans*/potato and *Colletotrichum lindemuthianum*/French bean interactions, cell death invariably precedes the onset of phytoalexin accumulation (Sato *et al.*, 1971; Bailey *et al.*, 1980; see also chapter 8). Nevertheless, it is possible that some of the phytotoxic effects of phytoalexins may limit the development of obligate, biotrophic parasites. In pea, wilting of leaves infected with powdery mildew may be caused by pisatin accumulation (Shiraishi *et al.*, 1975). The presence of phytoalexins within virus infected leaves might also render cells unsuitable for viral replication (Bailey *et al.*, 1976*b*).

Animal toxicity

Two quite separate considerations arise with respect to animal toxicity. Firstly, there is the possibility that phytoalexin accumulation is an effective defensive component useful in protecting plants against animal pathogens, particularly insects (Jacobson, 1977; Russell *et al.*, 1978; Sutherland *et al.*, 1980), and nematodes (Kaplan *et al.*, 1980*a*; Veech, 1978*a*). Secondly, there may be toxic consequences to animals, including humans, consuming plant tissues containing phytoalexins (Kuć and Currier, 1976). Instances of acute poisoning occurring after ingestion of diseased plant tissues are of historical significance, though the toxic chemicals are usually of microbial, not plant, origin (Uraguchi and Yamazaki, 1978). Nevertheless, potentially serious problems might arise if new resistant crop varieties are selected on the basis of the ability to accumulate toxic chemicals when stressed.

Despite a comparatively modest literature concerning phytoalexin toxicity in

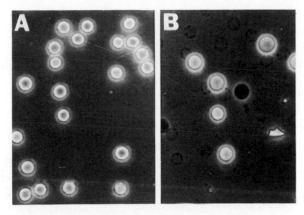

Figure 7.15 Observation of phytoalexin toxicity to animal cells by phase contrast microscopy. (A) Phase contrast photomicrograph of human red blood cells exposed to ethanol (0.5%) for twenty minutes. Note the bright, refractive appearance. (× 630).

(B) Phase contrast photomicrograph of human red blood cells ten minutes after treatment with phaseollin (45 μg/ml) in 0.5% ethanol). Note that many of the cells are no longer bright and refractive, and now appear as "ghosts" with small dark inclusions. (× 630).

Courtesy of C. A. Bull, Plant Biology Department, University of Hull, U.K. See Bull (1981).

animal systems, several different effects have been noted. One of the most striking concerns the observations that several isoflavonoid phytoalexins— including glyceollin, kievitone, medicarpin, pisatin and phaseollin (Bull, 1981; Oku *et al.*, 1976; VanEtten, 1972; VanEtten and Bateman, 1971)—will lyse red blood cells. The bright, refractive appearance of human erythrocytes (Figure 7.15A) changed considerably after a ten minute exposure to phaseollin (45 μg/ml) (Figure 7.15B); haemoglobin was lost and the individual cells exhibited something of the echinocytic change noted using pisatin (Oku *et al.*, 1976). Haemolysis is often more rapid, being almost complete in ovine erythrocytes five minutes after treatment with 30 μg phaseollin/ml (VanEtten and Bateman, 1971).

Apart from lysis of red blood cells, reports exist citing numerous and, apparently, miscellaneous, toxic effects of phytoalexins. Ipomeamarone (Uritani and Oshima, 1965) and pisatin (Oku *et al.*, 1976) repressed respiration in isolated rat liver mitochondria; ipomeamarone inhibited electron transport and oxidative phosphorylation, whereas pisatin served as an uncoupler.

Nor is it only isolated systems that are affected; whole animals are also vulnerable. Ipomeamarone derivatives, 4-ipomeanol, 1-ipomeanol, ipomeamine and 1,4-ipomeaidol, produced pulmonary oedema, congestion, and even death, in mice (Boyd *et al.*, 1972; Kuć and Currier, 1976). Liver and kidney damage may also arise. Concern is justified by the occurrence of these compounds, which are not destroyed by heat, in slightly blemished roots of sweet potato available for human consumption. The most serious danger, both in terms of the amounts occurring in plant tissues and its toxicity, seems to be associated with 4-ipomeanol.

Xanthotoxin (8-methoxypsoralen) has been cited as a phytoalexin in parsnip root (Johnson *et al.*, 1973). This compound also occurs in diseased celery tissues along with a related furanocoumarin, 4,5′,8-trimethylpsoralen (Scheel *et al.*, 1963). Both chemicals gave rise to blistering cutaneous disorders in humans and rabbits when applied to skin in the presence of sunlight or ultraviolet irradiation and this can represent a hazard to field workers harvesting celery. Coumestrol (Lyon and Wood, 1975), occurs in several legumes and often accumulates after infection (Loper *et al.*, 1967). This compound possesses oestrogenic activity in mouse uterine bioassays and may lead to infertility or other reproductive problems in sheep grazing on alfalfa or certain clovers. Several structurally-related coumestans, flavones and isoflavones—though not all strictly phyto-alexins—may present similar problems (Bickoff *et al.*, 1967; VanEtten and Pueppke, 1976). Consideration of mammalian reproductive abnormalities arising as a result of phytoalexin accumulation in forage legumes merits more intensive study.

Vestitol has been reported to be an active feeding deterrent for *Costelytra zealandica*, the subterranean larva of which is a serious agricultural pest in New Zealand (Russell *et al.*, 1978). A more extensive survey of isoflavonoids as insect feeding deterrents (Sutherland *et al.*, 1980) revealed several with activity. It is interesting to note that those compounds found to be effective, and those which were not, correspond closely to their respective activities against bacteria (Pankhurst and Biggs, 1980; Wyman and VanEtten, 1978). The possibility that plants may utilize the same chemicals to deter insects as well as inhibit micro-organisms merits closer attention (Sutherland *et al.*, 1980).

Low concentrations ($<22\,\mu g/ml$) of glyceollin isomers adversely affected the motility of nematode larvae of *Meloidogyne incognita*; the effect appeared nematistatic, not nematicidal (Kaplan *et al.*, 1980*a,b*). Differential sensitivity broadly comparable to that found with fungi was evidenced by the lack of effect of glyceollin on *M. javanica*. Oxygen uptake by the sensitive nematode was inhibited when phytoalexin was added to larval suspensions. Veech (1978*b*) has noted the toxicity of terpenoid aldehydes, including gossypol, to nematodes. Phaseollin, phaseollidin and phaseollinisoflavan killed water snails and brine shrimps (Bailey and Skipp, 1978; VanEtten and Pueppke, 1976). Insecticidal and piscicidal activities for structurally-related isoflavonoids have been reported (Fukami and Nakajima, 1971; Venkataraman, 1962).

Diverse toxicity to animals is also well represented by gossypol and its derivatives (Kuć and Lisker, 1978; see also chapter 3). Gossypol was toxic to both cotton bollworm and tobacco budworm larvae when introduced in their diet (Lukefahr and Martin, 1966). Considerable mortality occurred, less than 30 % of larvae reaching the pupal stage when the diet contained 0.2 % gossypol. Many non-ruminants, including rats, cats, dogs, rabbits, poultry and swine, may suffer chronic toxicity upon ingestion of gossypol or several closely related pigments in cottonseed meals (Berardi and Goldblatt, 1980). Pathological symptoms varied depending upon the animal species, though depressed appetite and loss of body weight were common. Other effects noted included diarrhoea, cardiac irregularity, lowering of haemoglobin and blood cell count, haemorrhage and congestion and oedema of many organs; death sometimes resulted. Ruminants were not adversely affected by gossypol, though young calves, in which the rumen is not fully functional, were vulnerable. Utilization of cotton-seed meal in less developed countries to alleviate protein shortages in human diets will depend upon clarifying the toxicological aspects, though present findings do not suggest serious problems. An additional consideration might, however, be its effect on fertility. Gossypol has been shown to function as a male contraceptive, apparently blocking sperm formation (Berardi and Goldblatt, 1980).

Despite the considerable volume of research on gossypol, a mechanism by which it might cause tissue damage in animals has not yet been established. The accumulation of fluid in body cavities would be consistent with membrane dysfunction, affecting permeability (Berardi and Goldblatt, 1980). The haemolysis noted above for several of the isoflavonoids would also be consistent with impaired membrane function. Likewise, the action of pisatin as a respiratory uncoupler in rat liver mitochondria might be mediated by the disruption of mitochondrial membranes. All in all, however, there seems insufficient data to propose a definitive site of action in animal cells.

One other point should be made—toxic effects of phytoalexins in animal systems occur at concentrations similar to those employed in fungitoxic assays (VanEtten and Pueppke, 1976). Animal cells are neither exceptionally sensitive nor particularly resistant, and the affinities and sensitivities of sites of action in animals and fungi are, therefore, likely to be similar.

Summary

Much has been learned about the toxicity of phytoalexins. Generalizations about the present state of knowledge—and its limitations—are summarized below.

(i) *Phytoalexins are biocides.* The notion of phytoalexins as exclusively antifungal chemicals is no longer accepted, nor are they thought to be selectively fungitoxic, for the amounts required to express antifungal activity are comparable to those used to inhibit other biological classes. Bacteria, higher plants and animals are all susceptible to the action of individual phytoalexins. Cell death as well as non-lethal inhibitions occur, indicating the biocidal as well as biostatic capabilities of phytoalexins.

(ii) *Organisms express differential sensitivity.* Despite the widespread biological activity of phytoalexins, individual organisms may vary greatly in their sensitivity to a given phytoalexin. Broad distinctions may sometimes be possible, as with the apparent vulnerability of Gram-positive bacteria (Gnanamanickam and Mansfield, 1981; Gnanamanickam and Smith, 1980). Closely-related species may also show quite different responses; for instance species of the fungal genus *Fusarium* (VanEtten and Stein, 1978) and of the nematode genus *Meloidogyne* (Kaplan *et al.*, 1980*b*). Not enough is known concerning tolerance mechanisms governing differential sensitivity; variability between organisms may depend on a number of factors, such as permeability characteristics, detoxification, differential affinities at sites of action or compensatory pathways.

(*iii*) *Phytoalexins are multi-site toxicants.*   Such a conclusion is perhaps based upon a too liberal generalization among several chemical classes. The very wide range of organisms detrimentally affected by phytoalexins is certainly suggestive of non-specific toxicity. The diverse physiological and cytological changes evident soon after treatment, the comparatively high concentrations needed for inhibition, indiscriminate binding and the occurrence of detoxification as a common means of tolerance likewise favour multi-site toxicants. This does not however constitute proof, as even a compound capable of binding to many cellular sites might only exert toxicity through one specific action. Considering the great diversity of phytoalexin structures, a single mode of action is unlikely. Competitive inhibition experiments utilizing phytoalexins from different chemical classes might throw some interesting light on this aspect.

(*iv*) *Membranes appear to be important sites of action.*   Phytoalexins are capable of causing membrane dysfunction, and sometimes physical rupture, in quite different organisms, those with and without cell walls. Damage to the cell membrane is consistent with the leakage, substantial loss of dry weight and lysis recorded. Since the plasmalemma would be the first membrane system encountered by an exogenous chemical, plasmalemma dysfunction may reflect a primary mode of action of phytoalexins.

Not all workers would agree with such a conclusion. Kaplan *et al.* (1980*b*) believe that glyceollin may have a specific intracellular site of action in electron transport and Hargreaves (1980) feels that mis-functioning of the tonoplast, rather than the plasmalemma, is important to phaseollin's phytotoxicity. However, multi-site toxicity, as proposed for phytoalexins, would presuppose interference with many intracellular systems. The possibility of defining primary, sensitive sites may be difficult. Certainly, more work needs to be undertaken with minimum inhibitory concentrations of phytoalexins, and records made of the earliest physiological and cytological changes. The considerable amounts of pure compounds needed to pursue mode of action studies remains a serious limitation.

The possibility of functional specificity also merits consideration. Many phytoalexins may inhibit a particular cellular function, perhaps the ability to control the passage of solutes across a membrane, but not just at one site, such as the tonoplast. Rather, all cellular membrane systems would be vulnerable, leading to widespread cellular dysfunction.

(v) *A toxic barrier* in vivo? This is a strong possibility, though far from established fact. Refinements in bioassay systems will be required to establish the toxic potentials of phytoalexins in infected plants. Information will also be needed concerning micro-site concentrations *in vivo* as well as to clarify whether contact between the phytoalexin and microorganism actually occurs. Research to elucidate the relevance of phytoalexin accumulation in plant tissues will need to pursue several avenues, but few more important than accurate determinations of *in vivo* toxicity.

Acknowledgements

Journal paper no. 81-11-26 of the Kentucky Agricultural Experiment Station, Lexington, Kentucky 40546.

The author is indebted to Christopher A. Bull, David E. Matthews, Malcolm R. Siegel and Hans D. VanEtten, all of whom read the manuscript in draft form and provided many valuable suggestions for its improvement. Any errors remaining are the responsibility of the author. Thanks are due to those who provided photographs for this chapter, to Judith M. Harrer and Raymond L. Mernaugh who helped prepare the illustrations and to Margaret Vest for preparation of the typescript.

REFERENCES

Albersheim, P. and Valent, B. S. (1978) Host-pathogen interactions in plants. *J. Cell Biol.*, **78**, 627–643.

Bailey, J. A. (1970) Pisatin production by tissue cultures of *Pisum sativum* L. *J. Gen. Microbiol.*, **61**, 409–415.

Bailey, J. A. and Burden, R. S. (1973) Biochemical changes and phytoalexin accumulation in *Phaseolus vulgaris* following cellular browning caused by tobacco necrosis virus. *Physiol. Plant Pathol.*, **3**, 171–177.

Bailey, J. A., Carter, G. A. and Skipp, R. A. (1976a) The use and interpretation of bioassays for fungitoxicity of phytoalexins in agar media. *Physiol. Plant Pathol.*, **8**, 189–194.

Bailey, J. A. and Deverall, B. J. (1971) Formation and activity of phaseollin in the interaction between bean hypocotyls (*Phaseolus vulgaris*) and physiological races of *Colletotrichum lindemuthianum*. *Physiol. Plant Pathol.*, **1**, 435–449.

Bailey, J. A. and Rowell, P. M. (1980) Viability of *Colletotrichum lindemuthianum* in hypersensitive cells of *Phaseolus vulgaris*. *Physiol. Plant Pathol.*, **17**, 341–345.

Bailey, J. A., Rowell, P. M. and Arnold, G. M. (1980) The temporal relationship between host cell death, phytoalexin accumulation and fungal inhibition during hypersensitive reactions of *Phaseolus vulgaris* to *Colletotrichum lindemuthianum*. *Physiol. Plant Pathol.*, **17**, 329–339.

Bailey, J. A. and Skipp, R. A. (1978) Toxicity of phytoalexins. *Ann. appl. Biol.*, **89**, 354–358.

Bailey, J. A., Vincent, G. G. and Burden, R. S. (1976b) The antifungal activity of glutinosone and capsidiol and their accumulation in virus-infected tobacco species. *Physiol. Plant Pathol.*, **8**, 35–41.

Bartnicki-Garcia, S. and Lippman, E. (1972) The bursting tendency of hyphal tips of fungi: presumptive evidence for a delicate balance between wall synthesis and wall lysis in apical growth. *J. Gen. Microbiol.*, **73**, 487–500.

Bateman, D. F. and Basham, H. G. (1976) "Degradation of plant cell walls and membranes by microbial enzymes", in *Encyclopedia of Plant Physiology* 4, *Physiological Plant Pathology*, eds. R. Heitefuss and P. H. Williams, Springer-Verlag, 316–355.

Berardi, L. C. and Goldblatt, L. A. (1980) "Gossypol", in *Toxic Constituents of Plant Foodstuffs*, ed I. E. Liener, 2nd edition, Academic Press, 183–237.

Bickoff, S. M., Loper, G. M., Hanson, C. H., Graham, J. H., Witt, S. C. and Spencer, R. R. (1967) Effect of common leafspot on coumestans and flavones in alfalfa. *Crop Sci.*, **7**, 259–261.

Bowman, B. J., Mainzer, S. E., Allen, K. E. and Slayman, C. W. (1978) Effects of inhibitors on the plasma membrane and mitochondrial adenosine triphosphatases of *Neurospora crassa*. *Biochim. Biophys. Acta*, **512**, 13–28.

Boyd, M. R., Burka, L. T., Harris, T. M. and Wilson, B. J. (1973) Lung-toxic furanoterpenoids produced by sweet potatoes (*Ipomoea batatas*) following microbial infection. *Biochim. Biophys. Acta*, **337**, 184–195.

Bull, C. A. (1981) Studies on the fungitoxicity, and relevance to disease resistance, of the phytoalexin, kievitone. Ph.D. thesis, University of Hull, U.K.

Bull, C. A. and Smith, D. A. (1981) Pectic enzyme inhibition by the phytoalexin, kievitone. *Phytopathology*, **71**, 206 (abstr.).

Carter, G. A., Chamberlain, K. and Wain, R. L. (1978) Investigations on fungicides. XX. The fungitoxicity of analogues of the phytoalexin 2-(2'-methoxy-4'-hydroxyphenyl)-6-methoxy-benzofuran (vignafuran). *Ann. appl. Biol.*, **88**, 57–64.

Chang, C., Suzuki, A., Kumai, S. and Tamura, S. (1969) Chemical studies on "clover sickness". Part II. Biological functions of isoflavonoids and their related compounds. *Agr. Biol. Chem.*, **33**, 398–408.

Christou, T. (1962) Penetration and host-parasite relationships of *Rhizoctonia solani* in the bean plant. *Phytopathology*, **52**, 381–389.

Cruickshank, I. A. M. (1962) Studies on phytoalexins. IV. The antimicrobial spectrum of pisatin. *Aust. J. Biol. Sci.*, **15**, 147–159.

Cruickshank, I. A. M. and Perrin, D. R. (1961) Studies on phytoalexins. III. The isolation, assay, and general properties of a phytoalexin from *Pisum sativum* L. *Aust. J. Biol. Sci.*, **14**, 336–348.

Cruickshank, I. A. M. and Perrin, D. R. (1971) Studies on phytoalexins. XI. The induction, antimicrobial spectrum and chemical assay of phaseollin. *Phytopath Z.*, **70**, 209–229.

Daly, J. M. (1972) The use of near-isogenic lines in biochemical studies of the resistance of wheat to stem rust. *Phytopathology*, **62**, 392–400.

Dekker, J. (1977a) "Resistance", in *Systemic Fungicides*, ed. R. W. Marsh, 2nd edition, Longman, 176–197.

Dekker, J. (1977b) "Tolerance and the mode of action of fungicides", in *Proceedings of the British Crop Protection Conference*, British Crop Protection Council, 689–697.

Delp, C. J. (1980) Coping with resistance to plant disease. *Pl. Disease*, **64**, 652–657.

Deverall, B. J. (1969) Biochemical changes in infection droplets containing spores of *Botrytis* spp. incubated in the seed cavities of pods of bean (*Vicia faba* L.). *Ann. appl. Biol.*, **59**, 375–387.

Deverall, B. J. and Rogers, P. M. (1972) The effect of pH and composition of test solutions on the inhibitory activity of wyerone acid towards germination of fungal spores. *Ann. appl. Biol.*, **72**, 301–305.

Dixon, R. A. and Fuller, K. W. (1977) Characterization of components from culture filtrates of *Botrytis cinerea* which stimulate phaseollin biosynthesis in *Phaseolus vulgaris* cell suspension cultures. *Physiol. Plant Pathol.*, **11**, 287–296.

Duzcek, L. J. and Higgins, V. J. (1976) Effect of treatment with the phytoalexins medicarpin and maackiain on fungal growth *in vitro* and *in vivo*. *Can. J. Bot.*, **54**, 2620–2629.

Fuchs, A., de Vries, F. W. and Platero Sanz, M. (1980) The mechanism of pisatin degradation by *Fusarium oxysporum* f. sp. *pisi*. *Physiol. Plant Pathol.*, **16**, 119–133.

Fukami, H. and Nakajima, M. (1971) "Rotenone and the rotenoids", in *Naturally Occurring Insecticides*, eds. M. Jacobsen and D. G. Crosby, Marcel Dekker, 71–97.

Glasby, J. S. (1976) *Encyclopaedia of Antibiotics*, John Wiley, 372.

Glazener, J. A. and VanEtten, H. D. (1978) Phytotoxicity of phaseollin to, and alteration of phaseollin by, cell suspension cultures of *Phaseolus vulgaris*. *Phytopathology*, **68**, 111–117.

Gnanamanickam, S. S. and Mansfield, J. W. (1981) Selective toxicity of wyerone and other phytoalexins to Gram-positive bacteria. *Phytochemistry*, **20**, 997–1000.

Gnanamanickam, S. S. and Patil, S. S. (1977) Accumulation of antibacterial isoflavonoids in hypersensitively responding bean leaf tissues inoculated with *Pseudomonas phaseolicola*. *Physiol. Plant Pathol.*, **10**, 159–168.

Gnanamanickam, S. S. and Smith, D. A. (1980) Selective toxicity of isoflavonoid phytoalexins to Gram-positive bacteria. *Phytopathology*, **70**, 894–896.

Gordon, M. A., Lapa, E. W., Fitter, M. S. and Lindsay, M. (1980) Susceptibility of zoopathogenic fungi to phytoalexins. *Antimicrob. Ag. Chemotherapy*, **17**, 120–123.

Grisebach, H. and Ebel, J. (1978) Phytoalexins, chemical defense substances of higher plants. *Angew. Chem. Int. Ed. Engl.*, **17**, 635–647.

Harborne, J. B. and Ingham, J. L. (1978) "Biochemical aspects of the coevolution of higher plants with their fungal parasites", in *Biochemical Aspects of Plant and Animal Coevolution*, ed. J. B. Harborne, Academic Press, 343–405.

Hargreaves, J. A. (1980) A possible mechanism for the phytotoxicity of the phytoalexin phaseollin. *Physiol. Plant Pathol.*, **16**, 351–347.

Hargreaves, J. A., Mansfield, J. W. and Rossall, S. (1977) Changes in phytoalexin concentrations in tissues of the broad bean plant (*Vicia faba* L.) following inoculation with species of *Botrytis*. *Physiol. Plant Pathol.*, **11**, 227–242.

Harris, J. E. and Dennis, C. (1976) Antifungal activity of post-infectional metabolites from potato tubers. *Physiol. Plant Pathol.*, **9**, 155–165.

Harris, J. E. and Dennis, C. (1977) The effect of post-infectional potato tuber metabolites and surfactants on zoospores of Oomycetes. *Physiol. Plant Pathol.*, **11**, 163–169.

Hart, J. H. and Shrimpton, D. M. (1979) Role of stilbenes in resistance of wood to decay. *Phytopathology*, **69**, 1138–1143.

Higgins, V. J. (1969) Comparative abilities of *Stemphylium botryosum* and other fungi to reduce and degrade a phytoalexin from alfalfa. Ph.D. thesis, Cornell University, Ithaca, N.Y., U.S.A.

Higgins, V. J. (1972) Role of the phytoalexin medicarpin in three leaf spot diseases of alfalfa. *Physiol. Plant Pathol.*, **2**, 289–300.

Higgins, V. J. (1978) The effect of some pterocarpanoid phytoalexins on germ tube elongation of *Stemphylium botryosum*. *Phytopathology*, **68**, 338–345.

Higgins, V. J. and Millar, R. L. (1968) Phytoalexin production by alfalfa in response to infection by *Colletotrichum phomoides*, *Helminthosporium turcicum*, *Stemphylium loti* and *S. botryosum*. *Phytopathology*, **58**, 1377–1383.

Hodgkin, T. and Lyon, G. D. (1979) Inhibition of *Solanum* pollen germination *in vitro* by the phytoalexin rishitin. *Ann. Bot.*, **44**, 253–255.

Hunter, R. E. (1974) Inactivation of pectic enzymes by polyphenols in cotton seedlings of different ages infected with *Rhizoctonia solani*, *Physiol. Plant Pathol.*, **4**, 151–159.

Hutson, R. A. and Mansfield, J. W. (1980) A genetical approach to the analysis of mechanisms of pathogenicity in *Botrytis/Vicia faba* interactions. *Physiol. Plant Pathol.*, **17**, 309–317.

Ingham, J. L. (1976a) Induced isoflavonoids from fungus-infected stems of pigeon pea (*Cajanus cajan*). *Z. Naturforsch.*, **31C**, 504–508.

Ingham, J. L. (1976b) Induced and constitutive isoflavonoids from stems of chickpeas (*Cicer arietinum* L.) inoculated with spores of *Helminthosporium carbonum* Ullstrup. *Phytopath. Z.*, **87**, 353–367.

Ingham, J. L., Keen, N. T. and Hymowitz, T. (1977) A new isoflavone phytoalexin from fungus-inoculated stems of *Glycine wightii*. *Phytochemistry*, **16**, 1943–1946.

Ishiguri, Y., Tomiyama, K., Murai, A., Katsui, N. and Masamune, T. (1978) Toxicity of rishitin, rishitin-M-1 and rishitin-M-2 to *Phytophthora infestans* and potato tissue. *Ann. Phytopath. Soc. Japan*, **44**, 52–56.

Jacobson, M. (1977) "Isolation and identification of toxic agents from plants", in *Host Plant Resistance to Pests*, ed. P. A. Hedin, American Chemical Society, 153–164.

Johnson, C., Brannon, D. R. and Kuć, J. (1973) Xanthotoxin: a phytoalexin of *Pastinaca sativa* root. *Phytochemistry*, **12**, 2961–2962.

Johnson, G., Maag, D. D., Johnson, D. K. and Thomas, R. D. (1976) The possible role of phytoalexins in the resistance of sugarbeet (*Beta vulgaris*) to *Cercospora beticola*. *Physiol. Plant Pathol.*, **8**, 225–230.

Kaplan, D. T., Keen, N. T. and Thomason, I. J. (1980a) Association of glyceollin with the incompatible response of soybean roots to *Meloidogyne incognita*. *Physiol. Plant Pathol.*, **16**, 309–318.

Kaplan, D. T., Keen, N. T. and Thomason, I. J. (1980b) Studies on the mode of action of glyceollin in soybean. Incompatibility to the root knot nematode, *Meloidogyne incognita*. *Physiol. Plant Pathol.*, **16**, 319–325.

Keen, N. T. and Kennedy, B. W. (1974) Hydroxyphaseollin and related isoflavonoids in the hypersensitive resistance reaction of soybeans to *Pseudomonas glycinea*. *Physiol. Plant Pathol.*, **4**, 173–185.

Keen, N. T., Sims, J. J., Erwin, D. C., Rice, E. and Partridge, J. E. (1971) 6a-Hydroxyphaseollin: an antifungal chemical induced in soybean hypocotyls by *Phytophthora megasperma* var. *sojae*. *Phytopathology*, **61**, 1084–1089.

Kenning, L. A. and Hanchey, P. (1980) Ultrastructure of lesion formation in *Rhizoctonia*-infected bean hypocotyls. *Phytopathology*, **70**, 998–1004.

Klarman, W. L. and Sanford, J. B. (1968) Isolation and purification of an antifungal principle from infected soybeans. *Life Sci.*, **7**, 1095–1103.

Kojima, M., Takeuchi, A. and Uritani, I. (1979) "Differential growth response of various fungal strains to divalent cations and phytoalexins", in *Recognition and Specificity in Plant Host-Parasite Interactions*, eds. J. M. Daly and I. Uritani, University Park Press, Baltimore, 335–349.

Kuć, J. and Currier, W. (1976) "Phytoalexins, plants and human health", in *Mycotoxins and Other Fungal Related Food Problems*, ed. J. R. Rodricks, American Chemical Society, 356–368.

Kuć, J. and Lisker, N. (1978) "Terpenoids and their role in wounded and infected plant storage tissue", in *Biochemistry of Wounded Plant Tissue*, ed. G. Kahl, de Gruyter, Berlin, 203–242.

Kuhn, P. J. (1979) Studies on the metabolism of the phytoalexin, kievitone, by *Fusarium solani* f. sp. *phaseoli*. Ph.D. thesis, University of Hull, U.K.

Loman, A. A. (1970) Bioassays of fungi isolated from *Pinus contorta* var. *latifolia* with pinosylvin, pinosylvinmonomethyl ether, pinobanksin, and pinocembrin. *Can. J. Bot.*, **48**, 1303–1308.

Loomis, W. D. (1974) "Overcoming problems of phenolics and quinones in the isolation of plant enzymes and organelles", in *Methods in Enzymology*, eds. S. Fleischer and L. Packer, Academic Press, **XXXI**, 528–544.

Loper, G. M., Hanson, C. H. and Graham, J. H. (1967) Coumestrol content of alfalfa as affected by selection for resistance to foliar diseases. *Crop Sci.*, **7**, 189–192.

Lukefahr, M. M. and Martin, D. F. (1966) Cotton plant pigments as a source of resistance to the bollworm and tobacco budworm. *J. Econ. Entomol.*, **69**, 176–179.

Lukens, R. J. (1971) *Chemistry of Fungicidal Action*, Springer-Verlag, 136.

Lyon, F. M. and Wood, R. K. S. (1975) Production of phaseollin, coumestrol and related compounds in bean leaves inoculated with *Pseudomonas* spp. *Physiol. Plant Pathol.*, **6**, 117–124.

Lyon, G. D. (1978) Attenuation by divalent cations of the effect of the phytoalexin rishitin on *Erwinia carotovora* var. *atroseptica*. *J. Gen. Microbiol.*, **109**, 5–10.

Lyon, G. D. (1980) Evidence that the toxic effect of rishitin may be due to membrane damage. *J. Exp. Bot.*, **31**, 957–966.

Lyon, G. D. and Bayliss, C. E. (1975) The effect of rishitin on *Erwinia carotovora* var. *atroseptica* and other bacteria. *Physiol. Plant Pathol.*, **6**, 177–186.

Lyon, G. D. and Mayo, M. A. (1978) The phytoalexin rishitin affects the viability of isolated plant protoplasts. *Phytopath. Z.*, **92**, 294–304.

Lyr, H. (1977) "Mechanism of action of fungicides", in *Plant Disease, Vol. I*, eds. J. G. Horsfall and E. B. Cowling, Academic Press, 239–261.

Mansfield, J. W. (1980) "Mechanisms of resistance to *Botrytis*", in *The Biology of Botrytis*, eds. J. R. Coley-Smith, W. R. Jarvis and K. Verhoeff, Academic Press, 181–218.

Mansfield, J. W., Hargreaves, J. A. and Boyle, F. C. (1974) Phytoalexin production by live cells in broad bean leaves infected with *Botrytis cinerea*. *Nature*, **252**, 316–317.

McCance, D. J. and Drysdale, R. B. (1975) Production of tomatine and rishitin by tomato plants. *Physiol. Plant Pathol.*, **7**, 221–230.

Misato, T. and Kakiki, K. (1977) "Inhibition of fungal cell wall synthesis and cell membrane function", in *Antifungal Compounds, Vol. 2*, eds. M. R. Siegel and H. D. Sisler, Marcel Dekker, New York, 277–300.

Müller, K. O. (1958) Relationships between phytoalexin output and the number of infections involved. *Nature*, **182**, 167–168.

Müller, K. O. and Börger, H. (1940) Experimentelle Untersuchungen über die *Phytophthora*-Resistenz der Kartoffel. Zugleich ein Beitrag zum Problem der "Erwobenen Resistenz", in *Pflanzenreich. Arbeiten aus der biologischen Reichsamstalt für Land und Forstwirtschaft*, **23**, 189–231.

Oku, H., Ouchi, S., Shiraishi, T., Utsumi, K. and Seno, S. (1976) Toxicity of a phytoalexin, pisatin, to mammalian cells. *Proc. J. Acad.*, **52**, 33–36.

Pankhurst, C. E. and Biggs, D. R. (1980) Sensitivity of *Rhizobium* to selected isoflavonoids. *Can. J. Microbiol.*, **26**, 542–545.

Paxton, J. D. (1980) A new working definition of the term "phytoalexin". *Pl. Disease*, **64**, 734.

Perrin, D. R., Biggs, D. R. and Cruickshank, I. A. M. (1974) Phaseollidin, a phytoalexin from *Phaseolus vulgaris*: isolation, physicochemical properties and antifungal activity. *Aust. J. Chem.*, **27**, 1607–1611.

Perrin, D. R. and Cruickshank, I. A. M. (1969) The antifungal activity of pterocarpans towards *Monilinia fructicola*. *Phytochemistry*, **8**, 971–978.

Pueppke, S. G. and VanEtten, H. D. (1974) Pisatin accumulation and lesion development in peas infected with *Aphanomyces euteiches*, *Fusarium solani* f. sp. *pisi*, or *Rhizoctonia solani*. *Phytopathology*, **64**, 1433–1440.

Pueppke, S. G. and VanEtten, H. D. (1976) The relation between pisatin and the development of *Aphanomyces euteiches* in diseased *Pisum sativum*. *Phytopathology*, **66**, 1174–1185.

Ravisé, A. and Kirkiacharian, B. S. (1976) Effects of the structure of phenolic compounds on the inhibition of *Phytophthora parasitica* and on lytic enzymes. *Phytopath. Z.*, **85**, 74–85.

Ravisé, A., Kirkacharian, B. S., Chopin, J. and Kunesch, G. (1980) Composés phénoliques et analogues structuraux de phytoalexines: influence des structures et des substituants sur l'inhibition *in vitro* de micromycètes et d'enzymes lytiques, Proc. 18ᵉᵐᵉ Colloque de la Société Française de Phytopathologie, paper 3, in *Ann. Phytopathol.*, in press.

Rossall, S. and Mansfield, J. W. (1978) The activity of wyerone acid against *Botrytis*. *Ann. appl. Biol.*, **89**, 359–362.

Rossall, S., Mansfield, J. W. and Hutson, R. A. (1980) Death of *Botrytis cinerea* and *B. fabae* following exposure to wyerone derivatives *in vitro* and during infection development in broad bean leaves. *Physiol. Plant Pathol.*, **16**, 135–146.

Russell, G. B., Sutherland, O. R. W., Hutchins, R. F. N. and Christmas, P. E. (1978) Vestitol: a phytoalexin with insect feeding-deterrent activity, *J. Chem. Ecol.*, **4**, 571–579.

Sato, N., Kitasawa, K. and Tomiyama, K. (1971) The role of rishitin in localizing hyphae of *Phytophthora infestans* in infection sites at the cut surface of potato tubers. *Physiol. Plant Pathol.*, **1**, 289–295.

Scheel, L. D., Perone, V. B., Larkin, R. L. and Kupel, R. E. (1963) The isolation and characterization of two phototoxic furanocoumarins (psoralens) from diseased celery. *Biochemistry*, **2**, 1127–1131.

Shiraishi, T., Oku, H., Isono, M. and Ouchi, S. (1975) The injurious effect of pisatin on the plasma membrane of pea. *Plant and Cell Physiol.*, **16**, 939–942.

Siegel, M. R. (1975) "Biological and biochemical considerations in the use of selective and nonselective fungicides", in *Pesticide Selectivity*, ed. J. C. Street, Marcel Dekker, 21–46.

Skipp, R. A. and Bailey, J. A. (1976) The effect of phaseollin on the growth of *Colletotrichum lindemuthianum* in bioassays designed to measure fungitoxicity. *Physiol. Plant Pathol.*, **9**, 253–263.

Skipp, R. A. and Bailey, J. A. (1977) The fungitoxicity of isoflavonoid phytoalexins measured using different types of bioassay. *Physiol. Plant Pathol.*, **11**, 101–112.

Skipp, R. A. and Carter, G. A. (1978) Adaptation of fungi to isoflavonoid phytoalexins. *Ann. appl. Biol.*, **89**, 366–369.

Skipp, R. A., Selby, C. and Bailey, J. A. (1977) Toxic effects of phaseollin on plant cells. *Physiol. Plant Pathol.*, **10**, 221–227.

Slayman, C. L. and VanEtten, H. D. (1974) Are certain pterocarpanoid phytoalexins and steroid hormones inhibitors of membrane ATP-ase? *Pl. Physiol., Suppl.*, **54**, 24 (Abstr.).

Smith, D. A. (1976) Some effects of the phytoalexin, kievitone, on the vegetative growth of *Aphanomyces euteiches, Rhizoctonia solani* and *Fusarium solani* f. sp. *phaseoli*. *Physiol. Plant Pathol.*, **9**, 45–55.

Smith, D. A. (1978) Observations on the fungitoxicity of the phytoalexin, kievitone. *Phytopathology*, **58**, 81–87.

Smith, D. A. and Bull, C. A. (1978) "Kievitone—a membranolytic phytoalexin?" in *Proceedings of the Third International Congress of Plant Pathology*, Parey, Berlin, 245.

Smith, D. A. and Ingham, J. L. (1980) "Legumes, fungal pathogens, and phytoalexins", in *Advances in Legume Science*, eds. R. J. Summerfield and A. H. Bunting, Her Majesty's Stationery Office, London, 207–223.

Smith, D. A., VanEtten, H. D. and Bateman, D. F. (1973) The principal antifungal component of "substance II" isolated from *Rhizoctonia* infected bean tissues. *Physiol. Plant Pathol.*, **3**, 179–186.

Smith, D. A., VanEtten, H. D. and Bateman, D. F. (1975) Accumulation of phytoalexins in *Phaseolus vulgaris* hypocotyls following infection by *Rhizoctonia solani*. *Physiol. Plant Pathol.*, **5**, 51–64.

Sprang, M. L. and VanEtten, H. D. (1978) Effects of pisatin on *Fusarium solani* f. sp. *pisi* and f. sp. *cucurbitae*. *Phytopathology News*, **12**, 131 (Abstr.).

Sutherland, O. R. W., Russell, G. B., Biggs, D. R. and Lane, G. A. (1980) Insect feeding-deterrent activity of phytoalexin isoflavonoids. *Biochem. Syst. and Ecol.*, **8**, 73–75.

Tegtmeier, K. J. and VanEtten, H. D. (1979) Genetic analysis of sexuality, phytoalexin sensitivity and virulence of *Nectria haematococca* MP VI (*Fusarium solani*). *Phytopathology*, **69**, 1047 (Abstr.).

Tomiyama, K., Sakuma, T., Ishizaka, N., Sato, N., Katsui, N., Takasugi, M. and Masamune, T. (1968) A new antifungal substance isolated from resistant potato tuber tissues infected by pathogens. *Phytopathology*, **58**, 115–116.

Uritani, I. and Oshima, K. (1965) Effects of ipomeamarone on respiratory enzyme systems in mitochondria. *Agr. Biol. Chem.*, **29**, 641–648.

Uraguchi, K. and Yamazaki, M. (1978) *Toxicology, Biochemistry and Pathology of Mycotoxins*. John Wiley, 288.

Van den Heuvel, J. and VanEtten, H. D. (1973) Detoxification of phaseollin by *Fusarium solani* f. sp. *phaseoli*. *Physiol. Plant Pathol.*, **3**, 327–339.

VanEtten, H. D. (1972) Antifungal and hemolytic activities of four pterocarpan phytoalexins. *Phytopathology*, **62**, 795 (Abstr.).

VanEtten, H. D. (1973) Differential sensitivity of fungi to pisatin and to phaseollin. *Phytopathology*, **63**, 1477–1482.

VanEtten, H. D. (1976) Antifungal activity of pterocarpans and other selected isoflavonoids. *Phytochemistry*, **15**, 655–659.

VanEtten, H. D. and Barz, W. (1980) Some properties of the pisatin demethylase activity of *Nectria haematococca* (*Fusarium solani*) mating population. VI. *Phytopathology*, **70**, 470 (Abstr.).

VanEtten, H. D. and Bateman, D. F. (1971) Studies on the mode of action of the phytoalexin phaseollin. *Phytopathology*, **61**, 1363–1372.

VanEtten, H. D., Matthews, P. S., Tegtmeier, K. J., Dietert, M. F. and Stein, J. I. (1980) The association of pisatin tolerance and demethylation with virulence on pea in *Nectria haematococca*. *Physiol. Plant Pathol.*, **16**, 257–268.

VanEtten, H. D., Maxwell, D. P. and Bateman, D. F. (1967) Lesion maturation, fungal development, and distribution of endopolygalacturonase and cellulase in *Rhizoctonia*-infected bean hypocotyl tissue. *Phytopathology*, **57**, 121–126.

VanEtten, H. D. and Pueppke, S. G. (1976) "Isoflavonoid phytoalexins", in *Biochemical Aspects of Plant-Parasite Relationships*, eds. J. Friend and D. R. Threlfall, Academic Press, 239–289.

VanEtten, H. D. and Stein, J. I. (1978) Differential response of *Fusarium solani* isolates to pisatin and phaseollin. *Phytopathology*, **68**, 1276–1283.

Van Sumere, C. F., Albrecht, J., Dedonder, A., De Pooter, H. and Pé, I. (1975) "Plant proteins and phenolics", in *The Chemistry and Biochemistry of Plant Proteins*, eds. J. B. Harborne and C. F. Van Sumere, Academic Press, 211–264.

Vasyukova, N. I., Davydova, M. A., Shcherbakova, L. A. and Ozeretskovskaya, O. L. (1977) Phytosterols as a factor protecting the potato phytophthorosis pathogen from the action of phytoalexins. *Dok. Acad. Nauk. SSSR*, **235**, 216–219.

Veech, J. A. (1978a) An apparent relationship between methoxy-substituted terpenoid aldehydes and the resistance of cotton to *Meloidogyne incognita*. *Nematologica*, **24**, 81–87.

Veech, J. A. (1978b) The toxicity of terpenoid aldehydes to nematodes. *J. Nematol.*, **10**, 301 (Abstr.).

Venkataraman, K. (1962) "Methods for determining the structures of flavonoid compounds", in *The Chemistry of Flavonoid Compounds*, ed. T. A. Geissman, Macmillan, 70–106.

Ward, E. W. B., Unwin, C. H. and Stoessl, A. (1974) Postinfectional inhibitors from plants. XIII. Fungitoxicity of the phytoalexin, capsidiol and related sesquiterpenes. *Can. J. Bot.*, **52**, 2481–2488.

Ward, E. W. B., Unwin, C. H. and Stoessl, A. (1975) Postinfectional inhibitors from plants. XV. Antifungal activity of the phytoalexin orchinol and related phenanthrenes and stilbenes. *Can. J. Bot.*, **53**, 964–971.

Wyman, J. G. and VanEtten, H. D. (1978) Antibacterial activity of selected isoflavonoids. *Phytopathology*, **68**, 583–589.

8 The role of phytoalexins in disease resistance

J. W. MANSFIELD

To the plant pathologist the undoubted *in vitro* antimicrobial activity of phytoalexins raises the possibility that the compounds directly inhibit the growth of microorganisms in plants and thereby regulate host/parasite interactions. This chapter assesses the role of phytoalexins in the resistance of plants to microbial colonization.

Resistance to fungi

It is clear from the arguments and examples presented in reviews by Kuć (1976), Deverall (1977) and Bailey (1981) that the involvement of phytoalexins in disease resistance is not restricted to a particular interaction between genotypes, for example race-specific or non-host resistance, but is more closely associated with morphologically similar types of response. The majority of data supporting a role for phytoalexin accumulation as the cause of the cessation of fungal growth in resistant plants comes from interactions in which resistance is expressed following penetration and is associated with the necrosis of plant cells. Such a local lesion response, often described as a hypersensitive reaction, can occur during the expression of non-host, race-specific or race non-specific resistance (*sensu* Scott *et al.*, 1980). It is important to recognize, however, that resistance may be expressed at other stages of infection development and need not be associated with necrosis. The following sections summarize the evidence available concerning the contribution of phytoalexins to the restriction of fungal growth at different stages of colonization.

Inhibition on plant surfaces

Fungal spores often fail to germinate following their deposition on leaf surfaces (Blakeman, 1973; Preece, 1976). A striking example of this concerns the behaviour of saprophytes in the phyllosphere. Dickinson (1967) and Hollomon (1967), have described the increased growth of epiphytic fungi coincident with the onset of senescence. The ability to produce phytoalexins declines during senescence (Cruickshank and Perrin, 1965; Bailey, 1969) and it has been proposed that fungal growth on young leaves may be restricted by phytoalexins produced by underlying cells in response to fungal metabolites diffusing from germinating spores (Bailey, 1969; Last and Warren, 1972). However, the limited evidence available does not support this attractive hypothesis. Thus, Mansfield *et al.* (1975) found that germination of the saprophytes *Aureobasidium pullulans*, *Cladosporium herbarum* and *Epicoccum nigrum* on pea leaves did not induce formation of the phytoalexin pisatin. Results of Rossall and Mansfield (1980) also failed to implicate the accumulation of phytoalexins as the cause of the inhibition of germination and germ tube growth of *Botrytis cinerea* and *B. elliptica* on the surface of broad bean leaves. The apparent lack of influence of phytoalexins on fungal development in the phyllosphere may be explained if the cuticle acts as a barrier preventing the diffusion to underlying cells of compounds eliciting phytoalexin biosynthesis.

As parts of the root system are devoid of cuticle, the same argument cannot be applied to the growth of microorganisms around roots. Bowen (1979) stated that the root's contribution to the rhizosphere included "... a 'blur' of exudates leaking from cells, active secretions of low molecular weight compounds and of plant mucilages, lysates from autolysing epidermal cells and sloughed root cap and epidermal cells". Mucilages and autolysing epidermal cells might be expected to contain elicitors of phytoalexin biosynthesis in many plants (Ayers *et al.*, 1976; Hargreaves and Bailey, 1978; see also chapter 9). The additional contribution of elicitors from the rhizosphere microflora would also be likely to induce phytoalexin production in underlying root cells. However, the possible influence of phytoalexins on the rhizosphere microflora has not received detailed examination.

Inhibition during attempted penetration into plant cells

Resistance to fungi is frequently expressed by the failure of infection hyphae to penetrate into or through plant cell walls (Aist, 1976; Ride, 1978; Heath, 1980). Various types of deposit (papillae) have been found to accumulate within living cells beneath sites of attempted penetration. It has been suggested that papilla formation and other localized changes in cell wall structure including lignification (Vance *et al.*, 1980) and silicification (Heath, 1980) may provide purely physical barriers to the continued progress of invading hyphae. Recently, however, Toyoda *et al.* (1978) isolated a fungitoxic flavonoid (which may be considered a phytoalexin) from papillae formed in resistant barley leaves in response to *Erysiphe graminis* f. sp. *hordei*. Localization and isolation of the flavonoid was greatly facilitated by its bright yellow fluorescence under UV radiation. It is possible that other phytoalexins may also be incorporated into papillae or cell walls, thereby producing a localized, fungitoxic barrier to penetration.

Consideration of the presence of phytoalexins in papillae raises the recurrent problem of the technical difficulties involved in attempting to determine the presence and concentrations of phytoalexins at micro-sites within infected tissues. Available histochemical techniques are not sufficiently specific and the use of fluorescence microscopy (which can be applied to only a few groups of phytoalexins) may not be sufficiently sensitive. The detection of fluorescing phytoalexins *in vivo* is also subject to interference by the presence of UV-absorbing substances (Mansfield et al., 1974). The answer may lie in the use of phytoalexin-specific antibodies labelled with ferritin which may allow localization of the inhibitors at the ultrastructural level, as discussed by Knox and Clarke (1978).

Inhibition after penetration

Following penetration of resistant plants, fungal growth may be restricted at one of a number of different sites: (a) within the partially degraded walls of epidermal cells (for example, *Botrytis* spp. in non-host plants); (b) intracellularly, either within the epidermis (*Colletotrichum* spp. in non-host plants or resistant cultivars) or in mesophyll cells (restricted development of haustoria of rust fungi); (c) in intercellular spaces (*Cladosporium fulvum* in resistant tomato leaves) and (d) within xylem vessels (*Verticillium* and *Fusarium* spp. in wilt resistant plants). In order to prove whether or not inhibition of hyphal growth at these sites is caused by phytoalexins, it would be necessary to measure the concentrations of inhibitors to which hyphae are exposed at the time they stop growing and also to examine the activity of what may be a mixture of phytoalexins at the site of exposure. An essential requirement for obtaining such a direct proof is that the biochemical changes occurring within infected tissues must be closely associated with the biology of infection development. Interactions which have been the subject of coordinated biochemical and microscopical studies will now be discussed.

Botrytis spp. and *Vicia faba*.   The production of phytoalexins by tissues of *Vicia faba* in response to infection by *Botrytis* spp. had been examined for several years before attention was paid to the precise timing of phytoalexin accumulation and the cessation of fungal growth during resistant reactions (Deverall and Vessey, 1969; Mansfield and Deverall, 1974; Mansfield, 1980). Recent microscopical studies have established the following sequence of events during the development of infections in detached, field-grown broad bean leaves inoculated with conidia on their abaxial surface. The pathogen, *Botrytis fabae*, penetrates the cuticle between 5 and 6h after inoculation and grows rapidly through epidermal cell walls which swell in advance of fungal colonization. Formation of infection hyphae within the epidermal wall causes the death of numerous cells around penetration points. When high inoculum concentrations are used, areas of dead cells soon coalesce and the fungus ramifies through the dead epidermal and mesophyll tissues producing black lesions which spread rapidly away from sites of inoculation. By contrast, if inoculum droplets contain few spores, the growth of infection hyphae may often be restricted in the epidermis within the localized lesions which develop. Non-pathogenic species such as *B. cinerea*, *B. elliptica* and *B. tulipae* also produce infection hyphae within the swollen walls of epidermal cells but they kill far fewer cells than *B. fabae* during the early stages of infection and their hyphae stop growing by 12 h after inoculation (Figure 8.1; Mansfield, 1980; Rossall *et al.*, 1978; Mansfield and Richardson, 1981).

In an attempt to determine the concentrations of phytoalexins to which infection hyphae might be exposed within the epidermis, Dr Rossall and I (Rossall, 1978 and unpublished observations) have measured fungal growth and the concentrations of wyerone and wyerone acid in epidermal strips collected from sites inoculated with *B. cinerea* and low or high numbers of *B. fabae* conidia. Strips were comparatively easy to collect from leaves undergoing resistant reactions but became increasingly difficult to recover from sites inoculated with high concentrations of *B. fabae* because of the extensive maceration of tissue even within the first 9 h after inoculation. Strips from these sites were often heavily contaminated with mesophyll cells. The phytoalexins were analysed by TLC, not by HPLC, and therefore the yields recorded in Figure 8.1 include a small percentage of their dihydro analogues which possess similar fungitoxicity (Mansfield *et al.*, 1980).

During resistant reactions, the onset of phytoalexin accumulation preceded the restriction of fungal growth, and both wyerone acid and wyerone continued to accumulate throughout the course of the experiment. By contrast, however, in tissues inoculated with high numbers of *B. fabae* spores, an initial increase in phytoalexin concentrations was followed by a decrease in concentration as the epidermis and mesophyll tissues which were collected became totally necrotic and were colonized by the chocolate spot pathogen (Figure 8.1).

It is clear from Figure 8.1 a and b that phytoalexin accumulation within the epidermis occurs at the right time to explain the inhibition of fungal growth during resistant reactions. However, we do not know if infection hyphae growing within epidermal walls are exposed to the phytoalexins, or if these compounds are active at the site of exposure.

Microspectrofluorimetric studies have demonstrated that wyerone acid and related furanoacetylenes are synthesized by live cells at the edge of necrotic lesions (Mansfield *et al.*, 1974; Mansfield, 1980). The phytoalexins may be released into the lesion containing the invading fungus either by diffusion from living phytoalexin-producing cells, or following cell death and loss of membrane integrity. As wyerone acid is found in inoculum droplets it seems probable that this polar molecule diffuses freely throughout the lesion and is therefore very likely to reach intramural hyphae. By contrast, wyerone does not diffuse into droplets and appears to be bound to cell walls within infected tissues (Hargreaves *et al.*, 1977). Unless this phytoalexin is bound to the walls which

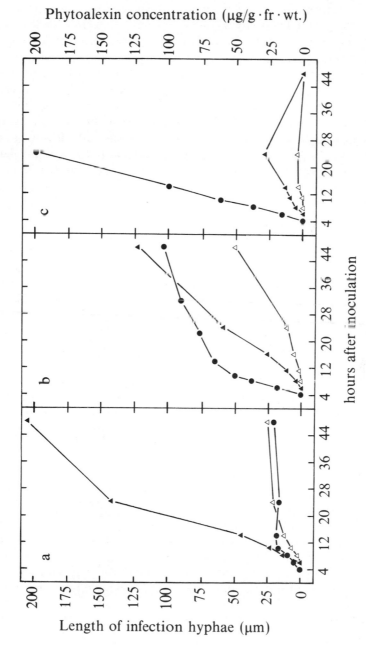

Figure 8.1 Relationship between growth of infection hyphae (●) and the accumulation of wyerone (△) and wyerone acid (▲) in the abaxial epidermis of broad bean leaves inoculated with (a) *Botrytis cinerea* (2000 conidia/droplet); (b) *B. fabae* (40 conidia/droplet); (c) *B. fabae* (2000 conidia droplet). From Rossall (1978) and unpublished data.

contain infection hyphae, the invading fungus is unlikely to be exposed to this inhibitor during the early stages of leaf infection.

Wyerone derivatives are normally highly fungitoxic, but the activity of wyerone acid, the predominant phytoalexin from leaves, is pH-dependent. The activity of the acid increases with decreasing pH, an effect probably caused by the greater ability of the fully protonated molecule to enter fungal cells. At pH 4.0, in a sucrose-casamino acids nutrient solution, the ED_{50}, MID and lethal concentrations of wyerone acid towards germ-tubes recorded after incubation for 18 h at 18°C, are 3.5, 15 and 35 µg/ml for *B. cinerea* and 6.8, 25 and 55 µg/ml for *B. fabae* (Deverall and Rogers, 1972; Hargreaves *et al.*, 1977; Rossall *et al.*, 1980). The release of galacturonic acid from bean cell walls attacked by pectic enzymes produced by *Botrytis* spp. causes a decrease in the pH of inoculum droplets often to values less than 4.5 during infection development (Deverall, 1967; Rossall, 1978). It seems probable, therefore, that the pH around infection

(a) *B. cinerea*

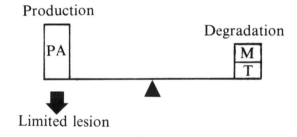

(b) *B. fabae*

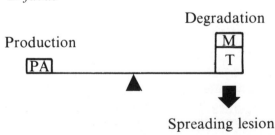

Figure 8.2 Interaction between *Botrytis* and leaf and pod tissues of *Vicia faba* expressed as a balance between phytoalexin production (PA) by the plant and phytoalexin degradation by the fungus. The degradation weighting incorporates both metabolic capacity to detoxify the inhibitors (M) and a tolerance factor (T) which accounts for the differential sensitivity of *B. cinerea* and *B. fabae* to wyerone derivatives. From Mansfield (1980).

hyphae within partially degraded cell walls is less than in overlying droplets and certainly sufficiently low to render any wyerone acid present highly antifungal.

Bearing in mind the circumstantial evidence described in the preceding paragraphs, it seems reasonable to conclude that phytoalexin accumulation restricts the growth of infection hyphae during resistant reactions. The changes in phytoalexin concentrations during the formation of spreading lesions suggest that, under suitable conditions, *B. fabae* is able to metabolize and detoxify the inhibitors to which it is exposed in leaf and pod tissues and thereby to prevent their accumulation to fungitoxic concentrations around invading hyphae (Figure 8.1, and Mansfield and Deverall, 1974; Hargreaves, *et al.*, 1977; Mansfield, 1980). Differentiation between *B. fabae* and species of *Botrytis* avirulent towards *V. faba* has been attributed to the greater tolerance of the pathogen to wyerone derivatives which allows *B. fabae* to detoxify the inhibitors (Rossall and Mansfield, 1978), and also to the greater ability of *B. fabae* to kill cells during the early stages of infection development, before phytoalexins accumulate. By killing numerous cells quickly, *B. fabae* reduces the synthetic activity of infected tissue and thus suppresses phytoalexin production, particularly if inoculum droplets contain high numbers of spores (Mansfield and Hutson, 1980). The interaction between *Botrytis* and leaves of *V. faba* has therefore been envisaged as a balance between phytoalexin production by the plant and phytoalexin degradation by the fungus (Figure 8.2). Modification of any of the factors contributing to this balance could affect the outcome of the interaction. Thus, decreasing spore numbers leads to reduced development of *B. fabae*, probably by reducing the numbers of plant cells killed at inoculation sites and thereby increasing phytoalexin production and accumulation around invading hyphae (Mansfield and Hutson, 1980).

Colletotrichum lindemuthianum and French bean. *Colletotrichum lindemuthianum*, the cause of an anthracnose disease, exists as several physiological races which can be differentiated by their reaction with various cultivars of *Phaseolus vulgaris*. Most studies comparing mechanisms of resistance and susceptibility have been carried out with hypocotyls. Spores of *C. lindemuthianum* germinate within 48 h of inoculation and produce similar numbers of appressoria on resistant and susceptible plants. Differences between cultivars become evident only when infection hyphae penetrate underlying epidermal cells. In excised hypocotyls of susceptible cultivars, developing intracellular hyphae cause no observable host response and grow between the cell wall and plasmalemma of epidermal and cortical cells. At temperatures below 20°C this biotrophic growth can continue for several days, for example, for more than 5 days at 16°C, leading to extensive colonization of hypocotyl tissues (Skipp and Deverall, 1972; Bailey *et al.*, 1980; Bailey, 1981). After this period of biotrophy, infected cells die, the tissue collapses and a brown lesion is produced. Considerable necrotrophic fungal growth may occur and rotting may become extensive, particularly when tissues are incubated at lower temperatures or inoculated at numerous sites along the length of the hypocotyl so that the brown lesions may coalesce as they appear (Rahe, 1973; Bailey, 1981). When inoculation sites are widely spaced, lesions often become limited (Rahe, 1973). Lesion limitation may also result when hypocotyls are transferred to between 22 and 25°C after a biotrophic relationship has been established by previous incubation at 16°C (Bailey, 1974). In resistant hypocotyls the initially infected cell, and perhaps one or two adjacent cells, die and turn brown soon after infection, appearing as scattered flecks beneath inoculum droplets. Growth of incompatible races is usually restricted to the initially penetrated cells; the process often referred to as hypersensitivity (Skipp and Deverall, 1972; Elliston *et al.*, 1976; Landes and Hoffman, 1979).

Phaseollin is the main phytoalexin produced by French bean hypocotyls challenged by *C. lindemuthianum* (Bailey, 1974). Early studies showed that the accumulation of phaseollin was associated with cell death and browning in both resistant and susceptible hypocotyls. Thus, the phytoalexin was not detected within symptomless tissue during the biotrophic period of colonization by a compatible race, but accumulated after the death of infected tissues and formation of lesions. At sites where anthracnose lesions coalesced, very low concentrations of phaseollin (less than $5 \mu g/g$ fresh wt) were recovered, whereas much greater concentrations, ca. $300 \mu g/g$, were found where isolated lesions became limited. Phaseollin accumulated rapidly following the death of cells during the hypersensitive reaction to incompatible races (Bailey and Deverall, 1971; Rahe, 1973; Bailey, 1974). The elegant experiments of Hargreaves and Bailey (1978) have demonstrated that phaseollin and other isoflavonoid phyto-alexins are probably synthesized in tissues around necrotic cells and are adsorbed and accumulate in the dead tissue containing intracellular hyphae. On the basis of the assumption that phytoalexins accumulated in dead tissue, Bailey and Deverall (1971) had earlier estimated that the concentration of phaseollin within cells which underwent hypersensitive reactions to *C. lindemuthianum* approached $3000 \mu g/ml$, 4 days after inoculation; almost 300 times the MID concentration for activity against germ-tube growth by each of the three races of the fungus tested *in vitro*. However, because the intracellular hyphae produced were too small to determine precisely when growth became inhibited during the hypersensitive reaction doubts remained concerning the role of phaseollin in restricting fungal growth.

More recently Bailey *et al.* (1980) have utilized the resistance induced by transfer of infected hypocotyls from 16 to 25°C to examine in detail the timing of fungal growth inhibition, host cell death and phytoalexin accumulation. Hypocotyls were inoculated with a compatible race of the fungus and incubated at 16°C for 3 days, allowing the formation of biotrophic infection hyphae, then transferred to 25°C. Incubation at 25°C led to premature death of invaded cells and the inhibition of intracellular hyphae. Fungal growth during this induced resistant response was greater than during normal hypersensitivity, enabling measurements to be made of the lengths of intracellular hyphae throughout the resistance process, from the time of transfer to 25°C until cessation of fungal

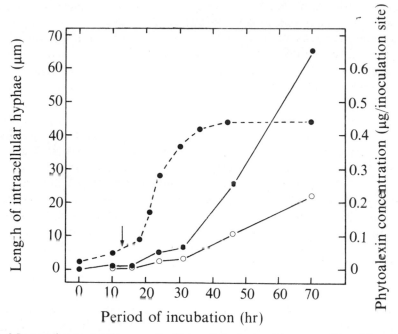

Figure 8.3 Relationship between growth of intracellular hyphae (● · · · · · · ●) and the accumulation of phaseollin (●———●) and phaseollinisoflavan (○) at inoculation sites in French bean hypocotyls inoculated with a compatible race of *Colletotrichum lindemuthianum*, incubated at 16°C for 72 h (=0 on x axis) and then placed at 25°C to induce a resistant reaction. The arrow indicates when cellular necrosis became visible. Adapted and redrawn from Bailey *et al.* (1980).

growth. Results obtained are summarized in Figure 8.3 and indicate that death of infected cells occurred several hours before phytoalexins formed at inoculation sites and that inhibition of hyphal growth occurred shortly after the phytoalexins began to accumulate. These findings are consistent with the view that the accumulated phytoalexins cause the restriction of fungal growth during the resistance of beans to *C. lindemuthianum*. Conversely, the unrestricted colonization of hypocotyls occurs when the fungus avoids exposure to the phytoalexins by establishing a prolonged biotrophic phase of growth. This leads to the formation of coalescing lesions, within and around which cells remain alive for insufficient time to synthesize the amounts of inhibitors required to reach fungitoxic concentrations.

Phytophthora infestans and potato tissue. The successful colonization of potato tissues by compatible races of *P. infestans* also involves the establishment of a biotrophic phase of infection during which the fungus grows intracellularly through host cells and may also produce haustoria within cells adjacent to intercellular hyphae. The timing and extent of necrosis during susceptible reactions to *P. infestans* has not been as well defined as in bean hypocotyls invaded by compatible hyphae of *Colletotrichum lindemuthianum*, but it always follows biotrophic colonization by the late-blight fungus.

Race-specific resistance of potato foliage and tubers conferred by the presence of R-genes is expressed by the hypersensitive reaction of penetrated cells to invading hyphae of incompatible races of *P. infestans*. Growth of incompatible hyphae is restricted following death of resistant cells which occurs soon after penetration (Müller, 1959; Kitazawa and Tomiyama, 1970; Sato *et al.*, 1971; Shimony and Friend, 1975, 1976; Hohl and Stössel, 1976; Hohl and Suter, 1976).

Sato *et al.* (1971) were the first workers to coordinate measurements of hyphal growth and phytoalexin accumulation within infected tissues. Their most detailed studies were carried out using cut surfaces of the leaf petiole cortex of cv. Rishiri, which possesses the R_1 gene and high field resistance, inoculated with zoospores of the incompatible race 0 or the compatible race 1. Under these conditions penetration occurred 1 to 2.5 h after inoculation. With race 0, approximately 50% of the invaded cortical cells had died by 2 h after inoculation and the remainder were dead within 3 h. Most hyphae caused death and subsequent browning of single cells which contained the restricted hyphae, but at 20 to 30% of invasion sites, intracellular hyphae also grew into the adjacent cell causing its death and browning. The invasion and death of three cells was rarely observed. Compatible hyphae ramified through the tissue without killing plant cells, for at least 2 days after inoculation. No rishitin was detected in petioles inoculated with the compatible race (Sato *et al.*, 1971) but accumulation of the phytoalexin began 7 h after inoculation with race 0 and clearly preceded the restriction of hyphal growth.

The localization of rishitin within responding petioles has not been examined, but results of experiments with tuber tissue demonstrated that rishitin accumulated within brown cells undergoing the hypersensitive response and reached local concentrations greater than 100 μg/ml which were completely inhibitory to zoosporangial germ tubes *in vitro* (Ishisaka *et al.*, 1969; Sato and Tomiyama, 1969; Sato *et al.*, 1971).

These data strongly suggest that rishitin is the primary cause of the inhibition of fungal growth. In potato tissue undergoing the hypersensitive reaction to *P. infestans*, levels of rishitin and other sesquiterpenoid stress metabolites often drop markedly by the fourth day after inoculation (Kuć, 1975). Unless fungicidal levels of the phytoalexins are reached at all infection sites, inhibition of further fungal growth may be due to lignification and also the accumulation of oxidized phenolics within dead cells (Sato *et al.*, 1971; see also chapter 3).

Phytophthora megasperma var. sojae and soybean.   *Phytophthora megasperma* var. *sojae* (*P. megasperma* f. sp. *glycinea* of Kuan and Erwin, 1980) causes root and stem rot of susceptible soybean cultivars. Inoculation procedures developed for screening for resistance involve either root inoculation, in which whole plants are grown in soil infested with *P. megasperma*, or hypocotyl inoculation in which mycelium is inserted into artificial wounds. The wound inoculation technique produces very consistent results and has been used extensively for studies on the role of phytoalexins in race-specific resistance conferred by the presence of dominant major genes in soybean cultivars. Recently, zoosporial inocula applied to intact hypocotyls have also been used for experiments with this system (Keeling, 1976; Bruegger and Keen, 1977; Ward et al., 1979).

Susceptible hypocotyls inoculated with mycelium or zoospores develop water-soaked, spreading lesions within 24 h after inoculation. Resistance is expressed by the equally rapid appearance of dark brown lesions which remain localized to inoculation sites and within which fungal growth is restricted. The resistant response has been described as hypersensitivity by several workers but it can involve far more widespread necrosis of infected tissue and more extensive invasion by incompatible hyphae of P. megaspermu than is observed in race-specific resistance of bean hypocotyls to Colletotrichum lindemuthianum or potato tissue to P. infestans. The limited information available indicates that soybean tissues at inoculation sites may die more rapidly following invasion by incompatible than by compatible hyphae of P. megasperma, but such differences between the interactions are not very great, because mixtures of living and dead invaded cells occur within tissues undergoing resistant or susceptible reactions (Keen, 1971; Yoshikawa et al., 1978; Stössel et al., 1980).

Yoshikawa et al. (1978) have examined the role of the phytoalexin glyccollin (a mixture of three isomeric 6a-hydroxy pterocarpans) in the resistance of soybean hypocotyls to P. megasperma. They used the wound inoculation technique and studied the responses of Harosoy (susceptible) and Harosoy 63 (resistant) hypocotyls to race 1 of the fungus. Changes in phytoalexin concentrations during infection were also examined in Harosoy 63 hypocotyls treated with the protein synthesis inhibitor blasticidin S which renders them completely susceptible to stem rot (Yoshikawa, 1978).

Initial time-course experiments, in which glyceollin levels were determined in 1 cm long segments of whole hypocotyls which contained the inoculation site at the centre, demonstrated that glyceollin accumulated at a greater rate and reached higher levels in resistant than in susceptible tissues. Rapid accumulation of the phytoalexin coincided with the restriction of fungal growth occurring between 9 and 10 h after inoculation, but levels recorded in resistant hypocotyls did not exceed ED_{50} values against mycelial growth (c. 100 μg/ml) until after 16 h (Figure 8.4).

In an attempt to determine the concentrations of glyceollin around invading hyphae, Yoshikawa et al. (1978) also measured phytoalexin levels in 0.25 mm thick serial sections prepared from inoculation sites, so that the distribution of glyceollin within seven layers of tissues was examined. Analysis showed that ED_{90} values, c. 200 μg/ml, were present as early as 8 h after inoculation in the first tissue layer of resistant hypocotyls where many invading hyphae were present (Table 8.1). Levels close to the ED_{50} value were also detected in the second layer where only a few hyphae were present, but glyceollin was not detected in deeper tissue layers which had not been invaded by the fungus.

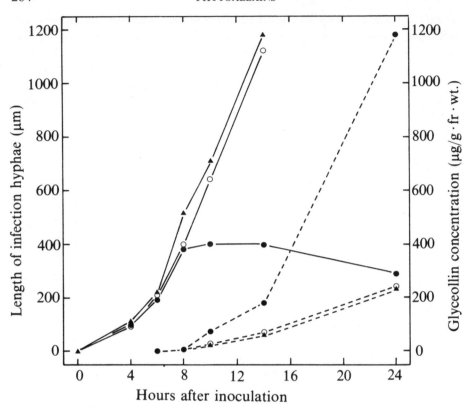

Figure 8.4 Relationship between fungal growth (solid lines) and glyceollin accumulation (broken lines) in hypocotyls of soybean cv. Harosoy (▲) and Harosoy 63 treated (○) or untreated (●) with blasticidin S (0.5 μg/ml) inoculated with *Phytophthora megasperma* var. *sojae*. Adapted and redrawn from Yoshikawa *et al.* (1978).

Accumulation was similarly localized at 9 and 10 h after inoculation. Although the time when fungal growth ceased in resistant hypocotyls varied slightly in repeated experiments, it always coincided with the time when high levels of localized glyceollin accumulation occurred. Interestingly, 24 h after inoculation, glyceollin levels exceeding the ED_{90} value were also recorded in susceptible Harosoy or blasticidin S-treated Harosoy 63 hypocotyls, especially in the tissue layers close to the hypocotyl surface. However, the glyceollin concentrations within newly invaded tissue layers in susceptible hypocotyls did not exceed the ED_{90} value at any time after inoculation, whereas, as we have seen, those in resistant hypocotyls reached the ED_{90} value after 8 h. The extremely thorough and detailed work of Yoshikawa and colleagues provides convincing evidence that the accumulation of glyceollin causes the cessation of fungal growth in resistant soybean hypocotyls.

Table 8.1 Glyceollin concentrations in tissue sections of soybean hypocotyls of Harosoy (H) and Harosoy 63 treated (H63-BcS) and untreated (H63) with blasticidin S (0.5 μg/ml) at various times after inoculation with race 1 of *Phytophthora megasperma* var. *sojae*. Modified from Yoshikawa *et al.* (1978).

Time after inoculation (h)	Host[a]	Glyceollin (μg/g fr. wt.) in tissue layers (0.25 mm thick)						
		1st	2nd	3rd	4th	5th	6th	7th
8	H63	364**[b]	75*[b]	<10	<10	0	0	0
	H63-BcS	10**	10*	0	0	0	0	0
	H	10**	18*	0	0	0	0	0
10	H63	752**	261**	73	20	0	0	0
	H63-BcS	104**	80**	60*	17	0	<10	0
	H	20**	18**	19*	25	0	<10	<10
24	H63	3889**	1820**	515	117	71	<10	0
	H63-BcS	690**	697**	591**	443**	223**	143**	60**
	H	557**	420**	273**	272**	155**	135**	87**

[a] H63 hypocotyls were resistant, H63-BcS and H were susceptible to colonization.
[b] * indicates slight colonization by the tips of advancing hyphae, ** indicates that the tissue layer was extensively colonized.

Other interactions.   The principal features of other interactions, upon which less detailed coordinated biochemical and microscopical studies have been performed, are summarized in Table 8.2. In each of these cases it has been suggested that phytoalexin accumulation causes the inhibition of invading hyphae. Only fungicide-induced resistance, vascular wilt and rust infections will be considered in more detail here, but the reader is urged to examine thoroughly the evidence for phytoalexin involvement in the other systems listed in Table 8.2.

The enhanced accumulation of phytoalexins in plants treated with systemic fungicides has been described in detail by Cartwright *et al.* (1980) and by Ward *et al.* (1981). Studies on the mode of action of the systemic fungicide 2,2-dichloro-3,3-dimethyl-cyclopropane carboxylic acid (WL 28325) which is specific for rice blast, led to the first characterization of phytoalexins from a member of the Gramineae; the momilactones A and B from *Oryza sativa* (Cartwright *et al.*, 1977, 1981). The cyclopropane derivative possesses little activity against the rice blast fungus (*Pyricularia oryzae*) *in vitro*. It appears to act *in vivo* by enhancing the host's ability to produce the momilactones. Following direct penetration of untreated leaves, *P. oryzae* initially grows intracellularly through the epidermis and mesophyll without causing cell death, but in fungicide-treated tissues infection hyphae grow into the epidermis only slightly before death and browning of the invaded host cell occurs. Cartwright *et al.* (1980) found that death of infection hyphae following this "fungicide-induced" hypersensitive reaction was associated with the accumulation of the momilactones in and also around inoculation sites, whereas the phytoalexins were detected in only trace

Table 8.2 Further examples of interactions examined by coordinated biochemical and microscopical studies

Fungus	Plant	Tissue	Resistance	Main site of restriction of hyphae	Principal phytoalexin	Key reference
Phytophthora infestans and Sclerotinia fructicola	French bean	Pod	Non-host	Intracellular[a]	Phaseollin[b]	Müller (1958)
Various	Alfalfa	Leaf	Non-host	Inter- and intracellular	Medicarpin[b]	Higgins and Millar (1968)
Ascochyta pisi and Mycosphaerella pinodes	Pea	Leaf	Race non-specific	Inter- and intracellular	Pisatin	Heath and Wood (1971)
Phytophthora drechsleri	Safflower	Stem	Race specific	Inter- and intracellular	Safynol	Allen and Thomas (1971)
Phytophthora infestans	Pepper	Fruit	Non-host	Inter- and intracellular	Capsidiol	Jones, Unwin and Ward (1975)
Rhizoctonia solani	French bean	Hypocotyl	Race non-specific	Inter- and intracellular	Kievitone	Smith et al. (1975)
Ceratocystis fimbriata	Sweet potato	Tuber	Non-host	Intracellular	Ipomeamarone	Kojima and Uritani (1976)
Phytophthora vignae	Cowpea	Hypocotyl	Race specific	Inter- and intracellular	Kievitone	Partridge and Keen (1976)
Melampsora lini	Flax	Leaf	Race specific	Intracellular	Coniferyl alcohol	Keen and Littlefield (1979)
Botrytis and Sclerotinia spp.	Clover	Leaf	Non-host	Inter- and intracellular	Maackiain	Macfoy and Smith (1979)
Cladosporium fulvum	Tomato	Leaf	Race specific	Intercellular	Unknown	De Wit and Flach (1979)
Erysiphe graminis	Pea	Leaf	Non-host	Intracellular	Pisatin	Oku et al. (1979)
Piricularia oryzae	Rice	Leaf	Fungicide induced	Inter- and intracellular	Momilactone A	Cartwright et al. (1980)
Fusarium oxysporum and Verticillium albo-atrum	Tomato	Stem	Race specific	In xylem vessels	Unknown terpenoid	Hutson and Smith (1980)
Phakopsora pachyrhizi	Soybean	Leaf	Race specific	Inter- and intracellular	Glyceollin	Keogh et al. (1980)
Plasmopara viticola	Frost grape	Leaf	Race non-specific	Inter- and intracellular	ε-viniferin	Langcake (1981)
Botrytis spp.	Narcissus	Bulb	Non-host and race non-specific	Inter- and intracellular	7-hydroxyflavan	O'Neill (1981)
Phytophthora megasperma var. sojae	Soybean	Hypocotyl	Fungicide induced	Inter- and intracellular	Glyceollin	Ward et al. (1981)

[a] Intracellular includes hyphae growing within cell walls.
[b] Probable identity of the uncharacterized phytoalexin studied.

amounts in untreated, infected tissue. The biochemical basis for the enhancement of momilactone production by WL 28325 is unclear but it is the first compound whose systemic fungicidal activity can be attributed solely to its ability to activate natural resistance mechanisms.

Enhanced phytoalexin accumulation has also been reported in response to inoculation with *Phytophthora megasperma* in soybean hypocotyls treated with the systemic fungicide Ridomil (DL-methyl *N*-(2,6-dimethylphenyl)-*N*-(2-methoxyacetyl) alaninate). The activity of this fungicide is also associated with an induced necrotic reaction to invading hyphae, lesions produced being indistinguishable from those formed on genetically resistant hypocotyls (Ward *et al.*, 1981). Ridomil is much more fungitoxic than WL 28325 *in vitro* and the enhanced glyceollin accumulation in Ridomil treated soybeans may therefore be caused by the release of phytoalexin elicitors from hyphae damaged by the fungicide. Ward *et al.* (1981) concluded that if Ridomil were to reach concentrations at infection sites that were less than fully inhibitory it seems quite possible that glyceollin accumulating there might make an important contribution to the prevention of disease development. They also pointed out that other reports have referred to hypersensitive symptoms during the control of fungal diseases by systemic fungicides and it is possible that associated enhanced phytoalexin production may be of general occurrence.

The involvement of phytoalexins in resistance to vascular wilts was first indicated by Bell (1969), who showed that cotton stems inoculated with spores of *Verticillium albo-atrum* accumulated gossypol and related compounds in the vascular system. Accumulation of the phytoalexins was greater in resistant than in susceptible varieties during the first few days after inoculation, but continued interaction between the susceptible host and the fungus led to increased total amounts in the susceptible cultivar. Bell argued that the rapid, localized accumulation of phytoalexins causes containment of hyphae in resistant varieties. Similar conclusions were reached by Tjamos and Smith (1974) and Hutson and Smith (1980) from their experiments on the role of phytoalexins in the resistance of tomato to *Fusarium oxysporum* f. sp. *lycopersici* and *V. albo-atrum*. They found that the accumulation of rishitin and other terpenoid phytoalexins was closely associated with vascular browning. Hutson and Smith (1980) utilized antimony pentachloride, a chromogenic reagent used for locating terpenoid compounds on TLC plates, as a histochemical stain for the detection of sites of terpenoid accumulation within infected tissues. In resistant plants, terpenoid accumulation appeared to be restricted to vessel walls and a few layers of xylem parenchyma cells around infected vessels. Tyloses also gave a very positive reaction to the stain. In infected susceptible plants reactions were much less intense and the terpenoids were more widespread in the xylem parenchyma. The production of phytoalexins by tyloses during resistant reactions may be particularly important as it would allow inhibitors to be secreted directly into the path of invading hyphae. The occlusion of vessels by tyloses, which Hutson and Smith (1980) found to be more frequent during resistant reactions, may also lead to the accumulation of phytoalexins to high concentrations within the blocked elements. Results of Mace (1978) working with *Verticillium* wilt of cotton have similar implications.

The hypersensitive necrotic reaction of resistant cells to infection with incompatible rust fungi has been considered to be sufficient *per se* to account for the restriction of growth of these biotrophic parasites (Stakman, 1915; Ingram, 1978). In flax and French bean leaves, however, phytoalexin accumulation may also contribute to fungal inhibition during necrotic reactions. Keen and Littlefield (1979) found that the fungitoxic compounds coniferyl alcohol and coniferyl aldehyde accumulated, following the necrosis of cells, at sites in flax leaves inoculated with incompatible races of *Melampsora lini* (see also Littlefield, 1973). The accumulation of phaseollin was also closely associated with the onset of necrosis in bean leaves of the resistant cultivar Imuna inoculated with *Uromyces appendiculatus* (Bailey and Ingham, 1971). Interestingly, Deverall (1977) using techniques similar to those applied by Bailey and Ingham (1971) and Keen and Littlefield (1979), failed to demonstrate phytoalexin accumulation in wheat leaves responding hypersensitively to *Puccinia graminis*. Cartwright and Russell (1981), however, have reported preliminary experiments indicating the accumulation of phytoalexins in wheat resistant to *Puccinia striiformis*.

It is clear from the preceding discussion that there are numerous papers which provide evidence, albeit circumstantial, supporting a primary role for phytoalexin accumulation in restricting fungal growth within necrotic tissue. Conversely, successful progressive colonization by virulent pathogens is often characterized by the failure of phytoalexins to reach or maintain fungitoxic concentrations around invading hyphae. An outstanding exception to this general pattern concerns the infection of pea seedlings by *Aphanomyces euteiches*. Pueppke and VanEtten (1974) found that *A. euteiches* was much more sensitive than other pea pathogens to pisatin in assays carried out against mycelium on agar or in liquid culture. On agar the ED_{50} value against *A. euteiches* was *c.* $40\,\mu g/ml$ and it was the only species tested which was completely inhibited by $100\,\mu g/ml$. Despite this sensitivity *in vitro*, the rapidly spreading lesions formed by the fungus in pea stems contained *c.* $800\,\mu g$ pisatin/ml. In an attempt to explain this apparent inconsistency Pueppke and VanEtten (1976) examined the histology of infection development, the location of pisatin within lesions, the sensitivity of *A. euteiches* to pisatin in several different media and the ability of the fungus to detoxify and metabolize the phytoalexin. Their results indicated that hyphae were almost certainly exposed to pisatin *in vivo*, that composition of the growing medium had little effect on sensitivity to the phytoalexin and that *A. euteiches* was unable to metabolize or become adapted to tolerate the phytoalexin *in vitro*. They therefore argued that data from their studies did not support available theories which offer mechanisms whereby compatible pathogens overcome phytoalexins. They recognized the possibility that plant components might modify the fungitoxicity of pisatin *in vivo* but rightly concluded from the data that the phytoalexin had no role in controlling infection by *A.*

Table 8.3 Interaction combinations between bacteria and plants and evidence for the involvement of phytoalexins in restricting bacterial multiplication. Adapted from Kelman and Sequeira (1972)

Types of bacterium/plant interaction	Plant response		
	Hypersensitive reaction	Typical disease symptoms	Phytoalexin involvement
Saprophytes/plants in general	−	−	−
Animal pathogens/plants in general	−	−	−
Plant pathogens/non-host plants	+ or −	− −	+ −
Pathogen/resistant cultivar of susceptible plant	+	−	+
Pathogen/susceptible plant which has become resistant due to environmental conditions	+[a]	−	+
Pathogen/susceptible plant	−	+	−

[a] Necrosis may be more extensive in this response than in other hypersensitive reactions.

euteiches. This work serves to re-emphasize the difficulty of comparing phyto-alexin activities *in vitro* and *in vivo*. But, as Smith (1978) has pointed out, with *Aphanomyces* and pisatin it may be the sensitivity *in vitro* which is anomalous rather than the apparent insensitivity *in vivo* (see also chapter 7). There are no reports, as yet, of such a striking and inexplicable discrepancy with other phytoalexins.

Resistance to bacteria

Interactions between bacteria and plants in which there is evidence for the involvement of phytoalexins in resistance are summarized in Table 8.3. Studies on the role of phytoalexins in resistance have been mainly concerned with the restriction of bacterial multiplication within intercellular spaces.

Pseudomonads and French bean or soybean

The most detailed studies of the involvement of phytoalexins in bacterial infections concern the French bean plant and the resistance of leaves of certain cultivars to halo blight caused by *Pseudomonas phaseolicola* and of pods to avirulent isolates of *P. syringae*.

The multiplication of compatible and incompatible races of *P. phaseolicola* in bean cv. Red Mexican and the timing of symptom appearance established by O'Brien and Wood (1973a) are given in Figure 8.5a. The compatible race 2 multiplies rapidly causing water-soaked lesions to develop between 2 and 4 days after inoculation; these lesions become brown and desiccated after 5 days. The incompatible race multiplies less rapidly and causes a hypersensitive reaction,

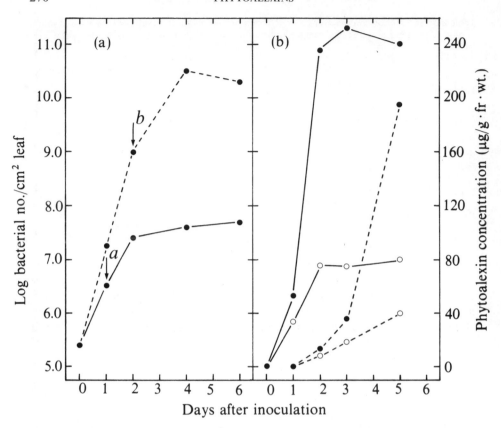

Figure 8.5 Relationship between (a) bacterial multiplication and (b) accumulation of phaseollin (●) and coumestrol (○) in leaves of French bean cv. Red Mexican inoculated with the incompatible race 1 (solid lines) and compatible race 2 (broken lines) of *Pseudomonas phaseolicola*. Arrows indicate earliest appearance of *a*, hypersensitive necrosis and *b*, water soaked lesions at inoculation sites. Adapted and redrawn from O'Brien and Wood (1973) and Lyon and Wood (1975).

inoculation sites collapsing to form localized desiccated brown lesions within 2 days. Collapse of tissue during the hypersensitive reaction is closely associated with the cessation of bacterial multiplication.

O'Brien and Wood (1973b) were the first to suggest the involvement of phytoalexins in resistance following their detection of the accumulation of antibacterial compounds in bean leaves undergoing hypersensitivity. In subsequent work they and Gnanamanickam and Patil (1977a) quantified changes in concentrations of isoflavonoids occurring during infection. Data for the accumulation of coumestrol and phaseollin taken from O'Brien and Wood (1973a) and Lyon and Wood (1975) are given in Figure 8.5b. Gnanamanickam and Patil (1977a,b) confirmed these results and found that phaseollinisoflavan and kievitone accumulated to concentrations similar to those recorded for phaseollin. In both resistant and susceptible reactions phytoalexin accumulation closely followed necrosis of plant cells, the earlier appearance of necrotic lesions during the hypersensitive reaction being paralleled by earlier accumulation of isoflavonoids. Highest yields were recovered from tissue undergoing the hypersensitive reaction. Experiments with other pseudomonads showed that the stone-fruit pathogen *P. mors-prunorum* multiplied briefly, inducing a rapid hypersensitive reaction and phytoalexin accumulation, but the saprophyte *P. fluorescens* failed to multiply, cause symptoms or cause changes in isoflavonoid concentrations (O'Brien and Wood, 1973a,b; Lyon and Wood, 1975). Gnanamanickam and Patil (1977a,b) found that phaseotoxin, produced by *P. phaseolicola*, suppressed hypersensitive cell death and also phytoalexin accumulation.

In the experimental systems used by Lyon and Wood (1975) and Gnanamanickam and Patil (1977a,b) attached leaves were inoculated and incubated under low relative humidities, conditions under which the collapse and desiccation of tissue during the hypersensitive reaction, even in the absence of accumulation of inhibitory compounds, would be expected to restrict bacterial multiplication (Rudolph, 1976). Webster and Sequeira (1977), however, used a saturated atmosphere for the incubation of detached bean pods, thereby reducing the influence of desiccation on bacterial multiplication. In the pod system, virulent and avirulent isolates of *Pseudomonas syringae* multiplied equally rapidly during the first 2 days after inoculation and caused water-soaked flecking in infected tissues. The small lesions caused by the avirulent isolate turned brown during the next 12 h and did not enlarge, whereas those caused by the virulent isolate remained water-soaked, coalesced and spread into surrounding uninoculated tissue. Changes in lesion appearance were followed closely by changes in bacterial populations, multiplication of the avirulent isolate being restricted. Ethyl acetate extracts from tissues undergoing the resistant reaction became highly inhibitory to both strains of *P. syringae* at the time of restriction of bacterial multiplication within infected pods, but extracts from tissue developing spreading lesions, or healthy pods, lacked inhibitory activity. Results obtained with these crude extracts provide convincing evidence for the involvement of phytoalexins in resistance. The inhibitor was a phenolic compound (λ_{max} 286 nm) but was not identified. The antibacterial activity of tissue extracts was closely paralleled by the concentrations of phaseollin therein, but Webster and Sequeira found that, when purified, this isoflavonoid caused no significant effect on the multiplication of *P. syringae*.

The results of bioassays carried out by Webster and Sequeira (1977) raise the problem of conflicting evidence on the antibacterial activity of compounds (particularly isoflavonoids) initially recognized as phytoalexins because of their fungitoxic properties. Thus although Lyon and Wood (1975) found phaseollin to be inactive, coumestrol was antibacterial in their assays and must therefore be considered to be a phytoalexin despite its lack of fungitoxicity (Sherwood *et al.*, 1970). Gnanamanickam and Patil (1977), using a different assay technique, found that coumestrol, kievitone, phaseollin and phaseollinisoflavan all inhibited *P. phaseolicola*. Thus, although opinions differ on the relative importance of individual compounds, there is general agreement that antibacterial phytoalexins accumulate during resistant reactions of French beans to plant pathogenic bacteria. These phytoalexins are not involved in the inhibition of the saprophyte *P. fluorescens* in bean tissue (Lyon and Wood, 1975).

The response of soybean leaves to compatible and incompatible races of *Pseudomonas glycinea* is closely comparable to that of French bean leaves to *P. phaseolicola*. Keen and Kennedy (1974) found that the hypersensitive reaction in soybean caused by incompatible races of *P. glycinea*, or the non-pathogen *P. lachrymans*, was associated with rapid accumulation of glyceollin, coumestrol, daidzein and sojagol. Glyceollin and coumestrol were antibacterial *in vitro*. Little accumulation of the isoflavonoids was found during susceptible reactions.

Erwinia spp. and potato

Potato tubers incubated in nitrogen containing 1 % or less oxygen are highly susceptible to rotting by *Erwinia carotovora* var. *atroseptica* or *E. carotovora* var. *carotovora*. Rots formed during incubation in air are less extensive and are often restricted as firm, dark lesions, contrasting with the soft, cream-coloured rots which occur when tubers are depleted of oxygen (Lund and Nicholls, 1970). Lyon (1972) found that rishitin and phytuberin accumulated within the restricted rots formed in tubers maintained in air. Phytoalexin accumulation was largely confined to the firm rotted tissue (Table 8.4). Lyon *et al.* (1975) also found an inverse correlation between the volume of rot formed in air and rishitin concentration after inoculation with *E. carotovora* var. *atroseptica*.

Rishitin was the major cause of the considerable antibacterial activity of aqueous methanol extracts of rotted tubers; phytuberin was comparatively inactive against *E. carotovora* var. *atroseptica* (Lyon and Bayliss, 1975). Thus, rishitin accumulation may be an important factor contributing to the degree of

Table 8.4 Rishitin and phytuberin concentrations in tubers inoculated with *Erwinia carotovora* var. *atroseptica* and incubated at 11°C in air. Modified from Lyon (1972)

Time after inoculation (days)	Tissue	Phytoalexin concentration ($\mu g/g\,fr.\ wt.$)	
		Rishitin	Phytuberin
7	Black rot	665	92
	0–3 mm from rot	21.3	4.5
	Healthy (> 2 cm from rot)	< 0.2	< 0.2
14	Black rot	1220	165
	0–3 mm from rot	11.8	1.2
	Healthy (> 2 cm from rot)	< 0.2	< 0.2

resistance of tubers to this organism. Additional factors which may contribute to resistance are oxidized polyphenols (Lovrekovich *et al.*, 1967) and the formation of a barrier of suberized cells (Fox *et al.*, 1971).

Other interactions

The accumulation of unidentified antibacterial compounds during resistant reactions has been reported for a number of bacteria/plant combinations (see Rudolph, 1976). An interesting example concerns the response of pepper leaves to *Xanthomonas vesicatoria*, which was studied by Stall and Cook (1968). Their work is particularly important as they extracted inhibitors from intercellular spaces, the sites of bacterial multiplication, using the technique devised by Klement (1965). Phytoalexins accumulated more rapidly and reached higher concentrations in extracts from leaves undergoing the hypersensitive resistant reaction than the susceptible response.

Extraction of intercellular fluids is a difficult technique, particularly when dealing with tissue undergoing confluent necrosis, but it merits further application as it allows the recovery of compounds to which bacteria will almost certainly be exposed within the infected plant. Use of the technique would help to clarify the proposed role of phytoalexins in the resistance of French bean and soybean, as recent work suggests that isoflavonoid phytoalexins may be adsorbed on to dead plant cells and in consequence may not occur in intercellular spaces at concentrations sufficient to prevent bacterial multiplication (Hargreaves and Bailey, 1978). The technique would also be suitable for the examination of phytoalexin involvement in the phenomenon of localized bacteriostasis in cotton leaves resistant to *Xanthomonas malvacearum*, elegantly described by Essenberg *et al.* (1979).

Resistance to nematodes

Keen and co-workers have studied the resistance of legume roots to nematodes. Lima bean roots exhibit a hypersensitive resistant response to *Pratylenchus scribneri*. Rich *et al.* (1977) found that tissue bearing necrotic lesions caused by attempted feeding of the nematode in the epidermis and cortex accumulated the fluorescent isoflavonoids coumestrol and psoralidin. These compounds were present only in low concentrations in uninoculated roots. Coumestrol inhibited the motility of *P. scribneri* at concentrations less than those found within infected roots leading Rich *et al.* (1977) to conclude that induced accumulation of the coumestan phytoalexin is probably the chemical basis for the resistance of Lima bean roots to *P. scribneri*. An interesting feature of their work is that they were unable to detect antifungal compounds in extracts of Lima bean roots challenged by the nematode, although other parts of the plant are known to accumulate the fungitoxic phytoalexins phaseollin and kievitone following fungal infection (Cruickshank and Perrin, 1971; Keen, 1976). Neither fungitoxic nor nematistatic compounds accumulated in roots of the susceptible *Phaseolus vulgaris* during their colonization by *P. scribneri* (Rich *et al.* 1977).

The role of glyceollin in the expression of varietal resistance of soybean cv. Centennial to the root knot nematode *Meloidogyne incognita* has also been examined. In the susceptible cultivar Pickett 71, second stage larvae migrate to the root protostele without causing symptoms in the cortex, during the first day after soil inoculation with eggs. Cells associated with the head regions of larvae present in the stele undergo mitosis or enlarge, becoming multinucleate, and form characteristic giant cells by 6 days after inoculation. Despite the rapid multiplication of *M. incognita* within infected roots there is very little necrosis of plant cells during the susceptible reaction. Second stage larvae also migrate to the protostele of resistant roots, but do not induce giant cell formation; instead, cells around larval heads usually die and turn brown. *Meloidogyne incognita* larvae fail to multiply or migrate from areas of dead cells (Kaplan *et al.*, 1979).

Kaplan *et al.* (1980) found that necrosis of cells was closely followed by increases in glyceollin concentrations; levels greater than $60 \mu g/g$ fresh wt being recovered from resistant roots, but less than $15 \mu g/g$ from roots undergoing a susceptible reaction. Dissection experiments showed that glyceollin accumulation was localized to stelar tissue where cellular browning was most frequent. Glyceollin was found to be nematistatic rather than nematicidal to *M. incognita*, the ED_{50} for inhibition of motility after 24 h being $11 \mu g/ml$. The striking effects of glyceollin on motility may help to explain the observed failure of larvae to migrate within responding resistant roots. Kaplan *et al.* (1980) concluded that glyceollin accumulation is the cause of the failure of *M. incognita* to multiply in roots of soybean cv. Centennial. By contrast, they found that glyceollin did not accumulate in Centennial roots during invasion by the compatible *M. javanica* and that this nematode was highly tolerant of glyceollin.

Resistance to viruses

Antifungal phytoalexins accumulate during the production of local necrotic lesions by viruses in leaves of legumes and *Nicotiana* spp., but they are absent from systemically infected plants (Klarman and Hammerschlag, 1972; Bailey, 1973; Bailey *et al.*, 1976). There have been few attempts to determine if phytoalexins suppress viral replication and thereby restrict lesion size. Klarman and Hammerschlag (1972) found that incubation of tobacco necrosis virus (TNV) in a soybean leaf extract containing low concentrations of glyceollin had no effect on viral infectivity. They suggested, however, that the presence of the phytoalexin in tissues immediately surrounding lesions might indirectly render them unsuitable for further virus multiplication. Glyceollin was not translocated and was not involved in systemic protection against TNV afforded by the prior inoculation of soybean leaves with the virus (Klarman and Hammerschlag, 1972).

Cross-protection phenomena

There have been many reports of plants becoming resistant to microbial pathogens following previous exposure to related avirulent strains or non-pathogenic species (Matta, 1971; Kelman and Sequeira, 1972; Kuć, 1976; Kuć and Caruso, 1977; Sequeira, 1979). The term "cross-protection" has often been used to refer to this phenomenon (Skipp and Deverall, 1973).

The protection of tuber tissue against *Phytophthora infestans* led Müller (1958, 1959) to propose the phytoalexin theory of disease resistance. In their initial experiments Müller and Börger (1940) demonstrated that a compatible race of *P. infestans* was unable to colonize tissue previously inoculated with zoospores of an incompatible race of the late blight fungus. The protective effect was highly localized and Müller and Börger (1940) suggested that it was caused by the production of fungitoxic substances (phytoalexins) by potato cells responding to the initial inoculation. The phytoalexins produced were considered not only to prevent the growth of incompatible hyphae but also to inhibit subsequently the development of the compatible race on the protected tissue. As previously discussed, the available evidence suggests that accumulation of the phytoalexin rishitin is the main cause of the restriction of incompatible hyphae within hypersensitively responding cells (Sato *et al.*, 1971), but the precise role of rishitin and other phytoalexins in cross protection has not been studied in the same detail. Results obtained by Varns and Kuć (1971) led them to suggest that protection was not caused entirely by the direct effects of rishitin or phytuberin produced in response to the initial challenge, but by an induced alteration in the response of cells to the compatible race. Affected cells appeared to respond to the second inoculation in a hypersensitive manner, thereby triggering further synthesis and accumulation of the phytoalexins which restricted fungal growth. Interestingly, Varns and Kuć (1971) also found that prior inoculation with the compatible race suppressed the hypersensitive reaction and associated phyto-alexin accumulation in response to incompatible races or homogenates of the fungus (see chapter 3).

A more direct role for phytoalexins has been proposed to explain protection against *Verticillium albo-atrum* in cotton induced by prior inoculation with heat-inhibited or heat-killed conidia of the fungus. The accumulation of gossypol and related compounds found in xylem exudates and vascular tissues was considered by Bell and Pressley (1969) to be sufficient to prevent subsequent colonization by the pathogen. Svoboda and Paxton (1972) also concluded that phytoalexin accumulation was the cause of the local protection of soybean against *Phytophthora megasperma* induced by inoculation with incompatible races of the fungus or the non-pathogen *P. cactorum*. By contrast, Deverall *et al.* (1979) failed to detect significant accumulation of phytoalexins in wheat roots protected against *Gaeumannomyces graminis* var. *tritici* by prior inoculation with *G. graminis* var. *graminis*, and suggested that enhanced lignification might be involved in this cross-protection reaction.

Cross-protection against *Colletotrichum lindemuthianum* in French bean hypocotyls has been studied extensively (see Kuć and Caruso, 1977; Skipp and Deverall, 1973). Two types of cross-protection can be achieved; local protection, when challenge inocula are applied to the same sites as the initial infection, and systemic protection, when sites of inducer and challenge are separated on the bean hypocotyl. Systemic protection appears to operate only on etiolated hypocotyls (Berard *et al.*, 1972; Skipp and Deverall, 1973). Both types of protection are induced by fungi causing hypersensitivity, for example, incompatible races of *C. lindemuthianum*, or *C. lagenarium*. Local, but not systemic protection is induced by *C. trifolii* which does not cause lesion formation as it fails to penetrate into epidermal cells. In protected tissues, inhibition of the compatible race occurs at a later stage of infection development than in genetically resistant hypocotyls, but it is similarly associated with necrosis and browning of invaded plant cells (Skipp and Deverall, 1973; Elliston *et al.*, 1976).

The isoflavonoid phytoalexins from *Phaseolus vulgaris* were found by Elliston *et al.* (1977) to accumulate only at inoculation sites bearing brown cells; they were not detected at sites inoculated with *C. trifolii* or in systemically protected tissues until the challenge inoculation had caused the necrotic response. Although the accumulation of phytoalexins in bean hypocotyls in response to certain inducer inoculations may therefore directly help to explain local cross-protection, it clearly does not, in itself, explain either systemic protection or the local protection induced by *C. trifolii*. Protection in the latter cases seems to involve the "sensitization" of cells so that they respond to the challenge inoculum by undergoing a necrotic reaction which leads to the highly localized accumulation of phytoalexins. The further significance of the work on protection against bean anthracnose is that it has demonstrated a separation of phytoalexin accumulation from some unknown factor which "sensitizes" susceptible tissues. Thus, although phytoalexin accumulation represents the expression of resistance and inhibits fungal growth, other factors determine whether or not the resistance mechanism is expressed.

Currently, the most striking examples of systemic cross-protection occur in the cucurbits. For example, inoculation of one leaf of a cucumber plant with *Colletotrichum lagenarium*, TNV, or *Pseudomonas lachrymans* induced systemic resistance to anthracnose caused by *C. lagenarium* in other leaves (Jenns and Kuć, 1980). The involvement of phytoalexins in these interactions has received little attention. No phytoalexins have been characterized from cucumber but Kuć and Caruso (1977) reported the detection of a chloroform soluble inhibitor produced in tissue bearing limited lesions caused by *C. lagenarium*. In contrast to the behaviour of *C. lindemuthianum* in protected French bean tissue, *C. lagenarium* typically fails to penetrate into systemically protected leaves and in consequence does not cause lesion development (Richmond *et al.*, 1979). If phytoalexins are involved in protection phenomena in cucumber, the factors controlling their synthesis and also their sites of activity in the cucurbit must be quite different from those in French bean or potato. Kuć and Caruso (1977) have suggested that the accumulation of agglutinating factors may be a more important mechanism of resistance in cucurbits.

Cross protection against bacterial pathogens may involve the induced accumulation of phytoalexins. For example, Rathmell and Sequeira (1975) demonstrated the presence of a low molecular weight bacteriostatic compound in the intercellular fluids of tobacco leaves inoculated with heat killed cells of *Pseudomonas solanacearum*. They suggested that accumulation of this phytoalexin might explain the failure of compatible or incompatible bacteria to multiply rapidly and cause symptoms in tissues previously inoculated with heat killed cells (Sequeira and Hill, 1974).

There is no evidence to suggest that phytoalexins contribute to systemic protection against plant viruses (Sequeira, 1979).

Experimental approaches to the examination of the role of phytoalexins in resistance

Induced resistance and susceptibility

If phytoalexins are the determinants of resistance in a particular host/parasite interaction, treatments which induce resistance or susceptibility should affect, either directly or indirectly, the accumulation of phytoalexins. Experimental alteration of the plant's response therefore provides a means of testing the role of phytoalexins in resistance. Some examples of this type of experimentation have been described above, for example, cross protection phenomena, heat induced resistance to bean anthracnose (Figure 8.3) and the induction of susceptibility to *Phytophthora megasperma* in soybean hypocotyls treated with blasticidin S (Figure 8.4). A number of other examples also merit discussion.

The virulence of *Botrytis cinerea* to leaves of *Vicia faba* can be enhanced by the addition of certain nutrients to inoculum droplets. Most striking are the effects of pollen grains and wheat-germ extract which transform *B. cinerea* into an aggressive pathogen causing spreading lesions (Chou and Preece, 1968; Mansfield, 1972; Deramo, 1980). Pollen extracts decrease the sensitivity of *B. cinerea* to wyerone acid, the predominant phytoalexin from broad bean leaves (Deverall and Rogers, 1972) and this may alter the balance described in Figure 8.2 in favour of the invading fungus. Concentrations of wyerone acid decrease rapidly once the lesions formed by *B. cinerea* start to spread in response to added nutrients (Mansfield, 1972; Deramo, 1980), suggesting that loss of resistance is due to the failure of wyerone acid to reach inhibitory concentrations around invading hyphae.

Jerome and Müller (1958) found that heating French bean pods at 44°C for 2 h immediately prior to inoculation decreased their resistance to colonization by the normally avirulent fungi *Botrytis cinerea* and *Sclerotinia fructicola*. Treated pods were colonized and rotted by both fungi, whereas untreated pods developed only localized flecked lesions. Loss of resistance was reversible, recovery taking place after 3 days when pods were stored at 20°C. However, when treated pods were stored at 2°C no significant recovery took place. The degree of susceptibility of the pods following various treatments was correlated inversely with phytoalexin accumulation recorded in inoculum droplets collected one day after inoculation. Heat treatment also reduced the respiration rates (oxygen uptake) of pods, but effects on respiration were much less striking than those on resistance or phytoalexin production. Thus, heat treatment reduced phytoalexin production by 90–100% while reductions in oxygen uptake within similar tissues were only around 40%. Jerome and Müller (1958) concluded that the sub-lethal heat treatment suppressed phytoalexin biosynthesis. An alternative explanation is that heat treatment caused tissues to leak more stimulants into inoculum droplets, thereby promoting fungal growth and phytoalexin detoxification, and altering a balance similar to that depicted in Figure 8.2, in favour of the invading fungus. Whatever the mechanism involved, the experiments of Jerome and Müller (1958) clearly support phytoalexin accumulation as the cause of restriction of fungal growth in untreated pod tissue.

Heat induced susceptibility has been associated with suppression of phytoalexin accumulation in other host-parasite interactions, for example, *Phytophthora megasperma* and soybean (Keen, 1971; Ward *et al.*, 1981) and *Erysiphe graminis* and barley (Oku *et al.*, 1979). Conversely, the resistance of soybean and French bean to *P. megasperma* and *Colletotrichum lindemuthianum* respectively, induced by treatment of hypocotyls with UV radiation, is thought to be due directly to the accumulation of isoflavonoid phytoalexins in response to irradiation (Bridge and Klarman, 1973; Andebrhan and Wood, 1980).

Resistance of bean hypocotyls to *C. lindemuthianum* has also been induced by incubating inoculated tissues in an atmosphere of 15% CO_2. Resistance was associated with the formation of limited lesions and the accumulation of high concentrations of phaseollin within infected tissue (Arnold and Rahe, 1977). In addition to providing support for the role of phaseollin in resistance, an intriguing feature of this study is that point-freezing injuries on etiolated hypocotyls caused phaseollin accumulation in air, but not in 15% CO_2. These results indicate that the processes leading to the accumulation of phaseollin at infection sites and at injury sites may be regulated differently.

There has been relatively little use of selective antimetabolites for evaluating resistance mechanisms in plants (Vance *et al.*, 1976; Yoshikawa, 1978; Heath, 1980). The main reason for this is probably the lack of suitable inhibitors. Ideally, the inhibitor will affect some aspect of the plant but not pathogen metabolism. Yoshikawa (1978) found that these criteria were fulfilled by actinomycin D and blasticidin S when applied to soybean hypocotyls inoculated with *Phytophthora* spp. Treatment with the transcription inhibitor actinomycin D or the protein synthesis inhibitor blasticidin S at the time of inoculation suppressed completely the resistance of hypocotyls to incompatible races of *P. megasperma*. Actinomycin D and blasticidin S also diminished resistance when applied up to 4 and 6 h after inoculation respectively. Loss of resistance was very closely correlated with the reduced accumulation of glyceollin at infection sites. Results obtained indicate that the expression of resistance and glyceollin accumulation both require *de novo* messenger RNA and protein synthesis soon after infection. Treatment with the inhibitors also suppressed glyceollin accumulation and resistance following inoculation with *Phytophthora* spp., normally non-pathogens of soybean.

Genetical approaches

If mutant plants were available lacking only the ability to produce phytoalexins, tests on their susceptibility to microbial attack would allow an unequivocal conclusion to be drawn on the contribution of phytoalexins to resistance. Such a useful mutant might lack an enzyme necessary for one of the final steps in phytoalexin biosynthesis, for example dimethylallylpyrophosphate:trihydroxy-pterocarpan dimethylallyl transferase in soybean (Zähringer *et al.*, 1979). Similarly, experiments with bacterial or fungal mutants highly tolerant or highly sensitive to phytoalexins would also allow their involvement to be critically assessed. Unfortunately, no such mutants have been obtained. Although Hadwiger *et al.* (1976) obtained mutants of *Pisum sativum* which produced pisatin constitutively, resistance of the plants to disease was not examined. Clearly, this type of genetical approach merits more detailed examination. Recent developments in tissue and protoplast culture may facilitate the selection of suitable mutant plants (Ingram and Helgeson, 1980).

Genetical approaches have been applied to the analysis of the relationship between virulence and phytoalexin tolerance with *Nectria haematococca* on pea, and *Botrytis fabae* and *B. cinerea* on broad bean. VanEtten *et al.* (1980) examined the association of pisatin tolerance and demethylation with virulence on pea in *N. haematococca* mating population (MP) VI. The conidial state of *N. haematococca* which causes disease on pea is called *Fusarium solani* f. sp. *pisi* but *N. haematococca* can be found in a variety of habitats other than diseased pea tissue, suggesting that a great deal of natural variability exists. Studies on an isolate of *N. haematococca* pathogenic to pea revealed that it was highly tolerant of pisatin which accumulated within infected tissue, and was able to detoxify the phytoalexin by demethylation (VanEtten and Pueppke, 1976; VanEtten and Stein, 1978). A total of 59 isolates with diverse origins were examined and virulence was found to be closely correlated both with tolerance to and ability to demethylate pisatin. Crosses between isolates with widely differing characteristics further confirmed the close relationships between virulence and pisatin tolerance which were always associated with an ability to metabolize the phytoalexin (see also chapter 7).

A similar approach was used for the analysis of the relative importance of several factors of potential significance to the pathogenicities of *B. cinerea* and *B. fabae* towards broad bean leaves (Hutson and Mansfield, 1980). After six generations of single spore isolation, 15 pure strains of each species were selected and found to differ in their degree of pathogenicity, indicating that both species were heterokaryotic or heteroplasmic for virulence factors. The performance of each isolate was examined in tests of different factors including germination on the leaf surface, formation of infection hyphae, ability to kill epidermal cells soon after penetration, germ-tube growth *in vitro* and sensitivity to wyerone acid. The virulence of each species was more closely correlated with the numbers of epidermal cells killed between $7\frac{1}{2}$ and $8\frac{1}{2}$ h after inoculation than with other factors, but phytoalexin sensitivity also differentiated most isolates of *B. cinerea* and *B. fabae*. Cell killing ability was also closely correlated with the virulence of 52 possible mutants of *B. fabae* recovered after treatment with nitrosoguanidine. Mutants of *B. fabae*, differing widely in virulence and cell killing ability were found to be equally sensitive to wyerone acid but they were all less sensitive to the phytoalexin than isolates of *B. cinerea*. Hutson and Mansfield concluded that the ability of *B. fabae* to kill epidermal cells soon after penetration, and thereby to suppress phytoalexin accumulation, is the major determinant of the virulence of the species. However, as phytoalexin tolerance distinguished between most isolates of *B. cinerea* and *B. fabae* the possibility remains that a degree of tolerance to wyerone acid is an essential requirement for the pathogenicity of *B. fabae* to *Vicia faba*.

Concluding remarks

In many plants, notably members of the Leguminosae and Solanaceae, phyto-alexin accumulation *alone* provides a satisfactory explanation for the restriction of microbial growth during some resistant reactions. An outstanding question remaining concerns the ubiquity of phytoalexins throughout the plant kingdom. As I have already mentioned, cucumber and wheat are two examples of plants which do not produce phytoalexins easily detectable by methods which have proved successful for many other species (Deverall, 1977; Baker and Smith, 1977; Kuć and Caruso, 1977). In these and possibly other plants, the release of preformed inhibitors, lignification or silicification of infected tissue or the presence and accumulation of agglutinating factors may also contribute to the restriction of microorganisms (Friend, 1976; Schönbeck, 1976; Kojima and Uritani, 1978; Vance *et al.*, 1980; Heath, 1980). The concept of coordinated mechanisms of defence was recently proposed by Kuć and Caruso (1977) and Mansfield (1980). In some plants phytoalexins, as low molecular weight lipo-philic compounds synthesized from remote precursors, may play a com-paratively minor part in a coordinated response.

Where phytoalexin accumulation makes a major contribution to resistance, virulence is associated with the failure of the inhibitors to reach or maintain antimicrobial concentrations around the invading microorganism. As we have seen, phytoalexin accumulation almost invariably follows necrosis of isolated challenged cells during infection development, phytoalexins being synthesized

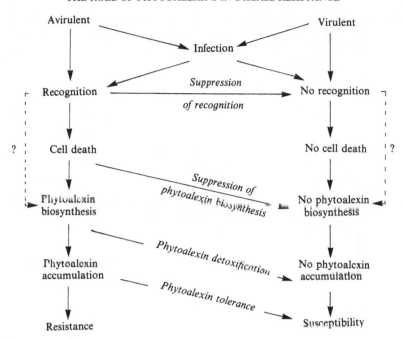

Figure 8.6 Factors affecting phytoalexin accumulation and disease resistance in plants following their inoculation with avirulent and virulent pathogens.

primarily by adjacent living tissues (see also chapter 9). The evidence therefore points to cell death being the trigger for subsequent phytoalexin biosynthesis and accumulation. In many examples of race-specific resistance, incompatible races seem to be recognized by invaded plant cells which then undergo a hypersensitive reaction. By contrast, cells fail to recognize the presence of compatible races which grow biotrophically at least during the early stages of infection. The concept of recognition is included in Figure 8.6 which summarizes the means by which virulent pathogens can avoid exposure to inhibitory concentrations of phytoalexins. In view of earlier discussions this figure is largely self-explanatory. I have, however, included an alternative route for triggering phytoalexin biosynthesis which avoids cell death, as it has been suggested (albeit in the absence of microscopical studies) that recognition may involve direct and specific elicitation of phytoalexin biosynthesis (Bruegger and Keen, 1979; Keen and Legrand, 1980). Mechanisms of phytoalexin accumulation are discussed more fully in chapter 9.

The development of mechanisms of disease resistance was undoubtedly of major importance during plant evolution. With the possible exception of Ginkgo (Christensen and Sproston, 1972) phytoalexins have not been reported from plants other than angiosperms (see chapter 4). Examination of the occurrence of phytoalexin production and other resistance processes in other gymnosperms and non-vascular plants should provide valuable clues to the role of phytoalexins in the evolution of plants and their parasites.

REFERENCES

Aist, J. R. (1976) Papillae and related wound plugs of plant cells. *Ann. Rev. Phytopathol.*, **14**, 145–163.

Allen, E. H. and Thomas, C. A. (1971) A second antifungal polyacetylene compound from *Phytophthora*-infected safflower. *Phytopathology*, **61**, 1107–1109.

Andebrhan, T. and Wood, R. K. S. (1980) The effects of ultraviolet radiation on the reaction of *Phaseolus vulgaris* to species of *Colletotrichum. Physiol. Plant Pathol.*, **17**, 105–110.

Arnold, R. M. and Rahe, J. E. (1977) Effects of 15% CO_2 on the accumulation of the phytoalexin phaseollin in *Phaseolus vulgaris* in response to mechanical injury and to infection by *Colletotrichum lindemuthianum. Can. J. Bot.*, **55**, 867–871.

Ayers, A. R., Ebel, J., Valent, B. and Albersheim, P. (1976) The fractionation and biological activity of an elicitor isolated from the mycelial walls of *Phytophthora megasperma* var. *sojae. Pl. Physiol.*, **57**, 760–765.

Bailey, J. A. (1969) Phytoalexin production by leaves of *Pisum sativum* in relation to senescence. *Ann. appl. Biol.*, **64**, 315–324.

Bailey, J. A. (1973) Production of antifungal compounds in cowpea (*Vigna sinensis*) and pea (*Pisum sativum*) after virus infection. *J. Gen. Microbiol.*, **75**, 119–123.

Bailey, J. A. (1974) The relationship between symptom expression and phytoalexin concentration in hypocotyls of *Phaseolus vulgaris* infected with *Colletotrichum lindemuthianum. Physiol. Plant Pathol.*, **4**, 477–488.

Bailey, J. A. (1981) "Physiological and biochemical events associated with the expression of resistance to disease", in *Active Defence Mechanisms in Plants*, ed. R. K. S. Wood, Plenum Press, New York and London, in press.

Bailey, J. A. and Deverall, B. J. (1971) Formation and activity of phaseollin in the interaction between bean hypocotyls (*Phaseolus vulgaris*) and physiological races of *Colletotrichum lindemuthianum. Physiol. Plant Pathol.*, **1**, 435–449.

Bailey, J. A. and Ingham, J. L. (1971) Phaseollin accumulation in bean (*Phaseolus vulgaris*) in response to infection by tobacco necrosis virus and the rust *Uromyces appendiculatus. Physiol. Plant Pathol.*, **1**, 451–456.

Bailey, J. A., Rowell, P. M. and Arnold, G. M. (1980) The temporal relationship between infected cell death, phytoalexin accumulation and the inhibition of hyphal development during resistance of *Phaseolus vulgaris* to *Colletotrichum lindemuthianum. Physiol. Plant Pathol.*, **17**, 329–339.

Bailey, J. A., Vincent, G. G. and Burden, R. S. (1976) The antifungal activity of glutinosone and capsidiol and their accumulation in virus-infected tobacco species. *Physiol. Plant Pathol.*, **8**, 35–41.

Baker, E. A. and Smith, I. M. (1977) Antifungal compounds in winter wheat resistant and susceptible to *Septoria nodorum. Ann. appl. Biol.*, **87**, 67–73.

Bell, A. A. (1969) Phytoalexin production and *Verticillium* wilt resistance in cotton. *Phytopathology*, **59**, 1119–1127.

Bell, A. A. and Pressley, J. T. (1969) Heat-inhibited or heat killed conidia of *Verticillium albo-atrum* induce disease resistance and phytoalexin synthesis in cotton. *Phytopathology*, **59**, 1147–1151.

Berard, D. F., Kuć, J. and Williams, E. B. (1972) A cultivar-specific protection factor from incompatible interactions of green bean with *Colletotrichum lindemuthianum. Physiol. Plant Pathol.*, **2**, 123–127.

Blakeman, J. P. (1973) The chemical environment of leaf surfaces with special reference to germination of pathogenic fungi. *Pestic. Sci.*, **4**, 575–588.

Bowen, G. D. (1979) "Integrated and experimental approaches to the study of growth of organisms around roots", in *Soil-borne Plant Pathogens*, eds. R. Schippers and W. Gams, Academic Press, London, 209–227.

Bridge, M. A. and Klarman, W. L. (1973) Soybean phytoalexin hydroxyphaseollin, induced by ultraviolet irradiation. *Phytopathology*, **63**, 606–609.

Bruegger, B. and Keen, N. T. (1977) "Phytoalexins and chemicals that elicit their production in plants", in *Host Plant Resistance to Pests*, ed. P. A. Hedin, American Chemical Society, Washington, D.C., 1–26.

Bruegger, B. B. and Keen, N. T. (1979) Specific elicitors of glyceollin accumulation in the *Pseudomonas glycinea*-soybean host-parasite system. *Physiol. Plant Pathol.*, **15**, 43–51.

Cartwright, D. W., Langcake, P., Pryce, R. J., Leworthy, D. P. and Ride, J. P. (1977) Chemical activation of host defence mechanisms as a basis for crop protection. *Nature*, **267**, 511–513.

Cartwright, D. W., Langcake, P., Pryce, R. J., Leworthy, D. P. and Ride, J. P. (1981) Isolation and characterization of phytoalexins from rice as momilactones A and B. *Phytochemistry*, **20**, 535–537.

Cartwright, D. W. and Russell, G. E. (1981) Possible involvement of phytoalexins in durable resistance of winter wheat to yellow rust. *Trans. Br. mycol. Soc.*, **76**, 323–325.

Cartwright, D. W., Langcake, P. and Ride, J. P. (1980) Phytoalexin production in rice and its enhancement by a dichlorocyclopropane fungicide. *Physiol. Plant Pathol.*, **17**, 259–267.

Chou, M. C. and Preece, T. F. (1968) The effect of pollen grains on infections caused by *Botrytis cinerea* Fr. *Ann. appl. Biol.*, **62**, 11–22.

Christensen, J. G. and Sproston, J. (1972) Phytoalexin production in *Ginkgo biloba* in relation to inhibition of fungal penetration. *Phytopathology*, **62**, 493–494 (abstract).

Cruickshank, I. A. M. and Perrin, D. R. (1965) Studies on phytoalexins. VII. The effect of some further factors on the formation, stability and localization of pisatin *in vivo*. *Aust. J. Biol. Sci.*, **18**, 817–828.

Cruickshank, I. A. M. and Perrin, D. R. (1971) Studies on phytoalexins. XI. The induction, antimicrobial spectrum and chemical assay of phaseollin. *Phytopath. Z.*, **70**, 209–229.

Deramo, A. (1980) Virulence enhancement of *Botrytis cinerea*. Ph.D. thesis, University of London.

Deverall, B. J. (1967) Biochemical changes in infection droplets containing spores of *Botrytis* spp. incubated in the seed cavities of pods of bean (*Vicia faba* L.). *Ann. appl. Biol.*, **59**, 375–387.

Deverall, B. J. (1977) *Defence Mechanisms of Plants*. Cambridge University Press.

Deverall, B. J and Rogers, P. M. (1972) The effect of pH and composition of test solutions on the inhibitory activity of wyerone acid towards germination of fungal spores. *Ann. appl. Biol.*, **72**, 301–305.

Deverall, B. J. and Vessey, J. C. (1969) Role of a phytoalexin in controlling lesion development in leaves of *Vicia faba* after infection by *Botrytis* spp. *Ann. appl. Biol.*, **63**, 444–458.

Deverall, B. J., Wong, P. T. W. and McLeod, S. (1979) Failure to implicate antifungal substances in cross-protection of wheat against take-all. *Trans. Br. mycol. Soc.*, **72**, 233–236.

De Wit, P. J. G. M. and Flach, W. (1979) Differential accumulation of phytoalexins in tomato leaves but not in fruits after inoculation with virulent and avirulent races of *Cladosporium fulvum*. *Physiol. Plant Pathol.*, **15**, 257–267.

Dickinson, C. H. (1967) Fungal colonization of *Pisum* leaves. *Can. J. Bot.*, **45**, 915–927.

Elliston, J., Kuć, J. and Williams, E. B. (1976) A comparative study of the development of compatible, incompatible and induced incompatible interactions between *Colletotrichum* spp. and *Phaseolus vulgaris*. *Phytopath. Z.*, **87**, 289–303.

Elliston, J., Kuć, J., Williams, E. B. and Rahe, J. E. (1977) Relationship of phytoalexin accumulation to local and systemic protection of bean against anthracnose. *Phytopath. Z.*, **88**, 114–130.

Essenberg, M., Hamilton, B., Cason, E. T. Jr., Brinkerhoff, L. A., Gholson, R. K. and Richardson, P. E. (1979) Localized bacteriostasis indicated by water dispersal of colonies of *Xanthomonas malvacearum* within immune cotton leaves. *Physiol. Plant Pathol.*, **15**, 69–78.

Fox, R. T. V., Manners, J. G. and Myers, A. (1971) Ultrastructure of entry and spread of *Erwinia carotovora* var. *atroseptica* into potato tubers. *Potato Res.*, **14**, 61–73.

Friend, J. (1976) "Lignification in infected tissue", in *Biochemical Aspects of Plant-Parasite Relationships*, eds. J. Friend and D. R. Threlfall, Academic Press, London, 291–303.

Gnanamanickam, S. S. and Patil, S. S. (1977a) Accumulation of antibacterial isoflavonoids in hypersensitively responding bean leaf tissues inoculated with *Pseudomonas phaseolicola*. *Physiol. Plant Pathol.*, **10**, 159–168.

Gnanamanickam, S. S. and Patil, S. S. (1977b) Phaseotoxin suppresses bacterially induced hypersensitive reaction and phytoalexin synthesis in bean cultivars. *Physiol. Plant Pathol.*, **10**, 169–179.

Hadwiger, L. A., Sander, C., Eddyvean, J. and Ralston, J. (1976) Sodiumazide induced mutants of peas that accumulate pisatin. *Phytopathology*, **66**, 629–630.

Hargreaves, J. A. and Bailey, J. A. (1978) Phytoalexin production by hypocotyls of *Phaseolus vulgaris* in response to constitutive metabolites released by damaged cells. *Physiol. Plant Pathol.*, 13, 89–100.

Hargreaves, J. A., Mansfield, J. W. and Rossall, S. (1977) Changes in phytoalexin concentrations in tissues of the broad bean plant (*Vicia faba* L.) following inoculation with species of *Botrytis*. *Physiol. Plant Pathol.*, 11, 227–242.

Heath, M. C. (1980). Reactions of nonsuscepts to fungal pathogens. *Ann. Rev. Phytopathol.*, 18, 211–236.

Heath, M. C. and Wood, R. K. S. (1971) Role of inhibitors of fungal growth in the limitation of leaf spots caused by *Ascochyta pisi* and *Mycosphaerella pinodes. Ann. Bot.*, 35, 475–491.

Higgins, V. J. and Millar, R. L. (1968) Phytoalexin production by alfalfa in response to infection by *Colletotrichum phomoides, Helminthosporium turcicum, Stemphylium loti* and *S. botryosum. Phytopathology*, 58, 1377–1383.

Hohl, M. R. and Stössel, P. (1976) Host-parasite interfaces in a resistant and a susceptible cultivar of *Solanum tuberosum* inoculated with *Phytophthora infestans*: tuber tissue. *Can. J. Bot.*, 54, 900–912.

Hohl, H. and Suter, E. (1976) Host-parasite interfaces in a resistant and a susceptible cultivar of *Solanum tuberosum* inoculated with *Phytophthora infestans:* leaf tissue. *Can. J. Bot.*, 54, 1956–1970.

Hollomon, D. W. (1967) Observations on the phylloplane flora of potatoes. *Eur. Potato J.*, 10, 53–61.

Hutson, R. A. and Mansfield, J. W. (1980) A genetical approach to the analysis of mechanisms of pathogenicity in *Botrytis/Vicia faba* interactions. *Physiol. Plant Pathol.*, 17, 309–317.

Hutson, R. A. and Smith, I. M. (1980) Phytoalexins and tyloses in tomato cultivars infected with *Fusarium oxysporum* f. sp. *lycopersici* or *Verticillium albo-atrum. Physiol. Plant Pathol.*, 17, 245–257.

Ingram, D. S. (1978) Cell death and resistance to biotrophs. *Ann. appl. Biol.*, 89, 291–295.

Ingram, D. S. and Helgeson, J. P. (eds.) (1980) *Tissue Culture Methods for Plant Pathologists.* Blackwell Scientific Publications, Oxford.

Ishisaka, N., Tomiyama, K., Katsui, N., Murai, A. and Masamune, T. (1969) Biological activities of rishitin, an antifungal compound isolated from diseased potato tubers, and its derivatives. *Pl. Cell Physiol.*, 10, 185–192.

Jenns, A. E. and Kuć, J. (1980) Characteristics of anthracnose resistance induced by localized infection of cucumber with tobacco necrosis virus. *Physiol. Plant Pathol.*, 17, 81–91.

Jerome, S. M. R. and Müller, K. O. (1958) Studies on phytoalexins. II. Influence of temperature on resistance of *Phaseolus vulgaris* towards *Sclerotinia fructicola* with reference to phytoalexin output. *Aust. J. Biol. Sci.*, 11, 301–314.

Jones, D. R., Unwin, C. H. and Ward, E. W. B. (1975) The significance of capsidiol induction in pepper fruit during an incompatible interaction with *Phytophthora infestans. Phytopathology*, 65, 1286–1288.

Kaplan, D. T., Keen, N. T. and Thomason, I. J. (1980) Association of glyceollin with the incompatible response of soybean roots to *Meloidogyne incognita. Physiol. Plant Pathol.*, 16, 309–318.

Kaplan, D. T., Thomason, I. J. and Van Gundy, S. D. (1979) Histological study of the compatible and incompatible interaction of soybeans and *Meloidogyne incognita. J. Nematol.*, 11, 338–343.

Keeling, B. L. (1976) A comparison of methods used to test soybeans for resistance to *Phytophthora megasperma* var. *sojae. Plant Dis. Rep.*, 60, 800–802.

Keen, N. T. (1971) Hydroxyphaseollin production by soybeans resistant and susceptible to *Phytophthora megasperma* var. *sojae. Physiol. Plant Pathol.*, 1, 265–275.

Keen, N. T. (1976) "Phytoalexins: a summary of points from contributions and discussions", in *Specificity in Plant Diseases*, eds. R. K. S. Wood and A. Graniti, Plenum Press, New York, 268–271.

Keen, N. T. and Kennedy, B. W. (1974) Hydroxyphaseollin and related isoflavonoids in the hypersensitive resistance response of soybean against *Pseudomonas glycinea. Physiol. Plant Pathol.*, 4, 173–185.

Keen, N. T. and Legrand, M. (1980) Surface glycoproteins: evidence that they may function as the race specific phytoalexin elicitors of *Phytophthora megasperma,* f. sp. *glycinea. Physiol. Plant Pathol.*, 17, 175–192.

Keen, N. T. and Littlefield, L. J. (1979) The possible association of phytoalexins with resistance gene expression in flax to *Melampsora lini. Physiol. Plant Pathol.*, 14, 265–280.

Kelman, A. and Sequeira, L. (1972) Resistance in plants to bacteria. *Proc. Royal Soc. Lond.* B., 181, 247–266.

Keogh, R. C., Deverall, B. J. and McLeod, S. (1980) Comparison of histological and physiological responses to *Phakopsora pachyrhizi* in resistant and susceptible soybean. *Trans. Br. mycol. Soc.*, **74**, 329–333.

Kitazawa, K. and Tomiyama, K. (1970) Microscopic observations of infection of potato cells by compatible and incompatible races of *Phytophthora infestans*. *Phytopath. Z.*, **66**, 317–324.

Klarman, W. L. and Hammerschlag, F. (1972) Production of the phytoalexin hydroxyphaseollin, in soybean leaves inoculated with tobacco necrosis virus. *Phytopathology*, **62**, 719–721.

Klement, Z. (1965) Method of obtaining fluid from the intercellular spaces of foliage and the fluid's merit as substrate for phytobacterial pathogens. *Phytopathology*, **55**, 1033–1034.

Knox, R. B. and Clarke, A. E. (1978) "Localization of proteins and glycoproteins by binding to labelled antibodies and lectins", in *Electron Microscopy and Cytochemistry of Plant Cells*, ed. J. L. Hall, Elsevier/North-Holland Biomedical Press, Amsterdam, 150–185.

Kojima, M. and Uritani, I. (1976) Possible involvement of furanoterpenoid phytoalexins in establishing host-parasite specificity between sweet potato and various strains of *Ceratocystis fimbriata. Physiol. Plant Pathol.*, **8**, 97–111.

Kojima, M. and Uritani, I. (1978) Studies on factor(s) in sweet potato which specifically inhibits germ tube growth of incompatible isolates of *Ceratocystis fimbriata. Pl. Cell Physiol.*, **19**, 71–81.

Kuan, T.-L. and Erwin, D. C. (1980) Formae speciales differentiation of *Phytophthora megasperma* isolates from soybean and alfalfa. *Phytopathology*, **70**, 333–338.

Kuć, J. (1975) Teratogenic constituents of potatoes. *Rec. Adv. Phytochem.*, **9**, 139–150.

Kuć, J. (1976) "Phytoalexins and the specificity of plant-parasite interaction", in *Specificity in Plant Disease*, eds. R. K. S. Wood and A. Graniti, Plenum Press, New York and London, 253–268.

Kuć, J. and Caruso, F. L. (1977) "Activated co-ordinated chemical defense against disease in plants", in *Host Plant Resistance to Pests*, ed. P. A. Hedin, American Chemical Society, Washington, D.C., 78–89.

Landes, M. and Hoffmann, G. M. (1979) Ultrahistological investigations of the interaction in compatible and incompatible systems in *Phaseolus vulgaris* and *Colletotrichum lindemuthianum. Phytopath. Z.*, **96**, 330–350.

Langcake, P. (1981) Disease resistance of *Vitis* spp. and the production of the stress metabolites resveratrol, ε-viniferin, α-viniferin and pterostilbene. *Physiol. Plant Pathol.*, **18**, 213–226.

Last, F. T. and Warren, R. C. (1972) Non-parasitic microbes colonizing green leaves: their form and functions. *Endeavour*, **31**, 143–150.

Littlefield, L. J. (1973) Histological evidence for diverse mechanisms of resistance to flax rust, *Melampsora lini* (Ehrenb.) Lev. *Physiol Plant Pathol.*, **3**, 241–247.

Lovrekovich, L., Lovrekovich, H. and Stahmann, M. A. (1967) Inhibition of phenol oxidation by *Erwinia carotovora* in potato tuber tissue and its significance in disease resistance. *Phytopathology*, **57**, 737–742.

Lund, B. M. and Nicholls, J. C. (1970) Factors influencing the soft rotting of potato tubers by bacteria. *Potato Res.*, **13**, 210–215.

Lyon, F. M. and Wood, R. K. S. (1975) Production of phaseollin, coumestrol and related compounds in bean leaves inoculated with *Pseudomonas* spp. *Physiol. Plant Pathol.*, **6**, 117–124.

Lyon, G. D. (1972) Occurrence of rishitin and phytuberin in potato tubers inoculated with *Erwinia carotovora* var. *atroseptica. Physiol. Plant Pathol.*, **2**, 411–416.

Lyon, G. D. and Bayliss, C. E. (1975) The effect of rishitin on *Erwinia carotovora* var. *atroseptica* and other bacteria. *Physiol. Plant Pathol.*, **6**, 177–186.

Lyon, G. D., Lund, B. M., Bayliss, C. E. and Wyatt, G. M. (1975) Resistance of potato tubers to *Erwinia carotovora* and formation of rishitin and phytuberin in infected tissue. *Physiol. Plant Pathol.*, **6**, 43–50.

Mace, M. E. (1978) Contribution of tyloses and terpenoid aldehyde phytoalexins to *Verticillium* wilt resistance in cotton. *Physiol. Plant Pathol.*, **12**, 1–11.

Macfoy, C. A. and Smith, I. M. (1979) Phytoalexin production and degradation in relation to resistance of clover leaves to *Sclerotinia* and *Botrytis* spp. *Physiol. Plant Pathol.*, **14**, 99–112.

Mansfield, J. W. (1972) Studies on the process of resistance of *Vicia faba* L. to infection by *Botrytis*. Ph.D. Thesis, University of London.

Mansfield, J. W. (1980) "Mechanisms of resistance to *Botrytis*", in *The Biology of Botrytis*, eds. J. R. Coley-Smith, W. R. Jarvis and K. Verhoeff, Academic Press, London, 181–218.

Mansfield, J. W. and Deverall, B. J. (1974) Changes in wyerone acid concentrations in leaves of *Vicia faba* after infection by *Botrytis cinerea* or *B. fabae. Ann. appl. Biol.*, **77**, 227–235.

Mansfield, J. W., Hargreaves, J. A. and Boyle, F. C. (1974) Phytoalexin production by live cells in broad bean leaves infected with *Botrytis cinerea. Nature*, **252**, 316–317.

Mansfield, J. W. and Hutson, R. A. (1980) Microscopical studies on fungal development and host responses in broad bean and tulip leaves inoculated with five species of *Botrytis. Physiol. Plant Pathol.*, **17**, 131–144.

Mansfield, J. W., Dix, N. J. and Perkins, A. (1975) Role of the phytoalexin pisatin in controlling saprophytic fungal growth on pea leaves. *Trans. Br. mycol. Soc.*, **64**, 507–511.

Mansfield, J. W., Porter, A. E. A. and Smallman, R. V. (1980) Dihydrowyerone derivatives as components of the furanoacetylenic phytoalexin response of tissues of *Vicia faba. Phytochemistry*, **19**, 1057–1061.

Mansfield, J. W. and Richardson, A. (1981) Ultrastructure of interactions between *Botrytis* species and broad bean leaves. *Physiol. Plant Pathol.*, in press.

Matta, A. (1971) Microbial penetration and immunization of uncongenial host plants. *Ann. Rev. Phytopathol.*, **9**, 387–410.

Müller, K. O. (1958) Studies on phytoalexins. I. The formation and immunological significance of phytoalexin produced by *Phaseolus vulgaris* in response to infections with *Sclerotinia fructicola* and *Phytophthora infestans. Aust. J. Biol. Sci.*, **11**, 275–300.

Müller, K. O. (1959) "Hypersensitivity", in *Plant Pathology I*, eds. J. G. Horsfall and A. E. Dimond, Academic Press, New York, 469–519.

Müller, K. O. and Börger, H. (1940) Experimentelle Untersuchungen uber die *Phytophthora*-Resistenz der Kartoffel zugleich ein Beitrag zum Problem der „erworbenen Resistenz" im Pflanzenreich. *Arb. Biol. Aust. Reichsanst (Berl.)*, **23**, 189–231.

O'Brien, F. and Wood, R. K. S. (1973*a*) Role of ammonia in infection of *Phaseolus vulgaris* by *Pseudomonas* spp. *Physiol. Plant Pathol.*, **3**, 315–325.

O'Brien, F. M. and Wood, R. K. S. (1973*b*) Anti-bacterial substances in hypersensitive responses induced by bacteria. *Nature*, **242**, 532–533.

Oku, H., Shiraishi, T. and Ouchi, S. (1979) "The role of phytoalexins in host-parasite specificity", in *Recognition and Specificity in Plant Host-Parasite Interactions*, eds. J. M. Daly and I. Uritani, Japan Scientific Societies Press, Tokyo, 317–333.

O'Neill, T. M. (1981) Narcissus smoulder; cause, epidemiology and host resistance. Ph.D. thesis, University of Stirling.

Partridge, J. E. and Keen, N. T. (1976) Association of phytoalexin kievitone with single-gene resistance of cowpeas to *Phytophthora vignae. Phytopathology*, **66**, 426–429.

Preece, T. F. (1976) "Some observations on leaf surfaces during the early stages of infection by fungi", in *Biochemical Aspects of Plant-Parasite Relationships*, eds. J. Friend and D. R. Threlfall, Academic Press, 1–10.

Pueppke, S. G. and VanEtten, H. D. (1974) Pisatin accumulation and lesion development in peas infected with *Aphanomyces euteiches, Fusarium solani* f. sp. *pisi*, or *Rhizoctonia solani. Phytopathology*, **64**, 1433–1440.

Pueppke, S. G. and VanEtten, H. D. (1976) The relation between pisatin and the development of *Aphanomyces euteiches* in diseased *Pisum sativum. Phytopathology*, **66**, 1174–1185.

Rahe, J. E. (1973) Occurrence and levels of the phytoalexin phaseollin in relation to delimitation of sites of infection of *Phaseolus vulgaris* by *Colletotrichum lindemuthianum. Can. J. Bot.*, **51**, 2423–2430.

Rathmell, W. G. and Sequeira, L. (1975) Induced resistance in tobacco leaves: the role of inhibitors of bacterial growth in the intercellular fluid. *Physiol. Plant Pathol.*, **5**, 65–73.

Rich, J. R., Keen, N. T. and Thomason, I. J. (1977) Association of coumestans with the hypersensitivity of lima bean roots to *Pratylenchus scribneri. Physiol. Plant Pathol.*, **10**, 105–116.

Richmond, S., Kuć, J. and Elliston, J. E. (1979) Penetration of cucumber leaves by *Colletotrichum lagenarium* is reduced in plants systemically protected by previous infection with the pathogen. *Physiol. Plant Pathol.*, **14**, 329–338.

Ride, J. P. (1978) The role of cell wall alterations in resistance to fungi. *Ann. appl. Biol.*, **89**, 302–306.

Rossall, S. (1978) The resistance of *Vicia faba* L. to infection by *Botrytis*. Ph.D. thesis, University of Stirling.

Rossall, S. and Mansfield, J. W. (1978) The activity of wyerone acid against *Botrytis. Ann. appl. Biol.*, **89**, 359–362.

Rossall, S. and Mansfield, J. W. (1980) Investigation of the causes of poor germination of *Botrytis* spp. on broad bean leaves (*Vicia faba* L.). *Physiol. Plant Pathol.*, **16**, 369–382.

Rossall, S., Mansfield, J. W. and Hutson, R. A. (1980) Death of *Botrytis cinerea* and *B. fabae* following exposure to wyerone derivatives *in vitro* and during infection development in broad bean leaves. *Physiol. Plant Pathol.*, **16**, 135–146.

Rudolph, K. (1976) "Models of interaction between higher plants and bacteria", in *Specificity in Plant Disease*, eds. R. K. S. Wood and A. Graniti, Plenum Press, New York and London, 109–126.

Sato, N., Kitazawa, K. and Tomiyama, K. (1971) The role of rishitin in localizing the invading hyphae of *Phytophthora infestans* in infection sites at the cut surface of potato tubers. *Physiol. Plant Pathol.*, **1**, 289–295.

Sato, N. and Tomiyama, K. (1969) Localized accumulation of rishitin in potato-tuber tissue infected by an incompatible race of *Phytophthora infestans*. *Ann. Phytopath. Soc. Japan*, **35**, 202–217.

Schönbeck, F. (1976) "Role of preformed factors in specificity", in *Specificity in Plant Diseases*, eds. R. K. S. Wood and A. Graniti, Plenum Press, New York and London, 237–252.

Scott, P. R., Johnson, R., Wolfe, M. S., Lowe, H. J. B. and Bennett, F. G. A. (1980) "Host-specificity in cereal parasites in relation to their control", in *Applied Biology*, *Vol. 5*, ed. T. H. Croaker, Academic Press, London, 349–407.

Sequeira, L. (1979) "Acquisition of systemic resistance by prior inoculation", in *Recognition and Specificity in Plant Host-Parasite Interactions*, eds. J. M. Daly and I. Uritani, Japan Scientific Societies Press, Tokyo, 231–251.

Sequeira, L. and Hill, L. M. (1974) Induced resistance in tobacco leaves: The growth of *Pseudomonas solanacearum* in protected tissues. *Physiol. Plant Pathol.*, **4**, 447–455.

Sherwood, R. T., Olah, A. F., Oleson, W. H. and Jones, E. E. (1970) Effect of disease and injury on accumulation of a flavonoid estrogen, coumestrol, in alfalfa. *Phytopathology*, **60**, 684–688.

Shimony, C. and Friend, J. (1975) Ultrastructure of the interaction between *Phytophthora infestans* and leaves of two cultivars of potato (*Solanum tuberosum* L.) orion and majestic. *New Phytol.*, **74**, 59–65.

Shimony, C. and Friend, J. (1976) Ultrastructure of the interaction between *Phytophthora infestans* and tuber slices of resistant and susceptible cultivars of potato (*Solanum tuberosum* L.) orion and majestic. *Israel J. Bot.*, **25**, 174–183.

Skipp, R. A. and Deverall, B. J. (1972) Relationships between fungal growth and host changes visible by light microscopy during infection of bean hypocotyls (*Phaseolus vulgaris*) susceptible and resistant to physiologic races of *Colletotrichum lindemuthianum*. *Physiol. Plant Pathol.*, **2**, 357–374.

Skipp, R. A. and Deverall, B. J. (1973) Studies on cross-protection in the anthracnose disease of bean. *Physiol. Plant Pathol.*, **3**, 299–314.

Smith, D. A., VanEtten, M. D. and Bateman, D. F. (1975) Accumulation of phytoalexins in *Phaseolus vulgaris* hypocotyls following infection by *Rhizoctonia solani*. *Physiol. Plant Pathol.*, **5**, 51–64.

Smith, I. M. (1978) The role of phytoalexins in resistance. *Ann. appl. Biol.*, **82**, 325–329.

Stakman, E. C. (1915) Relation between *Puccinia graminis* and plants highly resistant to its attack. *J. agric. Res.*, **4**, 193–200.

Stall, R. E. and Cook, A. A. (1968) Inhibition of *Xanthomonas vesicatoria* in extracts from hypersensitive and susceptible pepper leaves. *Phytopathology*, **58**, 1584–1587.

Stössel, P., Lazarovits, G. and Ward, E. W. B. (1980) Penetration and growth of compatible and incompatible races of *Phytophthora megasperma* var. *sojae* in soybean hypocotyl tissues differing in age. *Can. J. Bot.*, 2594–2601.

Svoboda, W. E. and Paxton, J. D. (1972) Phytoalexin production in locally cross-protected Harosoy and Harosoy-63 soybeans. *Phytopathology*, **62**, 1457–1460.

Tjamos, E. C. and Smith, I. M. (1974) The role of phytoalexins in the resistance of tomato to *Verticillium* wilt. *Physiol. Plant Pathol.*, **4**, 249–259.

Toyoda, H., Koga, H., Mayama, S. and Shishiyama, J. (1978) The isolation and chemical properties of yellow-fluorescent compound from incompatible barley leaves inoculated with *Erysiphe graminis hordei*. Abstract from *3rd International Congress of Plant Pathology, Munchen*, Paul Parcy, Berlin, 218.

Vance, C. P., Kirk, T. K. and Sherwood, R. T. (1980) Lignification as a mechanism of disease resistance. *Ann. Rev. Phytopathol.*, **18**, 259–288.

VanEtten, H. D., Matthews, P. S., Tegtmeier, K., Dietert, M. F. and Stein, J. I. (1980) The association of pisatin tolerance and demethylation with virulence on pea in *Nectria haematococca*. *Physiol. Plant Pathol.*, **16**, 257–268.

VanEtten, H. D. and Stein, J. I. (1978) Differential response of *Fusarium solani* isolates to pisatin and phaseollin. *Phytopathology*, **68**, 1276–1283.

Varns, J. L. and Kuć, J. (1971) Suppression of rishitin and phytuberin accumulation and hypersensitive response in potato by compatible races of *Phytophthora infestans*. *Phytopathology*, **61**, 178–181.

Ward, E. W. B., Lazarovits, G., Stössel, P., Barrie, S. D. and Unwin, C. H. (1981) Glyceollin production associated with control of *Phytophthora* rot of soybeans by the systemic fungicide, Ridomil. *Phytopathology*, in press.

Ward, E. W. B., Lazarovits, G., Unwin, C. H. and Buzzell, R. I. (1979) Hypocotyl reactions and glyceollin in soybeans inoculated with zoospores of *Phytophthora megasperma* var. *sojae*. *Phytopathology*, **69**, 951–955.

Webster, D. M. and Sequeira, L. (1977) Expression of resistance in bean pods to an incompatible isolate of *Pseudomonas syringae*. *Can. J. Bot.*, **55**, 2043–2052.

Yoshikawa, M. (1978) *De novo* messenger R.N.A. and protein synthesis are required for phytoalexin-mediated disease resistance in soybean hypocotyls. *Pl. Physiol.*, **61**, 314–317.

Yoshikawa, M., Yamauchi, K. and Masago, H. (1978) Glyceollin: its role in restricting fungal growth in resistant soybean hypocotyls infected with *Phytophthora megasperma* var. *sojae*. *Physiol. Plant Pathol.*, **12**, 73–82.

Zähringer, U., Ebel, J., Mulheirn, L. J., Lyne, R. L. and Grisebach, H. (1979) Induction of phytoalexin synthesis in soybean. Dimethylallylpyrophosphate : trihydroxypterocarpan dimethylallyl transferase from elicitor induced cotyledons. *FEBS Lett.*, **101**, 90–92.

9 Mechanisms of phytoalexin accumulation

J. A. BAILEY

Introduction

Phytoalexins are diverse compounds contributing to many of the major groups
of secondary metabolites, including simple phenolics, flavonoids, isoflavonoids,
stilbenes, terpenes and polyacetylenes. Elucidation of the mechanisms by which
these products accumulate in plants is an important challenge to plant scientists,
and represents a great opportunity for increased understanding of metabolic
control systems. This opportunity should not be obscured by the arguments
which still remain about the contribution of phytoalexins to resistance to
disease.

It will become clear that phytoalexins are produced by plants not only in
response to interactions with fungi, bacteria, viruses, nematodes and other living
organisms, but also following treatment with many chemicals, irradiation by
ultra-violet light and exposure to the products of microbial metabolism. These
metabolites were first termed "inducers" (Cruickshank and Perrin, 1968), but
the implications of this word for an involvement in gene transcription and
translation led Keen (1975) to propose the alternative word "elicitor" which he
originally used to describe agents that caused plant tissues to become resistant
to disease. However, because phytoalexins are an important aspect of resistance,
an elicitor has become accepted to denote any compound which "elicits"
accumulation of phytoalexins (see also Ward and Stoessl, 1976). In addition,
elicitors have been divided into those of a biological origin, *biotic elicitors*, and
those of a physical or chemical nature, *abiotic elicitors* (Yoshikawa, 1978). It
must be emphasized that this division, although convenient, may be entirely
artificial and does not necessarily reflect differences in their modes of action, nor
imply involvement in host-parasite interactions. A further category of elicitor
has been proposed to describe those obtained from plant tissues. These were first
described as *constitutive* (Hargreaves and Bailey, 1978) but more recently
essentially similar materials have been termed *endogenous* (Hahn *et al.*, 1981).

This chapter will close by considering possible mechanisms of phytoalexin accumulation and the involvement of elicitors in disease processes, but this will be preceded by discussion of the many ways in which phytoalexins can be produced, the techniques which yield greatest amounts, the distribution of phytoalexin synthesis and accumulation in affected tissues and how synthesis of these products begins.

Methods of inducing accumulation of phytoalexins

Fungi

The concept and early isolation of phytoalexins were based on experiments with fungi, and, as indicated in earlier chapters, research with these organisms has always been predominant. Results have shown that "fungi are particularly efficient inducing agents" (VanEtten and Pueppke, 1976).

Various drop-diffusate techniques (chapter 2) have been used to demonstrate the ability of many fungi to cause phytoalexin accumulation. Fungi which were effective on pea (Cruickshank and Perrin, 1963), French bean (Cruickshank and Perrin, 1971) and pepper (Stoessl *et al.*, 1974) include pathogens of these species, pathogens of other species and even some saprophytes. It is also evident from studies with this technique that a single fungus, e.g. *Helminthosporium carbonum*, which is a pathogen of maize and has been used extensively in the studies by Ingham (chapter 2), can cause the formation of different phytoalexins in many different species. One reservation regarding the isolation of phytoalexins from infection droplets is the possibility that not all the metabolites obtained are necessarily phytoalexins. Some could be normal fungal metabolites, whilst others may be the product of metabolism of a phytoalexin by the fungus. Thus demethylmedicarpin was isolated with medicarpin from *Trifolium repens*, but its presence in infection droplets was shown to be due to metabolism of medicarpin by *Monilinia fructicola* (Woodward, 1981). The concentrations of phytoalexins produced in diffusates vary considerably, e.g. from 10 to 200 μg/ml. However, if the volume of diffusate is large, as from pods of legumes or from fruits of solanaceous plants, the amounts of phytoalexins produced may be quite considerable. The yields from other tissues, notably leaves, are usually much less.

The drop-diffusate technique has been of great importance to the isolation and identification of many new phytoalexins. It was soon realized, however, that diffusate techniques were of little value when assessing the contribution of phytoalexins to disease resistance. As a result, research has been increasingly orientated toward studies with infected tissues. Many of these have been referred to earlier (chapter 8), but further consideration of some of this information is also essential for assessing methods of obtaining phytoalexins and for considering the mechanisms leading to phytoalexin production.

The structures of phytoalexins are determined solely by the plant producing them (Cruickshank, 1963 and see chapter 1). Thus tissues infected by different fungi generally contain a similar range of phytoalexins; a range which is analogous to that obtained from diffusates or after treatment with elicitors. For example, phaseollin, phaseollidin and phaseollinisoflavan were produced by French bean in response to infection by *Botrytis cinerea* (Van den Heuvel and Grootveld, 1980; Fraile *et al.*, 1980), *Colletotrichum lindemuthianum* (Bailey, 1974), *Fusarium solani* (VanEtten and Smith, 1975), *Rhizoctonia solani* (Smith *et al.*, 1975) or *Thielaviopsis basicola* (VanEtten and Smith, 1975). Similarly, medicarpin was isolated from *Medicago sativa* infected with *Colletotrichum phomoides, Stemphylium loti, S. botryosum, Phoma herbarum* or *Leptosphaeria briossiana* (Higgins, 1972) and glyceollins from soybean infected with *Phytophthora megasperma* var. *sojae* or *P. cactorum.* Rishitin was often isolated from potato tubers infected with *P. infestans* (Kuć, 1975; Tomiyama *et al.*, 1979) but was also obtained from tissue inoculated with other öomycetes, *Helminthosporium carbonum, Ceratocystis fimbriata* (Lisker and Kuć, 1977) or *Fusarium* spp. (Corsini and Pavek, 1980). Other, albeit less frequent, reports have indicated some interesting exceptions. Kievitone was the predominant phytoalexin in French bean tissues infected with *Rhizoctonia solani* (Smith *et al.*, 1975) and also occurred in tissues infected with *Colletotrichum lindemuthianum* (Bailey, 1974) or *Botrytis cinerea* (Fraile *et al.*, 1980). It was not, however, detected in lesions caused by *Fusarium solani* or *Thielaviopsis basicola* (VanEtten and Smith, 1975). Analysis of sesquiterpenes produced by several cultivars of potato infected with various isolates of *Phytophthora infestans, Fusarium avenaceum* or *Phoma exigua* also revealed differences in the phytoalexins produced. Solavetivone was produced most consistently, but rishitin, although present in tissue infected with most isolates of *P. infestans* was rarely present in tissues infected with *F. avenaceum* or *Phoma exigua* (Price *et al.*, 1976).

The quantities of phytoalexins in infected tissue also change during the expression of resistant reactions. Normally this is evident as a gradual or a rapid increase in concentration to a degree which is often maintained or only slightly reduced. Other patterns have, however, been reported. Smith *et al.* (1975) found that kievitone accumulated very rapidly in young lesions of *Rhizoctonia solani* on French bean stems. Within a few days the amounts of kievitone present were very much lower, whereas the concentrations of phaseollin had increased during this period. A similar reduction in the concentration of rishitin was observed in potato tubers following the hypersensitive reaction (Kuć, 1975).

As found with diffusates, infection of different plants by the same fungus can lead to formation of different phytoalexins. The amount of evidence is, however, restricted by shortage of fungi which can interact pathogenically with more than one species. As a consequence, the clearest examples come from infections with the ubiquitous necrotroph *Botrytis cinerea*; acetylenic compounds were formed by broad bean (Mansfield, 1980), isoflavonoids by French bean (Fraile *et al.*, 1980), 6-methoxymellein by carrot (Coxon *et al.*, 1973), resveratrol by vine (Langcake and Pryce, 1977) and capsidiol by sweet peppers (Stoessl *et al.*, 1974).

Phytoalexin production is not restricted to infections by facultative pathogens.

Mildew (Shiraishi *et al.*, 1976) and rust fungi can also be effective. *Uromyces appendiculatus* and *Phakopsora pachyrhizi* produced phytoalexins in French bean (Bailey and Ingham, 1971) and soybean (Keogh *et al.*, 1980) respectively, but only in cultivars where infection caused death and browning of some tissues. Phytoalexins were not produced by cultivars in which growth was entirely biotrophic. The absence of phytoalexins from tissues supporting biotrophic growth has also been described with French beans infected with *Colletotrichum lindemuthianum* (Bailey, 1981) and potatoes infected with *Phytophthora infestans* (Tomiyama *et al.*, 1979). These examples illustrate important exceptions to the view that fungal infections always lead to accumulation of phytoalexins. An essential process for the formation of phytoalexins appears to be death of infected tissues. Thus when growth of *C. lindemuthianum* and *P. infestans* became necrotrophic, phytoalexins were again produced. The importance of this observation will be discussed later.

Bacteria

Early surveys indicated that bacteria caused less phytoalexin accumulation than fungi. Several bacteria failed to cause pisatin to form in pea pods (Cruickshank and Perrin, 1963) and only small amounts of phaseollin were produced by pods of French bean (Cruickshank and Perrin, 1971). These results, based on diffusate techniques, may have discouraged further work, but in 1971 Stholasuta *et al.* reported large concentrations of phaseollin (226 μg/g tissue) in French bean leaves which had reacted hypersensitively to infiltration with an incompatible race of *Pseudomonas phaseolicola*. Further work showed that phaseollidin, phaseollinisoflavan, kievitone and coumestrol are also produced (Gnanamanickan and Patil, 1977). The non-pathogenic bacterium *P. morsprunorum* was also effective but phytoalexins were not obtained from leaves inoculated with the saprophyte *P. fluorescens* (Lyon and Wood, 1975). An earlier failure to obtain phytoalexins from leaves infiltrated with *P. morsprunorum* (Stholasuta *et al.*, 1971) may have been due to very rapid dehydration of entire inoculated leaves (see chapter 8). *Pseudomonas* spp. cause phytoalexins to accumulate in other plant species. Glyceollin and coumestrol, along with daidzein and sojagol, were produced by soybean infected with *P. glycinea* (Keen and Kennedy, 1974). As with infection of French beans inoculation with incompatible races of these pathogens produced more phytoalexins than compatible races.

The potential complexities of responses to bacteria are illustrated by studies with soybean leaves and cotyledons. Leaves inoculated with *P. pisi* produced three glyceollin isomers, isoformonenetin, glyceofuran, 9-*O*-methylglyceofuran and glyceocarpin (Ingham *et al.*, 1981). Cotyledons inoculated with *Erwinia carotovora* also produced the glyceollin isomers, but the major product was a trihydroxypterocarpan, glycinol (Weinstein *et al.*, 1981). Glycinol was not produced by soybean leaves.

Sesquiterpene phytoalexins are produced in bacteria-infected tissues of solanaceous plants. In 1972, Lyon reported that rishitin and phytuberin occurred at high concentrations (100 to 1000 µg/g tissue) in potato tubers colonized by *E. atroseptica*. These compounds were also present, with several other sesquiterpenes, in potatoes infected with *E. carotovora*. Rishitinol and lubimin were not detected, although they had been obtained from fungus-infected tissues (Coxon *et al.*, 1974). *E. carotovora* also caused capsidiol to form in pepper fruits (Ward *et al.*, 1973) and phytuberin was isolated from leaves of *Nicotiana tabacum* inoculated with *P. lachrymans* (Hammerschmidt and Kuć, 1979).

Viruses

The formation of virus-induced local lesions is associated with many biochemical changes, including production of phenolic compounds and pigments (Farkas and Kiraly, 1962). Antifungal activity was first reported using extracts of soybean leaves infected with tobacco necrosis virus (TNV) (Hammerschlag and Klarman, 1969). Subsequently phaseollin, phaseollidin, phaseollinisoflavan and kievitone were obtained from TNV-infected French beans (Burden *et al.*, 1972) and glyceollin from virus-infected soybean (Klarman and Hammerschlag, 1972). Since this work, virus-induced local lesions in several plants have been shown to be an excellent source of phytoalexins. Phaseollin, phaseollidin, kievitone and 2-O-methylphaseollidinisoflavan were obtained from infected *Vigna* spp. (Bailey, 1973; Preston, 1975), pisatin from pea (Bailey, 1973) and medicarpin from *Canavalia ensiformis* (Lampard, 1974). Similarly, several phytoalexins were produced by virus-infected tobaccos. Capsidiol, solavetivone, 3-hydroxy-solavetivone, solascone, phytuberin and phytuberol were obtained from TNV-infected *Nicotiana tabacum* (Bailey *et al.*, 1976; Uegaki *et al.*, 1980) and glutinosone from *N. glutinosa* infected with tobacco mosaic virus. Glutinosone was not obtained from *N. glutinosa* infected systemically with TNV (Bailey *et al.*, 1976).

The amounts of phytoalexins obtained from these sources can be large (10 to 500 µg/g tissue), particularly if the integrity of the infected tissues can be maintained for several days after symptoms appear. Thus virus-infected tissues have been useful not only for the isolation of some new phytoalexins, but also for the provision of quantities sufficient for studies on their metabolism and toxicity (Bailey and Skipp, 1978).

Animals

Infestation of plant tissues by weevils or nematodes leads to the production of phytoalexins. Ipomeamarone, ipomeamarol and dehydroipomeamarone were obtained from sweet potatoes infected with *Cylas formicarius* or *Euscepes postfasciatus* (Uritani *et al.*, 1975). Various coumestans, phaseollin or glyceollin occurred in lima bean (Rich *et al.*, 1977), French bean (Abawi *et al.*, 1971) or soybean (Kaplan *et al.*, 1980) infected with *Pratylenchus scribneri*, *P. penetrans* or *Meloidogyne incognita* respectively. *M. incognita* also caused several terpene aldehydes, including gossypol, to accumulate in cotton (Veech, 1978). In all these investigations the phytoalexins were concentrated within pigmented necrotic tissues.

Chemicals (*abiotic elicitors*)

Many chemicals cause phytoalexins to accumulate; the same chemical causing production of phytoalexins by different plant species. The information available is extremely extensive, and as a consequence only selected samples can be mentioned here. Further information is contained within the references below.

Salts of heavy metals, e.g. mercury and copper, induce accumulation of many phytoalexins. Isoflavonoids were produced by pea (Perrin and Cruickshank, 1965). French bean (Hargreaves, 1979) and soybean (Yoshikawa, 1978); wyerone and other acetylenes by *Vicia faba* (Mansfield *et al.*, 1980); 6-methoxymellein by carrot (Coxon *et al.*, 1973); furanoterpenes by sweet potato (Uritani *et al.*, 1960) and sesquiterpenes by potato (Tomiyama and Fukaya, 1975). The amounts of phytoalexins produced were sometimes small and in some investigations could not be obtained from chemically treated tissues (Lisker and Kuć, 1977; see chapter 3). It has been suggested that with potato tuber tissue, lack of accumulation may be a consequence of unsuitable storage conditions (Cheema and Haard, 1978).

Respiratory inhibitors, e.g. sodium iodoacetate, sodium fluoride, potassium cyanide and 2,4-dinitrophenol are also effective in producing isoflavonoids (Perrin and Cruickshank, 1965), sesquiterpenes (Cheema and Haard, 1978) and furanoterpenes (Uritani *et al.*, 1960). Several surfactants can also act as elicitors. Glyceollin was produced by soybeans treated with Triton X-100 or Nonidet P40 (Yoshikawa, 1978); furanoterpenes by sweet potato treated with sodium dodecyl sulphate (Oba *et al.*, 1976) and phaseollin, kievitone and licoisoflavone A by French bean treated with several phytotoxic Triton surfactants (Hargreaves, 1981). Finally, antibiotics, which have been used to assess whether formation of phytoalexins requires synthesis of new enzymes (see p. 362), are sometimes also elicitors *per se*. Effective compounds include actinomycin D, puromycin and cycloheximide (Schwochau and Hadwiger, 1968).

The chemicals discussed above affect legumes, sweet potatoes, carrots and perhaps potatoes. Many other chemicals have been shown to be elicitors, but their effects have only been investigated with pea. Hadwiger and his colleagues (see Hadwiger *et al.*, 1974; Sander and Hadwiger, 1979) have shown that pisatin was produced in response to phenothiazine-derived tranquillizers; various anti-histaminic, antiviral and antimalarial drugs; snake and bee venom and other basic polypeptides; compounds that intercalate with nucleic acids, e.g. acridine orange, ethidium bromide and 9-aminoacridine; sulphydryl compounds, e.g. dimercaptoethanol; amino acids, particularly DL-methionine, DL-norleucine, DL-norvaline and L-valine; and psoralen compounds when photoactivated. Several fungicides, e.g. triazin, triphenyltin and phenylmercuriacetate also elicited pisatin production in pea (Oku *et al.*, 1973), whilst others, e.g. maneb and benomyl were elicitors of glyceollin (Reilly and Klarman, 1980).

It has also been suggested that compounds with plant growth regulating properties, e.g. ethylene, indolyl 3-acetic acid, 2,4-dichlorophenoxyacetic acid and 2,4,5-trichlorophenoxyacetic acid are elicitors. Pisatin was obtained from wounded pea pods after exposure to ethylene (Chalutz and Stahmann, 1969) and phytuberin and phytuberol were produced in wounded tobacco leaves treated with ethrel. It is interesting to note that these wounds were also darkly pigmented (Uegaki *et al.*, 1980). The use of wounded tissue does not readily facilitate a distinction between ethylene (or other elicitors) instigating phytoalexin synthesis or modifying subsequent processes. Thus Chalutz *et al.* (1969) concluded that ethylene did not cause synthesis of 6-methoxymellein in carrot, but was essential for its subsequent accumulation. More recently Paradies *et al.* (1980) concluded that, in soybean, ethylene is only a consequence of elicitation and does not act on synthesis of glyceollin, nor does it affect the amounts which accumulate.

Physical agents

(i) *Wounds.* Wounding tissue, e.g. by cutting, bruising or pricking, rarely leads to the accumulation of phytoalexins. Phytoalexins could not be detected in water placed on pea pod tissue or on tobacco leaves which had been pricked with a sterile needle (Cruickshank and Perrin, 1963; Uegaki *et al.*, 1980). However, extract of leaves of *Vicia faba*, which had been bruised, became fungitoxic, although the identity of the active components, possibly acetylenic phytoalexins, was not established (Deverall and Vessey, 1969). These are examples of the few investigations which have set out to assess the effects of wounding. However, the effect of elicitor solutions has often been compared with that of water. In some of this work, Hadwiger and his colleagues repeatedly failed to obtain pisatin from water-treated pea pod tissue, whilst Albersheim's group usually obtained a little phytoalexin from water placed on cut soybean cotyledons, even when microbial contamination was prevented (Albersheim and Valent, 1978).

The analytical procedures used, e.g. UV spectroscopy of crude or purified extracts, and GLC, are appropriate for measuring quantities of phytoalexins generally greater than $10\,\mu g$; they would not necessarily demonstrate whether synthesis of phytoalexins had been induced in the absence of accumulation. Evidence that synthesis can occur in wounded tissues is available (e.g. Sakai *et al.*, 1979). This data and its significance to views on mechanisms of phytoalexin elicitation are discussed on p. 306.

(*ii*) *Partial freezing.* It is a general observation that phytoalexins are not present in extracts of uninfected or untreated tissues which have been obtained either before or after storage at −20°C and phytoalexins do not accumulate when completely frozen tissues are thawed and incubated (Rahe and Arnold, 1975). However, these workers found that phaseollin was produced in French bean hypocotyls which had been frozen by touching with solid carbon dioxide at separate sites along their length. Phaseollin was generally restricted to the wounded areas. This discovery was extended by the finding that several isoflavonoids were produced by hypocotyls which had been thawed and incubated after being partially frozen by placing at −20°C for 10 to 20 minutes or by contact with liquid nitrogen. These results are of significance to considerations of mechanisms of accumulation, see p. 306, but the yields from such experiments were small. A more likely technique for producing greater amounts might be partial freezing of cotyledons (Cain and Porter, 1979).

(*iii*) *Irradiation.* Short wavelength (254 nm) UV light is cytocidal or, at sub-lethal doses, mutagenic. It is another important method of producing phytoalexins in significant quantities. In 1971, Hadwiger and Schwochau obtained pisatin from irradiated pea pods and more recently several other isoflavonoids have been isolated from various legumes (Ingham and Dewick, 1980; Martin and Dewick, 1979; Munn and Drysdale, 1975; Reilly and Klarman, 1980; Weinstein *et al.*, 1981). The effect of UV light is not restricted to legumes. Irradiation (260 to 270 nm) was the major treatment for the production of pterostilbene and resveratrol from *Vitis vinifera* (Langcake and Pryce, 1977) and also caused rishitin and lubimin to form in potato tuber tissues (Cheem and Haard, 1978). Gamma irradiation caused pisatin to form in pea leaves (Hadwiger *et al.*, 1976).

Microbial metabolites

There have been many investigations aimed at isolating and characterizing the products of microbial metabolism which might be responsible for stimulating phytoalexin accumulation. That biotic elicitors could be obtained very easily was evident from demonstrations that culture filtrates, fungal cells killed by heat or partially purified cell walls were often as effective as the living organism (Albersheim and Valent, 1978; Cruickshank, 1980). It cannot, however, be assumed solely on the basis of their microbial origins that these biotic elicitors function in host-parasite interactions, although this possibility will be discussed later. At present the nature and activity of biotic elicitors will be considered. Evidence indicates that these compounds are of three types, peptides, glycopeptides and polysaccharides.

(*i*) *Peptides and glycopeptides.* The first purified biotic elicitor, monilicolin A, was extracted from mycelium of *Monilinia fructicola*, a pathogen of stone fruits

(Cruickshank and Perrin, 1968). It is a polypeptide, M wt approximately 8×10^3, which caused, at a concentration of 2×10^{-7} M (2 ppm), production of phaseollin by French bean (15 μg/ml pod diffusate). It was not effective on pea or broad bean (*Vicia faba*). Polysaccharides were not components of monilicolin A. Glycopeptide elicitors have, however, been isolated from several other fungi including *Cladosporium fulvum*, *Phytophthora megasperma* and *Rhizopus stolonifer*.

The elicitors from *C. fulvum* were effective on tomato, producing rishitin, on pea, producing pisatin and on soybean, producing glyceollin. They were not effective on potato or jack bean (De Wit and Roseboom, 1980). The elicitors are a group of heat stable polydisperse materials, M wt varying from 3×10^4 to 2.5×10^5, in which glucose, mannose and galactose are the carbohydrate moieties (De Wit and Kodde, 1981; Dow and Callow, 1979; Lazarovits *et al.*, 1979). Culture conditions affected the presence or absence of a galactofuranosyl derivative but these changes did not affect the elicitor activity. In contrast, the importance of the peptide chain was indicated by loss of activity after treatment with pronase.

Glycopeptide elicitors from *P. megasperma* caused glyceollin to form in soybean. Detailed characterizations of the glycoprotein elicitors isolated from cell walls of *P. megasperma* (Keen and Legrand, 1980; Wade and Albersheim, 1979) have not been reported. However, on the basis of their binding with concanavalin A, inactivation by periodate and no loss of activity with pronase, it was concluded that carbohydrates were the bases of activity (Keen and Legrand, 1980).

The glycoprotein elicitor from *R. stolonifer*, which stimulated formation of casbene synthase, a reaction used to indicate potential accumulation of casbene in *Ricinus communis*, has a M wt of 3.2×10^4 and requires both the carbohydrate and peptide moieties for activity. Highly purified preparations of the elicitor had polygalacturonase activity (Lee and West, 1981). Other microbial enzymes also act as elicitors. A protease, with endopeptidase activity, which caused benzoic acid to form in apple fruits, was isolated from *Nectria galligena* (Swinburne, 1975). Cell wall degrading enzymes from *Erwinia carotovora* (Albersheim *et al.*, 1981) and *Monilinia fructigena* (Hargreaves, personal communication) caused production of phytoalexins in soybean and French bean respectively. The concentrations of glycoproteins required for elicitation are commonly between 1 and 5 mg per ml; yields of phytoalexins from treated tissues are often low (less than 50 μg per g tissue), but this may reflect the tissues used rather than the activity of the elicitor.

(ii) *Polysaccharides.* In direct contrast to the isolation of glycoproteins from *P. megasperma*, Ayers *et al.* reported that carbohydrates were the only elicitors of glyceollin present in cultures of this fungus (see Albersheim and Valent, 1978). The carbohydrates were obtained from culture filtrates and also from cell walls after hydrolysis in hot trifluoroacetic acid, whereas the glycoproteins were obtained by hydrolysis of walls in cold alkali (Keen and Legrand, 1980). Analysis of the carbohydrates showed they were a heterogeneous group of heat-stable neutral polysaccharides, M wt 5×10^3 to 2×10^5; amino acids were not present. The polysaccharide elicitors had many common features; they were based on a β1,3-glucan, with branches at C-4 and C-6 and a small proportion of mannosyl residues were also present. A most remarkable feature of these elicitors and one which has encouraged the view that they are involved in host-parasite interactions (Albersheim and Valent, 1978), is that, like moni-licolin A, they are effective at low concentrations, i.e. 10^{-13} M; cotyledons of soybean receiving only 10 ng elicitor produced sufficient glyceollin to inhibit growth of *P. megasperma*. Activity of these elicitors appears to depend on their branched nature; simpler β1,3-glucans, e.g. laminarin, were much less effective. The elicitors from *P. megasperma* are similar to those isolated from *Colleto-trichum lindemuthianum* (Anderson, 1978) and from commercial extracts of *Saccharomyces cerevisiae* (Hahn and Albersheim, 1978). In addition to affecting soybean, the glucan elicitor from *P. megasperma* also caused production of rishitin by potato and isoflavonoids by French bean (Albersheim and Valent, 1978).

Chitosan, a β1,4 linked glucosamine derivative was the most active elicitor of pisatin obtained from cell walls of *Fusarium solani*. It was effective at a concentration of 3 μg/ml (Hadwiger and Beckmann, 1980). Many extracts with elicitor activity have been obtained from *Phytophthora infestans*. Activity has been associated with lipoidal materials, but often also with both polysaccharides and protein (see chapter 3). Recently, however, Bostock *et al.* (1981) have reported that the activity of these preparations was due to the presence of eicosapentaenoic and arachidonic acids. Biotic elicitors have been obtained from several other fungi including *Botrytis cinerea* (Dixon, 1980) and *Uromyces appendiculatus* (Hoppe *et al.*, 1980).

Elicitors from plants

Accumulation of phytoalexins in the absence of microorganisms and chemicals, especially after treatment with UV light or after partial freezing, suggests that exogenous materials are not always required to initiate phytoalexin synthesis, and it was as a consequence of studies with partially frozen French beans that Hargreaves and Bailey (1978) demonstrated the presence of water-soluble elicitors in extracts of plant tissues. Phaseollin and phaseollidin were produced by bean hypocotyls or suspension cells (Hargreaves and Selby, 1978) treated with juice expressed from frozen hypocotyls, with water-soluble materials from freeze-dried tissues or with juice obtained from tissues heated at 121°C for 20 minutes. The active components were not well characterized, nor was it shown that the active constituents of these different extracts were the same, although it was suggested that these elicitors were insoluble in organic solvents, had a M wt less than 10^4 and were heat stable. Because these materials were thought to originate within the tissues, they were termed constitutive elicitors. Constitutive elicitors were also found in pea tissues (Hargreaves, 1979). Hahn *et al.* (1981) have obtained essentially similar effects with hot water (121°C) extracts of either soybean tissues or purified cell walls or with walls of soybean, tobacco, sycamore and wheat treated with hot trifluoroacetic acid. These extracts caused the production of glyceollin in cut soybean cotyledons. The elicitor from acid-treated soybean cell walls was a pectic polysaccharide. It was suggested that the activity of the hot water extracts was of a similar nature, but experimental data supporting this was not presented. In this work these materials are referred to as endogenous elicitors. In more recent experiments, Lyon and Albersheim (personal communication) have obtained a heat labile elicitor from freeze-thawed soybean hypocotyls. This elicitor has enzymic properties and when applied to isolated soybean cell walls a heat stable elicitor was released. The heat stable elicitor may be akin to those obtained by Hahn *et al.* (1981). Lyon and Albersheim propose that the enzyme is released during freeze-thawing, and that it elicits glyceollin accumulation in soybean tissues by releasing the endogenous heat stable elicitor from the cell wall of the treated tissues.

The demonstration of elicitors in plant tissues appears to be an important discovery and one which requires further investigation. It is likely that this will provide further insight into the processes leading to synthesis of phytoalexins.

Optimizing phytoalexin accumulation

It is clear from the preceding discussion that phytoalexins can be obtained from plants in many ways. Substantial quantities of these compounds are required for their identification, for studies on their metabolism and toxicity and for assessing their potential as therapeutic agents against diseases of plants, and possibly animals, and as possible dangers to human and animal health. Lack of sufficient quantities has probably restricted many of these studies. Until chemical synthesis is possible, phytoalexins will continue to be obtained from natural sources. In order to optimize yields several parameters must be considered.

(*i*) *Plant species.*   Many phytoalexins are known from only one species and here there is clearly only one choice. However, it is increasingly evident that they can often be obtained from several related plants. It is then important to choose the plant likely to produce most of the phytoalexin which is required. For example, it is probable that kievitone could be more readily obtained from cowpea, where it is the major phytoalexin, than from French bean, where it is often a minor component.

(*ii*) *Plant tissue and environmental conditions.*   Phytoalexins have been obtained from most types of plant tissues and examples using leaves, stems, fruits and tubers have already been described. Roots also produce phytoalexins

(Sutherland *et al.*, 1980). Highest yields will be obtained, by diffusate techniques, from fruit cavities, e.g. of legumes and peppers, or directly from bulky tissues, e.g. stems or cotyledons, rather than leaves. Physiologically active tissues incubated with good aeration at between 15 and 30°C will produce most phytoalexin. Yields might be much reduced if older senescent tissues are used or if aeration is poor, perhaps because tissues are submerged in elicitor solutions. Plant cell suspensions have been used to study production of phytoalexins (Dixon, 1980). Unfortunately, the yields obtained are often low, but the development of these techniques could well prove successful, especially if particular stereoisomers are required.

(iii) Method of induction. Clear recommendations of the most appropriate inducing procedures are less easy, as possible methods may be limited by the facilities available within individual laboratories. As illustrated earlier, the range of phytoalexins produced is not usually affected by the treatment used, but this can affect the yield obtained. Pathogenic fungi, viruses, UV light, heavy metal salts and fungal extracts appear to offer greatest hope of success. If pathogens are used, tissue bearing symptoms reflecting partial death of tissues, e.g. limited

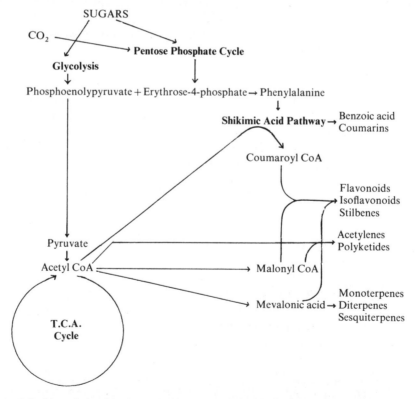

Figure 9.1 Interrelationships between primary metabolism and synthesis of phytoalexins.

lesions, will produce the most phytoalexin. It appears that all treatments need to cause some cell or tissue damage but this must be accompanied by residual healthy tissues. Extensive damage will lead to reduced yields.

Synthesis of phytoalexins

Diversity of phytoalexin structure implies a complex integration of synthetic organization. The interactions between primary and secondary metabolism and the sources of important substrates, e.g. acetate, cinnamate, phenylalanine and mevalonate are summarized in Figure 9.1. Details of the chemical conversions required for phytoalexin synthesis were discussed in chapter 5. The important question which now arises is how does this metabolism begin? Is it due to diversion of existing pathways or to activation of new pathways? To answer this the properties and activities of the enzymes catalysing the appropriate conversions need to be understood. Until recently there was not much information. However, research groups experienced in mechanisms of control of other biochemical processes have now reported some very relevant information. As yet, these reports are restricted almost entirely to work with isoflavonoid phyto alexins. Any generalization to all phytoalexins must bear in mind this serious limitation.

Accumulation of phytoalexins is accompanied by enzyme activities which are much greater than those in untreated tissues. Phenylalanine ammonia lyase (PAL), the enzyme which catalyses the conversion of phenylalanine to cinnamic acid, occupies a key position in the biosynthesis of many phenylpropanoid phytoalexins. Its activity and also those of cinnamic acid 4-hydroxylase, 4-coumarate:CoA ligase, and flavanone synthase (Figure 9.2) increase during

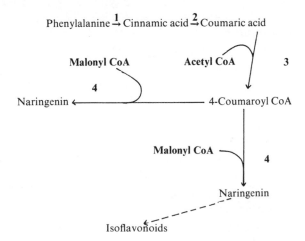

Figure 9.2 Biosynthetic pathway from phenylalanine to the flavanone naringenin. (1) phenylalanine ammonia lyase; (2) cinnamic acid 4-hydroxylase; (3) 4-coumaroyl CoA ligase; (4) flavanone synthase.

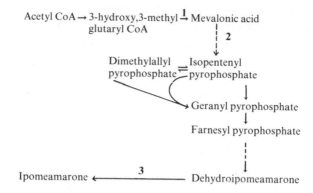

Figure 9.3 Biosynthetic pathway for synthesis of ipomeamarone. (1) 3-hydroxy,3-methyl
glutaryl CoA reductase; (2) pyrophosphomevalonate decarboxylase;
(3) furanosesquiterpene reductase.

accumulation of isoflavonoid phytoalexins (see Lamb *et al.*, 1980). These
enzymes act at a relatively early stage of biosynthesis and direct relationships
between enzyme activities and the eventual amounts of phytoalexins which
accumulate are difficult to establish. Nevertheless, production of isoflavonoid
phytoalexins does appear to be a consequence of a specific coordinated
activation of many enzymes (Rathmell and Bendall, 1971; Lamb *et al.*, 1980).

Increased enzyme activity is also associated with accumulation of furanoterpenes (Uritani *et al.*, 1976), sesquiterpenes (Corsini and Pavek, 1980) and
stilbenes (Fritzemeier and Kindl, 1981). These include enzymes directly responsible for the formation of the accumulated phytoalexins, e.g stilbene synthase
(Schoppner and Kindl, 1979) and furanosesquiterpene reductase (Inoue and
Uritani, 1980)—see Figure 9.3.

The mechanisms leading to increased enzyme activity have attracted much
attention. Early studies relied on the use of inhibitors of enzyme synthesis, e.g.
actinomycin D, cycloheximide and 6-methylpurine. These compounds prevented both the accumulation of isoflavonoids and the associated increases in
enzyme activities, particularly PAL, in various legumes. It was concluded that
production of these phytoalexins is mediated by synthesis of mRNA (Hadwiger
and Schwochau, 1971*b*; Yoshikawa *et al.*, 1978). Similar results were obtained
with sweet potatoes (Uritani *et al.*, 1960). However, as discussed by Biggs (1972),
reported increases in RNA content could not always be attributed solely to
mRNA and the antimetabolites used could not be relied on to show specific
effects on gene activation. They also affect aspects of primary metabolism, e.g.
glycolysis, and have also been shown to produce phytoalexins in the absence of
other elicitor treatments (see p. 294).

The production of phaseollin by suspension cells of French bean has been
used to re-examine this problem. Lamb and Dixon (1978) and Dixon and Lamb

(1979) found that phaseollin occurs at low concentrations in untreated cells, but its concentration increased greatly after treatment with culture filtrates of *Botrytis cinerea*, a cell wall elicitor from *Colletotrichum lindemuthianum* or denatured RNase. As previously, phaseollin accumulation was accompanied by increased PAL activity. The origin of this enzyme activity was examined by comparative density labelling. This technique distinguishes radioactively labelled and unlabelled enzymes, on the basis of their weight by separating them on density gradients of, for example, CsCl. The density label, ^2H_2O, is applied to tissue in the presence and absence of elicitor. Enzymes are extracted, subjected to density gradient centrifugation and the buoyant density of the enzymes is measured. PAL isolated from cells treated with elicitor was heavier than that from untreated cells, i.e. the enzyme had been synthesized in the presence of ^2H_2O. On the basis of these experiments it was concluded that *de novo* enzyme synthesis occurred within 2 h of treatment. Subsequent studies with KBr gradients have confirmed that new enzyme protein was produced and showed that the newly synthesized enzyme was more stable than that from untreated cells (Lamb *et al.*, 1980).

The application of other sensitive techniques is promising to extend quickly our knowledge of protein synthesis *de novo* and gene function. Immunoprecipitation of enzymes and measurements of mRNA synthesis, using a reticulocyte translation system, have already shown that, although a phytoalexin was not identified, the glucan elicitor from *Phytophthora megasperma* caused coordinate increases in the activities of PAL and 4-coumarate:CoA ligase in cells of parsley, *Petroselinium hortense*. Remarkably, synthesis of PAL occurred almost immediately (Hahlbrock *et al.*, 1981). Similar increases in the activity of PAL and chalcone synthase occurred in French bean cells treated with the elicitor from *C. lindemuthianum*. Again synthesis of these enzymes occurred immediately after treatment (Lamb, personal communication).

There is, therefore, good evidence that the increased activity of several enzymes believed to be involved in the synthesis of isoflavonoid phytoalexins in *Phaseolus vulgaris* is due primarily to *de novo* synthesis of enzymes, which occurs in response to both biotic (fungal cell walls) and abiotic elicitors (autoclaved RNase). The sources of the increased enzyme activities observed in other plants requires urgent consideration. Experiments with enzymes involved in later conversions, e.g. stilbene synthase, casbene synthase and furanosesquiterpene reductase, would be particularly interesting.

Location of phytoalexin synthesis and accumulation

The amounts of phytoalexins which accumulate have been assessed by chemical analysis of plant extracts. By excising affected tissues the distribution of phytoalexins can be investigated. However, location of synthesis requires more sensitive techniques, e.g. incorporation of a radiolabelled precursor. Equally relevant are studies on the distribution of activity of enzymes, particularly those responsible for the later metabolite conversions. Information, based on these procedures, is available concerning most of the major classes of phytoalexin.

Isoflavonoids

Phaseollin and related phytoalexins occur at high concentrations (300 μg to 3 mg/g tissue) in lesions excised from French bean hypocotyls (VanEtten and Pueppke, 1976), greatest concentrations being at the edge of lesions (Rahe, 1973). These phytoalexins were not detected in the healthy tissues surrounding the lesions. Lesions on leaves also contained large amounts of phytoalexins, but smaller quantities were also present in surrounding tissues and very occasionally in petioles of leaves bearing large spreading lesions (van den Heuvel and Grootveld, 1980). Glyceollin was concentrated (1 to 10 mg/g) in lesions in soybean hypocotyls, whilst much less was present in surrounding healthy tissues (Keen, 1971). Similarly, pisatin occurred to the greater extent in necrotic lesions on pea leaves (Heath and Wood, 1971), whilst it was restricted to the epidermis or first sub-epidermal layer of leaves infected with mildew (Oku et al., 1979). Infections with virus or treatment with UV-irradiation or chemicals also reveal that accumulation of these phytoalexins is usually localized in the treated tissues which also accumulated brown pigments and are considered to be dead. This is well illustrated by the accumulation of phenolic materials, including presumably the phytoalexins, in the dead outer cells of French bean cotyledons which had been treated with phytotoxic surfactants (Hargreaves, 1981).

Studies on the location of isoflavonoid synthesis have rarely been attempted. However, after confirming the restriction of phytoalexins to lesions, Rathmell and Bendall (1971) showed that enhanced PAL activity occurred throughout the entire hypocotyl. Similarly, whilst kievitone was isolated only from the outer tissues of cowpea affected by actinomycin D or cycloheximide or exposed to UV light, PAL activity increased in all the tissues (Munn and Drysdale, 1975).

Acetylenes

Accumulation of acetylenic phytoalexins is also associated with symptom expression. The highest concentrations of wyerone acid occurred in lesions on Vicia faba, particularly at lesion edges. This and related phytoalexins were present in trace amounts in surrounding healthy tissues. Fluorescence microscopy indicated that synthesis occurs in a narrow band of healthy cells immediately adjacent to the dead ones in the lesion (Mansfield et al., 1974; Mansfield, 1980).

Furanoterpenes

Ipomeamarone and several other furanoterpenes accumulated in the upper parts of sweet potato tissues infected with Ceratocystis fimbriata. Only small amounts were present in adjacent healthy cells (Imezaki and Uritani, 1964). This contrasts with the location of synthesis of these compounds which, as assessed by incorporation of ^{14}C-acetate or the enhanced activities of 3-hydroxy,3-methyl glutaryl CoA reductase, pyrophosphomevalonate decarboxylase and furanosesquiterpene reductase, occurred mostly in these underlying healthy cells (Imezaki et al., 1964; Oba et al., 1976; Uritani et al., 1976). Results obtained with furanosesquiterpene reductase are important as this enzyme catalyses the formation of ipomeamarone from dehydroipomeamarone (Inoue and Uritani, 1980)—see Figure 9.3.

Sesquiterpenes

Rishitin and lubimin occurred predominantly in the two upper pigmented and apparently dead cell layers of potato tuber tissue infected with *Phytophthora infestans* (Sato and Tomiyama, 1969; Varns et al., 1971) or treated with $HgCl_2$ or dibromochloropropane (Komai and Sato, 1974; Tomiyama and Fukaya, 1975; Varns et al., 1971). As in the other plants synthesis of phytoalexins appeared to occur in the cells beneath. The contribution of the surrounding cells to the eventual accumulation of phytoalexins was first indicated by demonstrations that changes in phenolic constituents occurred only in the upper 10 to 15 cell-layers and that resistance was not expressed if inoculated tissues were less than 10 cells thick (Sakai et al., 1967). Direct evidence for synthesis of rishitin in surrounding healthy cells was obtained by incorporation of radiolabelled precursors and its conversion to an insoluble 3,5-dinitro benzoate (Nakajima et al., 1975; Sakai et al., 1979). Other sesquiterpenes, e.g. capsidiol and glutinosone, accumulated at high concentrations in necrotic lesions formed in leaves of *Nicotiana* spp. following virus infection, but were not detected in the surrounding tissues (Bailey et al., 1976).

Mechanisms of phytoalexin synthesis and accumulation

So far this chapter has illustrated the many ways in which phytoalexins can be produced and some of the more efficient methods of obtaining them in quantities sufficient for further experimentation. The ability of different plant species to respond to the same stimuli by producing different phytoalexins has also been described and emphasis has been placed on the requirement for *de novo* protein synthesis. Finally, it has become evident that synthesis and accumulation of phytoalexins occur at different locations. Phytoalexins accumulate in the directly affected cells, which are often also pigmented and dead, whereas synthesis occurs in the adjacent healthy tissues. It is now necessary to consider how synthesis is activated and whether the various elicitor treatments act in similar or different ways.

The nature of elicitation during infection of plants by microorganisms is of particular importance to knowledge and possible future manipulation of resistance mechanisms. Effective elicitor treatments are very often associated with death or injury to plant tissues. Total tissue death does not lead to phytoalexin formation but when living and dead tissues exist together these compounds are usually present. Many abiotic elicitors are toxic and are effective only when used at concentrations exhibiting toxicity. This was apparent with heavy metal salts (Hargreaves, 1979; Tomiyama and Fukaya, 1975; Uritani *et al.*, 1960) and surfactants (Hargreaves, 1981), but toxic effects were also evident with de-natured RNase (Dixon and Bendall, 1978) various drugs and basic polypeptides (see Hadwiger *et al.*, 1974) and even mouse tumour cells (Teasdale *et al.*, 1974). The possible importance of cell death is further suggested by production of phytoalexins in lesions caused by viruses and in necrotrophic, but not bio-trophic, fungal infections (Deverall, 1977; Bailey, 1981).

Studies with wounded tissues provide direct evidence that phytoalexins can be synthesized and can sometimes accumulate as a consequence of cell injury. The number of reports available is small but the results appear highly signi-ficant. They have allowed a theory to be developed which might explain how synthesis and accumulation of phytoalexins occur. These proposals were based on experiments with French bean (Bailey, 1981), but there are indications that by slight modifications they can be extended to other plants. The relevant experiments have been alluded to before. They are those where phytoalexins accumulate in the absence of exogenous materials, i.e. in tissues wounded by cutting, slicing, bruising, partial freezing and exposure to short wavelength UV light. Phytoalexins did not accumulate in tissues which had been killed completely by freezing and thawing. Several isoflavonoids accumulated when only the outer layers were killed, i.e. when living and dead cells were allowed to interact. Phytoalexins were also formed if distinct living and dead (freeze-thawed) tissues were placed in contact with each other, when most accumulated in the dead tissues. This suggests that material emanating from the freeze-thawed tissue caused formation of phytoalexins in the living and that they subsequently moved from these into the dead tissues. The isolation of elicitors from frozen tissues of French bean and soybean (Hargreaves and Bailey, 1978; Lyon and Albersheim, personal communication) provides support for this view. The experiments with freeze-thawed tissues are also consistent with the differ-ential location of phytoalexin synthesis and accumulation which has been found in many plants. Finally, in addition to the studies with wounded tissues mentioned earlier, recent experiments with potato tissues indicate that synthesis of rishitin can be initiated solely by wounding. Sakai *et al.* (1979) have shown that synthesis of rishitin (but not its accumulation) occurred 30 minutes after cutting, whilst Ishiguri *et al.* (1978) found that wounding also induced formation of the enzymes responsible for metabolism of rishitin. These enzymes could not be detected in uninjured tissues.

The process of induction of phytoalexin synthesis in plant tissues can therefore be considered as follows. Healthy untreated cells contain within them material capable of initiating synthesis of appropriate enzymes, but they remain inactive. When cells are injured these materials become active and thus enzyme synthesis and subsequently phytoalexin synthesis occur in the injured cells and in the adjacent cells which remain healthy. It has been suggested that the action of these materials is facilitated by their release as a direct consequence of membrane dysfunction (Bailey, 1981) or by the action of enzymes which cleave the active products from the cell wall (Lyon and Albersheim, personal communication).

As emphasized throughout this book, the diversity of phytoalexin structure is remarkable. Nevertheless the processes outlined above would represent a unifying concept for phytoalexin synthesis and would allow considerable variation in the mechanisms which allow subsequent phytoalexin accumulation. It is now clear that some phytoalexins, but not necessarily all, can be metabolized by plant tissues. Most of the evidence relies on metabolism of added phytoalexins but, as indicated above, the enzymes responsible for metabolizing rishitin have been isolated from potato tissue (Ishiguri et al., 1978). In such situations, phytoalexins would be relatively readily metabolized, and their accumulation would require their removal from the effects of cellular metabolism. This could be achieved by adsorption to a suitable site, which for many phytoalexins, e.g. phaseollin, rishitin and ipomeamarone etc., would appear to be provided by dead cells. For other phytoalexins, e.g. pisatin, accumulation is not always associated with cell death, although evidence suggests a need for cell injury (Hargreaves, 1979). For such compounds, which may have different physical properties, different sites, e.g. cell walls, may be needed for adsorption. Alternatively, some phytoalexins may be metabolic end products and hence could accumulate in the cytoplasm. Diversity of this kind could allow explanations of the production of phytoalexins under somewhat exceptional conditions, e.g. pisatin in pea leaf discs floated on sucrose (Robinson and Wood, 1976). These processes would also be consistent with the formation of pisatin in the absence of demonstrable cell death, i.e. in pea pods treated with monilicolin A or low concentrations of $HgCl_2$ (Paxton et al., 1974).

The possible application of this theory to elicitation by chemicals has been referred to above and is discussed more fully by Bailey (1980). It may also provide an explanation for the occurrence of some secondary metabolites as phytoalexins and also as constitutents of heartwoods. Recent illustrations are provided by Vitis spp. and Pinus sylvestris. Stilbenes accumulate in these species; as phytoalexins in Vitis spp. (Langcake and Pryce, 1977) and also as components of their heartwoods. Recently the activity of the enzyme responsible for production of stilbenes, stilbene synthase, was shown to be induced in the leaves of both the Vitis spp. and P. sylvestris by exposure to UV light (Fritzemeier and Kindl, 1981; Schoppner and Kindl, 1979). Heartwoods are built up by the presence of dead xylem cells which occur adjacent to the living xylem parenchyma and phloem. When the xylem cells die, this could trigger synthetic activity in the adjacent cells and the products would subsequently accumulate by adsorption to the dead heartwood tissues.

Phytoalexin synthesis and accumulation during interactions with biotic elicitors

Biotic elicitors have been obtained from many fungi and a few bacteria and there have been many attempts to implicate these compounds in natural host-parasite interactions, including non-host resistance and more often race specific resistance, the highly specific interactions of host cultivars with different races of a pathogen. The simplest view, which is consistent with genetic analyses, suggests that fungi, particularly races which cannot effect disease, contain specific elicitors which cause accumulation of phytoalexins in resistant but not susceptible plants or cultivars (Keen, 1975; Keen and Legrand, 1980). The data supporting the existence of materials with such specificity remains equivocal (Ayers *et al.*, 1976). Alternatively, it has been suggested that elicitors are not specific, affecting equally all cultivars and also possibly non-host species (Albersheim and Valent, 1978; Albersheim *et al.*, 1981). In these proposals specificity is determined by the presence within pathogens of other materials. These may be specificity factors, which are thought to induce resistance in other ways (Wade and Albersheim, 1979) or specific suppressors, agents which prevent the action of the elicitor on susceptible cultivars (Oku *et al.*, 1977; Doke and Tomiyama, 1980; see chapter 3).

These suggestions consider production of elicitors as a property of the pathogen which is not influenced by reactions with the host. Other proposals, which have not encompassed cultivar specificity and which have been described as requiring "mutual" or "double induction" suggest that elicitors are produced by the pathogen in response to materials released during infection (see Cruickshank, 1980). In this context it is interesting that chitosan, an effective elicitor of pisatin, is released from hyphae of *Fusarium solani* in the presence of pea tissues (Hadwiger *et al.*, 1981) and a soluble elicitor has been obtained from insoluble purified walls of *Phytophthora infestans* after incubation with soybean tissues (Yoshikawa *et al.*, 1981).

All these proposals have emphasized the nature of biotic elicitors and their ability to cause phytoalexin accumulation. The physiological effects of these compounds and hence their modes of action have received less attention. It has been suggested that these elicitors act on cellular metabolism, by initiating or regulating phytoalexin synthesis (Cruickshank, 1980), possibly by modifying cellular DNA (Hadwiger and Beckman, 1980). This now seems unlikely, for there are several indications that biotic elicitors act through an intermediate response (see also Lewis, 1980). Stoessl (1980) has drawn attention to the findings that some of these elicitors can cause the formation of phytoalexins in different plants, e.g. isoflavonoids and sesquiterpenes in legume and solanaceous species (Cline *et al.*, 1978; De Wit and Roseboom, 1980). This requires activation of different biosynthetic routes (Figure 9.1) and it is highly unlikely that this could result by direct interaction of these elicitors with the different genes which control phytoalexin synthesis. It was suggested that elicitor action is exerted through a disturbance of plant metabolism at another site.

Comparisons of the effects of biotic and abiotic elicitors (Dixon and Lamb, 1979; Moesta and Grisebach, 1980) have also led to the conclusion that elicitation occurs through a common primary response.[1] In view of the preceding argument, the possibility that cell injury is the intermediate response must be explored.

Despite the dearth of relevant investigations there are some indications that biotic elicitors cause damage to plant cell membranes. The glycoprotein from *Cladosporium fulvum* caused ion leakage and death of tomato cells (Lazarovits and Higgins, 1979; Dow and Callow, 1979), the glucan from *Phytophthora megasperma* inhibited growth of suspension cells (Albersheim and Valent, 1978) and the cell wall elicitor from *P. infestans* was toxic to leaf cells of several plant species (Doke *et al.*, 1979). These results add experimental support to the suggestion that biotic elicitors act by causing cell injury and thus cause accumulation of phytoalexins as a consequence of the processes outlined earlier

Phytoalexin accumulation during host-parasite interactions

In recent years emphasis on biotic elicitors has overshadowed considerations of the physiological interactions between hosts and parasites. As a result the theories proposed concerning the involvement of these products in host-parasite interactions have often been based on information from situations, e.g. pod cavities or wounded tissues, where the interactions between host and parasite are quite different from those which occur with intact plant tissues. Thus when infections of soybean by incompatible and compatible races of *Phytophthora megasperma* were achieved through wounds in hypocotyls, resistance and susceptibility were considered to be determined by the different rates of fungal growth in the infected necrotic tissues. Phytoalexins accumulated more rapidly during resistant responses (Keen, 1971). The times when phytoalexins were produced were similar. Recently, however, Stössel *et al.* (1980) showed that on intact tissues a compatible race of *P. megasperma* is distinguished from an incompatible race by its ability to colonize tissue without killing cells, to establish biotrophy. Incompatible races caused early death of infected cells, i.e. hypersensitivity.

Similar processes occurred in the more extensively studied interactions between *P. infestans* and potato (Tomiyama *et al.*, 1979) and between *Colletotrichum lindemuthianum* and French bean (Bailey, 1981). In addition, information is available which allows the relationship between biotrophy, necrotrophy and phytoalexin synthesis and accumulation to be established. During both these

---

[1] Yoshikawa (1978) has concluded that biotic and abiotic elicitors have different modes of action; biotic elicitors stimulating synthesis of phytoalexins, whereas abiotic elicitors prevent their degradation. This assumes that the phytoalexins are continually metabolized by healthy cells, a view that is inconsistent with the occurrence of *de novo* protein synthesis. In addition, Moesta and Grisebach (1980) re-examined these effects and failed to demonstrate any differences between biotic and abiotic elicitors.

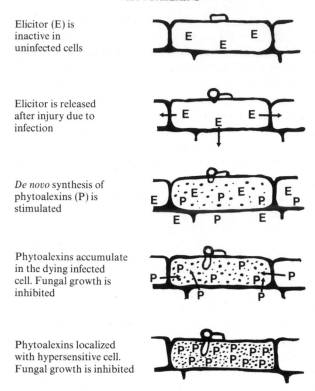

Elicitor (E) is
inactive in
uninfected cells

Elicitor is released
after injury due to
infection

De novo synthesis of
phytoalexins (P) is
stimulated

Phytoalexins accumulate
in the dying infected
cell. Fungal growth is
inhibited

Phytoalexins localized
with hypersensitive cell.
Fungal growth is inhibited

Figure 9.4   A schematic illustration of the role of constitutive elicitors in phytoalexin accumulation during a hypersensitive response of *P. vulgaris* to *C. lindemuthianum*. (Redrawn from Bailey, 1981).

interactions compatible races establish biotrophy. Biotrophy can last for several days and during this period phytoalexins are not produced. Thereafter, the infected tissues collapse and phytoalexins can then be detected (Bailey and Deverall, 1971; Ishizaki and Tomiyama, 1972). Their concentrations usually remain low and spreading lesions form, although under some circumstances, e.g. on French bean seedlings bearing healthy cotyledons the concentration of phytoalexins produced was much greater and the lesions then became restricted (Bailey and Rowell, unpublished results). By contrast, incompatible races induced early death of the infected cells and the early accumulation of phytoalexins. The results of studies with *P. infestans* and *Colletotrichum lindemuthianum* are clearly consistent with the traditional view of hypersensitivity, which is that the infected cell dies before the pathogen is restricted (Ingram, 1978). Investigations of host-parasite interactions provide no support for the view that restriction of the pathogen occurs first (Kiraly *et al.*, 1972). In addition, there is some evidence that phytoalexin formation occurs after infected cells die (Bailey *et al.*, 1980; Sato *et al.*, 1971) suggesting that phytoalexin accumulation in infected potatoes and French beans is a consequence of the death of infected cells.

To summarize, an interpretation consistent with the available data is that specificity, i.e. differentiation between races, is determined by the occurrence of biotrophy or nectrotrophy (see Ward and Stoessl, 1976; Doke and Tomiyama, 1980). Nectrophy, which follows infection by incompatible or compatible races and leads to hypersensitivity or lesion formation, is the trigger for the initiation of synthesis of phytoalexins. This synthesis is mediated through the action of elicitors released from infected cells as a result of membrane damage. The extent to which phytoalexins subsequently accumulate, which determines whether pathogen growth continues or is restricted, is controlled by the interaction of injured cells with the surrounding tissues. Thus, as illustrated in Figure 9.4, if only one or a few cells are injured, the surrounding cells can react strongly to the released elicitor and thus large quantities of phytoalexin are synthesized and their accumulation to high concentrations in the infected cells leads to the inhibition of pathogen growth (Bailey, 1981). When many cells are killed and extensive pathogen growth has already occurred, i.e. when biotrophic growth becomes necrotrophic the outcome is less predictable. The nectrotrophic phase of phytoalexin accumulation will depend on the ability of the pathogen to reduce the amounts of phytoalexin which accumulate. This could be achieved either by reducing the synthetic activity of the adjacent cells, perhaps by killing them, or by metabolizing the phytoalexins to less toxic products. This balance between the aggressiveness of the necrotrophic pathogen and the reactivity of the host will be complex, but it permits much flexibility and can readily explain the development of very variable symptoms, i.e. from small restricted lesions to completely rotted tissues.

Concluding remarks

Phytoalexins are antimicrobial secondary metabolites which can be regarded as part of a greater stimulation of new plant metabolism, which occurs when cells within a tissue are damaged, directly or by chemicals or by infection with microorganisms. Less attention has been paid to the many other metabolites which form coincidently with the phytoalexins, but which lack biological activity. Evolutionary pressures may have favoured the production of anti-microbial compounds by plants, but this may not have occurred in all species. Hence, a failure to obtain phytoalexins from some species may be a genuine reflection of the induced biochemical processes, rather than a result of problems associated with extraction procedures. The biochemical mechanisms controlling the synthesis and accumulation of these metabolites are only beginning to be understood. Nevertheless the importance of de novo protein synthesis to the production of isoflavonoid phytoalexins has been established and this now offers, possibly, the greatest opportunity for the study and eventual understanding of gene expression in plants.

Acknowledgements

I would like to thank John Hargreaves, Pat Rowell and Sarah Constantine for discussions during the preparation of this paper and Jackie Mizen for preparing the typescript.

REFERENCES

Abawi, G. S., VanEtten, H. D. and Mai, W. F. (1971) Phaseollin production induced by *Pratylenchus penetrans* in *Phaseolus vulgaris*. *J. Nematol.*, **3**, 301.

Albersheim, P., Darvill, A. G., McNeil, M., Valent, B. S., Hahn, M. G., Lyon, G., Sharp, J. K., Desjardins, A. E., Spellman, W. M., Ross, L. M., Robertson, B. K., Amen, P. and Franzen, L.-E. (1981) Structure and function of complex carbohydrates active in regulating plant-microbe interactions. *Pure Appl. Chem.*, **53**, 79–88.

Albersheim, P. and Valent, B. S. (1978) Host-pathogen interactions in plants. Plants, when exposed to oligosaccharides of fungal origin, defend themselves by accumulating antibiotics. *J. Cell Biol.*, **78**, 627–643.

Anderson, A. J. (1978) Isolation from three species of *Colletotrichum* of glucan-containing polysaccharides that elicit browning and phytoalexin production in bean. *Phytopathology*, **68**, 189–194.

Ayers, A. R., Ebel, J., Finelli, F., Berger, N. and Albersheim, P. (1976) Host-pathogen interactions. IX. Quantitative assays of elicitor activity and characterization of the elicitor present in the extracellular medium of cultures of *Phytophthora megasperma* var. *sojae*. *Plant Physiol.*, **57**, 751–759.

Bailey, J. A. (1973) Production of antifungal compounds in cowpea (*Vigna sinensis*) and pea (*Pisum sativum*) after virus infection. *J. Gen. Microbial.*, **75**, 119–123.

Bailey, J. A. (1974) The relationship between symptom expression and phytoalexin concentration in hypocotyls of *Phaseolus vulgaris* infected with *Colletotrichum lindemuthianum*. *Physiol. Plant Pathol.*, **4**, 477–488.

Bailey, J. A. (1980) Constitutive elicitors from *Phaseolus vulgaris*; a possible cause of phytoalexin accumulation. *Ann. Phytopathol.*, **12**, in press.

Bailey, J. A. (1981) "Physiological and biochemical events associated with the expression of resistance to disease", in *Active Defence Mechanisms in Plants*, ed. R. K. S. Wood, Plenum Press, London, in press.

Bailey, J. A. and Deverall, B. J. (1971) Formation and activity of phaseollin in the interaction between bean hypocotyls (*Phaseolus vulgaris*) and physiological races of *Colletotrichum lindemuthianum*. *Physiol. Plant Pathol.*, **1**, 435–449.

Bailey, J. A., Rowell, P. M. and Arnold, G. M. (1980) The temporal relationships between host cell death, phytoalexin accumulation and fungal inhibition during hypersensitive reactions of *Phaseolus vulgaris* to *Colletotrichum lindemuthianum*. *Physiol. Plant Pathol.*, **17**, 329–339.

Bailey, J. A. and Skipp, R. A. (1978) Toxicity of phytoalexins. *Ann. appl. Biol.*, **89**, 354–358.

Bailey, J. A., Vincent, G. G. and Burden, R. S. (1976) The antifungal activity of glutinosone and capsidiol and their accumulation in virus-infected tobacco species. *Physiol. Plant Pathol.*, **8**, 35–41.

Bailey, J. A. and Ingham, J. L. (1971) Phaseollin accumulation in bean (*Phaseolus vulgaris*) in response to infection by tobacco necrosis virus and the rust *Uromyces appendiculatus*. *Physiol. Plant Pathol.*, **1**, 451–456.

Biggs, D. R. (1972) Studies on phytoalexins. The relationship between actinomycin D and ribonucleic acid synthesis during the induction of phaseollin in the French bean (*Phaseolus vulgaris* L). *Plant Physiol.*, **50**, 660–666.

Bostock, R. M., Kuć, J. and Laine, R. A. (1981) Eicosapentaenoic and arachidonic acids from *Phytophthora infestans* elicit fungitoxic sesquiterpenes in potato. *Science*, **212**, 67–69.

Burden, R. S., Bailey, J. A. and Dawson, G. W. (1972) Structures of three new isoflavonoids from *Phaseolus vulgaris* infected with tobacco necrosis virus. *Tetrahedron Lett.*, 4175–4178.

Cain, R. O. and Porter, A. E. A. (1979) Biosynthesis of the phytoalexin wyerone in *Vicia faba*. *Phytochemistry*, **18**, 322–323.

Chalutz, E., DeVay, J. E. and Maxie, J. C. (1969) Ethylene-induced isocoumarin formation in carrot root tissues. *Plant Physiol.*, **44**, 235–241.

Chalutz, E. and Stahmann, M. A. (1969) Induction of pisatin by ethylene. *Phytopathology*, **59**, 1972–1973.

Cheema, A. S. and Haard, N. F. (1978) Induction of rishitin and lubimin in potato tuber discs by non-specific elicitors and the influence of storage conditions. *Physiol. Plant Pathol.*, **13**, 233–240.

Cline, K., Wade, M. and Albersheim, P. (1978) Host-pathogen interactions. XV. Fungal glucans which elicit phytoalexin accumulation in soybean also elicit the accumulation of phytoalexins in other plants. *Plant Physiol.*, **62**, 918–921.

Corsini, D. L. and Pavek, J. J. (1980) Phenylalanine ammonia lyase activity and fungitoxic metabolites produced by potato cultivars in response to *Fusarium* tuber rot. *Physiol. Plant Pathol.*, **16**, 63–72.

Coxon, D. T., Curtis, R. F., Price, K. R. and Levett, G. (1973) Abnormal metabolites produced by *Daucus carota* roots stored under conditions of stress. *Phytochemistry*, **12**, 1881–1885.

Coxon, D. T., Price, K. R., Howard, B., Osman, S. F., Kalan, E. B. and Zacharius, R. M. (1974) Two new vetispirane derivatives: stress metabolites from potato (*Solanum tuberosum*) tubers. *Tetrahedron Lett.*, 2921–2924.

Cruickshank, I. A. M. (1963) Phytoalexins. *Ann. Rev. Phytopathology*, **1**, 351–374.

Cruickshank, I. A. M. (1980) "Defenses triggered by the invader: Chemical defenses", in *Plant Disease*, Vol. V, eds. J. G. Horsfall and E. C. Cowling, Academic Press, New York and London.

Cruickshank, I. A. M. and Perrin, D. R. (1963) Studies on Phytoalexins, VI. Pisatin: the effect of some factors on its formation in *Pisum sativum* L., and the significance of pisatin in disease resistance. *Aust. J. Biol. Sci.*, **16**, 111–128.

Cruickshank, I. A. M. and Perrin, D. R. (1968) The isolation and partial characterization of monilicolin A, a polypeptide with phaseollin-inducing activity from *Monilinia fructicola. Life Sci.*, **7**, 449–458.

Cruickshank, I. A. M. and Perrin, D. R. (1971) Studies on phytoalexins. XI. The induction, antimicrobial spectrum and chemical assay of phaseollin. *Phytopathol. Z.*, **70**, 209–229.

Deverall, B. J. (1977) *Defence Mechanisms of Plants*. Cambridge University Press.

Deverall, B. J. and Vessey, J. C. (1969) Role of a phytoalexin in controlling lesion development in leaves of *Vicia faba* after infection by *Botrytis* spp. *Ann. appl. Biol.*, **63**, 449–458.

De Wit, P. J. G. M. and Kodde, E. (1981) Induction of polyacetylenic phytoalexins in *Lycopersicon esculentum* after inoculation with *Cladosporium fulvum* (syn. *Fulvia fulva*). *Physiol. Plant Pathol.*, **18**, 143–148.

De Wit, P. J. G. M. and Roseboom, P. H. M. (1980) Isolation, partial characterization and specificity of glycoprotein elicitors from culture filtrates, mycelium and cell walls of *Cladosporium fulvum* (syn. *Fulvia fulva*). *Physiol. Plant Path.*, **16**, 391–408.

Dixon, R. A. (1980) "Plant tissue culture methods in the study of phytoalexin induction", in *Tissue Culture Methods for Plant Pathologists*, eds. D. S. Ingram and J. P. Helgeson, 185–196.

Dixon, R. A. and Bendall, D. S. (1978) Changes in the levels of enzymes of phenylpropanoid and flavonoid synthesis during phaseollin production in cell suspension cultures of *Phaseolus vulgaris*. *Physiol. Plant Pathol.*, **13**, 295–306.

Dixon, R. A. and Lamb, C. J. (1979) Stimulation of *de novo* synthesis of L-phenylalanine ammonia lyase in relation to phytoalexin accumulation in *Colletotrichum lindemuthianum* elicitor-treated cell suspension cultures of French bean (*Phaseolus vulgaris*). *Biochim. Biophys. Acta*, **586**, 453–463.

Doke, N., Sakai, S. and Tomiyama, K. (1979) Hypersensitive reactivity of various host and nonhost plant leaves to cell wall components and soluble glucan isolated from *Phytophthora infestans*. *Ann. Phytopath. Soc. Japan*, **45**, 386–393.

Doke, N. and Tomiyama, K. (1980) Suppression of the hypersensitive response of potato tuber protoplasts to hyphal wall components by water soluble glucans isolated from *Phytophthora infestans*. *Physiol. Plant Pathol.*, **16**, 177–186.

Dow, J. M. and Callow, J. A. (1979a) Partial characterization of glycopeptides from culture filtrates of *Fulvia fulva* (Cooke) Ciferri (syn. *Cladosporium fulvum*), the tomato leaf mould pathogen. *J. Gen. Microbiol.*, **113**, 57–66.

Dow, J. M. and Callow, J. A. (1979b) Leakage of electrolytes from isolated leaf mesophyll cells of tomato induced by glycopeptides from culture filtrates of *Fulvia fulva* (Cooke) Ciferri (syn. *Cladosporium fulvum*). *Physiol. Plant Path.*, **15**, 27–34.

Farkas, G. L. and Kiraly, Z. (1962) Role of phenolic compounds in the physiology of plant diseases and disease resistance. *Phytopath. Z.*, **44**, 105–150.

Fraile, A., Garcia-Arenal, F. and Sagasta, E. M. (1980) Phytoalexin accumulation in bean (*Phaseolus vulgaris*) after infection with *Botrytis cinerea* and treatment with mercuric chloride. *Physiol. Plant Pathol.*, **16**, 9–18.

Fritzemeier, K.-H. and Kindl, H. (1981) Co-ordinate induction by UV light of stilbene synthase, phenylalanine ammonia lyase and cinnamate-4-hydroxylase in leaves of Vitaceae. *Planta.* **151**, 48–52.

Gnanamanickam, S. S. and Patil, S. S. (1977) Accumulation of antibacterial isoflavonoids in hypersensitively responding bean leaf tissues inoculated with *Pseudomonas phaseolicola. Physiol. Plant Pathol..* **10**, 159–168.

Hadwiger, L. A. and Beckman, J. M. (1980) Chitosan as a component of pea–*Fusarium solani* interactions. *Plant Physiol.,* **66**, 205–211.

Hadwiger, L. A., Beckmann, J. M. and Adams, M. J. (1981) Localization of fungal components in the pea–*Fusarium* interaction detected immunochemically with anti-chitosan and anti-fungal cell wall antisera. *Plant Physiol.,* **67**, 170–175.

Hadwiger, L. A., Jafri, A., von Broembsen, S. and Eddy, R., Jr. (1974) Mode of pisatin induction. Increased template activity and dye-binding capacity of chromatin isolated from polypeptide-treated pea pods. *Plant Physiol.,* **53**, 52–63.

Hadwiger, L. A., Sander, C., Eddyrean, J. and Ralston, J. (1976) Sodium azide-induced mutants of peas that accumulate pisatin. *Phytopathology,* **66**, 629–630.

Hadwiger, L. A. and Schwochau, M. E. (1971*a*) Ultraviolet light-induced formation of pisatin and phenylalanine ammonia lyase. *Plant Physiol.,* **47**, 588–590.

Hadwiger, L. A. and Schwochau, M. E. (1971*b*) Specificity of deoxyribonucleic acid intercalating compounds in the control of phenylalanine ammonia lyase and pisatin levels. *Plant Physiol.,* **47**, 346–351.

Hahlbrock, K., Lamb, C. J., Purwin, C., Ebel, J., Fautz, E. and Schafer, E. (1981) Rapid response of suspension-cultured parsley cells to the elicitor from *Phytophthora megasperma* var. *sojae. Plant Physiol.,* **67**, 768–773.

Hahn, M. G. and Albersheim, P. (1978) Host-pathogen interactions. XIV. Isolation and partial characterization of an elicitor from yeast extract. *Plant Physiol.,* **62**, 107–111.

Hahn, M. G., Darvill, A. G. and Albersheim, P. (1981) Host-pathogen interactions. XIX. The endogenous elicitor, a fragment of a plant cell wall polysaccharide that elicits phytoalexin accumulation in soybeans. *Plant Physiol.,* in press.

Hammerschlag, F. and Klarman, W. L. (1969) An antifungal principle produced by soybean plants inoculated with tobacco necrosis virus. *Phytopathology,* **59**, 1557.

Hammerschmidt, R. and Kuć, J. (1979) Isolation and identification of phytuberin from *Nicotiana tabacum* previously infiltrated with an incompatible bacterium. *Phytochemistry,* **18**, 874–875.

Hargreaves, J. A. (1979) Investigations into the mechanism of mercuric chloride stimulated phytoalexin accumulation in *Phaseolus vulgaris* and *Pisum sativum. Physiol. Plant Pathol.,* **15**, 279–287.

Hargreaves, J. A. (1981) Accumulation of phytoalexins in cotyledons of French bean (*Phaseolus vulgaris* L.) following treatment with triton (T-octyl-phenol polyethoxyethanol) surfactants. *New Phytol.,* **87**, 733–741.

Hargreaves, J. A. and Bailey, J. A. (1978) Phytoalexin production by hypocotyls of *Phaseolus vulgaris* in response to constitutive metabolites released by damaged cells. *Physiol. Plant Pathol.,* **13**, 89–100.

Hargreaves, J. A. and Selby, C. (1978) Phytoalexin formation in cell suspensions of *Phaseolus vulgaris* in response to an extract of bean hypocotyls. *Phytochemistry,* **17**, 1099–1102.

Heath, M. C. and Wood, R. K. S. (1971) Role of inhibitors of fungal growth in the limitation of leaf spots caused by *Ascochyla pisi* and *Mycosphaerella pinodes. Ann. Bot.,* **35**, 475–491.

Higgins, V. J. (1972) Role of the phytoalexin medicarpin in three leaf spot diseases of alfalfa. *Physiol. Plant Pathol.,* **2**, 289–300.

Hoppe, H. H., Humme, B. and Heitefuss, R. (1980) Elicitor induced accumulation of phytoalexins in healthy and rust infected leaves of *Phaseolus vulgaris. Phytopath. Z.,* **97**, 85–88.

Imezaki, H., Takei, S. and Uritani, I. (1964) Ipomeamarone accumulation and lipid metabolism in sweet potato infected by the black-rot fungus. II. Accumulation mechanism of ipomeamarone in the infected region, with special regard to the contribution of the non-infected tissue. *Plant Cell Physiol.,* **5**, 133–144.

Ingham, J. L. and Dewick, P. M. (1980) Sparticarpin: a pterocarpan phytoalexin from *Spartium junceum. Z. Naturforsch.,* **35c**, 197–200.

Ingham, J. L., Keen, N. T., Mulheirn, L. J. and Lyne, R. L. (1981) Inducibly-formed isoflavonoids from leaves of soybean (*Glycine max*). *Phytochemistry,* **20**, 795–798.

Ingram, D. S. (1978) Cell death and resistance to biotrophs. *Ann. appl. Biol.,* **89**, 291–295.

Inoue, H. and Uritani, I. (1980) Furanosesquiterpene reductase from fungal-inoculated sweet potato root tissue. *Agric. Biol. Chem.*, **44**, 2245–2248.

Ishiguri, Y., Tomiyama, K., Doke, N., Murai, A., Katsui, N., Yagihashi, F. and Masamune, T. (1978) Induction of rishitin-metabolizing activity in potato tuber tissue disks by wounding and identification of rishitin metabolites. *Phytopathology*, **68**, 720–725.

Ishizaki, N. and Tomiyama, K. (1972) Effect of wounding or infection by *Phytophthora infestans* on the contents of terpenoids in potato tubers. *Plant Cell Physiol.*, **13**, 1053–1063.

Kaplan, D. T., Keen, N. T. and Thomason, L. J. (1980) Association of glyceollin with the incompatible response of soybean roots to *Meloidogyne incognita*. *Physiol. Plant Pathol.*, **16**, 309–318.

Keen, N. T. (1971) Hydroxyphaseollin production by soybeans resistant and susceptible to *Phytophthora megasperma* var. *sojae*. *Physiol. Plant Pathol.*, **1**, 265–275.

Keen, N. T. (1975) Specific elicitors of plant phytoalexin production: determinants of race specificity in pathogens. *Science*, **187**, 74–75.

Keen, N. T. and Kennedy B. W. (1974) Hydroxyphaseollin and related isoflavonoids in the hypersensitive resistance response of soybean against *Pseudomonas glycinea*. *Physiol. Plant Pathol.*, **4**, 173–185.

Keen, N. T. and Legrand, M. (1980) Surface glycoproteins: evidence that they may function as the race specific phytoalexin elicitors of *Phytophthora megasperma* f. sp. *glycinea*. *Physiol. Plant Pathol.*, **17**, 175–192.

Keogh, R. C., Deverall, B. J. and McLeod, S. (1980) Comparison of histological and physiological responses to *Phakopsora pachyrhizi*. *Trans. Br. mycol. Soc.*, **74**, 329–333.

Kiraly, Z., Barna, B. and Ersek, T. (1972) Hypersensitivity as a consequence, not the cause, of plant resistance to infection. *Nature*, **239**, 456–458.

Klarman, W. L. and Hammerschlag, F. (1972) Production of the phytoalexin, hydroxyphaseollin, in soybean leaves inoculated with Tobacco Necrosis Virus. *Phytopathology*, **62**, 719–721.

Komai, K. and Sato, S. (1974) Induction of rishitin formation and its accumulation in the potato-tuber tissues by DBCP treatment. *J. Agric. Chem. Soc. Japan*, **48**, 599–604.

Kuć, J. (1975) Teratogenic constituents of potatoes. *Rec. Adv. Phytochem.*, **9**, 139–150.

Lamb, C. J. and Dixon, R. A. (1978) Stimulation of *de novo* synthesis of L-phenylalanine ammonia-lyase during induction of phytoalexin biosynthesis in cell suspension cultures of *Phaseolus vulgaris*. *FEBS Lett.*, **94**, 277–280.

Lamb, C. J., Lawton, M. A., Taylor, S. J. and Dixon, R. A. (1980) Elicitor modulation of phenylalanine ammonia-lyase in *Phaseolus vulgaris*. *Ann. Phytopathol.*, **12**, in press.

Lampard, J. F. (1974) Demethylhomopterocarpin: an antifungal compound in *Canavalia ensiformis* and *Vigna unguiculata* following infection. *Phytochemistry*, **13**, 291–292.

Langcake, P. and Pryce, R. J. (1977) The production of resveratrol and the viniferins by grapevines in response to ultra-violet irradiation. *Phytochemistry*, **16**, 1193–1196.

Lazarovits, G., Bhullar, B. S., Sugiyama, H. J. and Higgins, V. J. (1979) Purification and partial characterization of a glycoprotein toxin produced by *Cladosporium fulvum*. *Phytopathology*, **69**, 1062–1068.

Lazarovits, G. and Higgins, V. J. (1979) Biological activity and specificity of a toxin produced by *Cladosporium fulvum*. *Phytopathology*, **69**, 1056–1061.

Lee, S.-C. and West, C. A. (1981) Polygalacturonase from *Rhizopus stolonifer* is an elicitor of casbene synthase activity in Castor bean (*Rhicinus communis* L.) seedlings. *Plant Physiol.*, **67**, 633–639.

Lewis, D. H. (1980) Are there inter-relations between the metabolice role of boron, synthesis of phenolic phytoalexins and the germination of pollen? *New Phytologist*, **84**, 261–270.

Lisker, N. and Kuć, J. (1977) Elicitors of terpenoid accumulation in potato tuber slices. *Phytopathology*, **67**, 1356–1359.

Lyon, F. M. and Wood, R. K. S. (1975) Production of phaseollin, coumestrol and related compounds in bean leaves inoculated with *Pseudomonas* spp. *Physiol. Plant Pathol.*, **6**, 117–124.

Lyon, G. D. (1972) Occurrence of rishitin and phytuberin in potato tubers inoculated with *Erwinia carotovora* var. *atroseptica*. *Physiol. Plant Pathol.*, **2**, 411–416.

Mansfield, J. W. (1980) "Mechansims of resistance to *Botrytis*", in *The Biology of Botrytis*, eds. J. R. Coley-Smith, W. R. Jarvis and K. Verhoeff, Academic Press, London, 181–218.

Mansfield, J. W., Hargreaves, J. A. and Boyle, F. C. (1974) Phytoalexin production by live cells in broad bean leaves infected with *Botrytis cinerea*. *Nature*, **252**, 316–317.

Mansfield, J. W., Porter, A. E. A. and Smallman, R. V. (1980) Dihydrowyerone derivatives as components of the furanoacetylenic phytoalexin response of tissues of *Vicia faba*. *Phytochemistry*, **19**, 1057–1061.

Martin, M. and Dewick, P. M. (1979) Biosynthesis of the 2-arylbenzofuran phytoalexin vignafuran in *Vigna unguiculata*. *Phytochemistry*, **18**, 1309–1317.

Moesta, P. and Grisebach, H. (1980) Effects of biotic and abiotic elicitors on phytoalexin metabolism in soybean. *Nature*, **286**, 710–711.

Munn, C. B. and Drysdale, R. B. (1975) Kievitone production and phenylalanine ammonia-lyase activity in cowpea. *Phytochemistry*, **14**, 1303–1307.

Nakajima, T., Tomiyama, K. and Kinukawa, M. (1975) Distribution of rishitin and lubimin in potato tuber tissue infected by an incompatible race of *Phytophthora infestans* and the site where rishitin is synthesized. *Ann. Phytopath. Soc. Japan*, **41**, 49–55.

Oba, K., Tatematsu, H., Yamashita, K. and Uritani, I. (1976) Induction of furanoterpene production and formation of the enzyme system from mevalonate to ispentenyl pyrophosphate in sweet potato root tissue injured by *Ceratocystis fimbriata* and by toxic chemicals. *Plant Physiol.*, **58**, 51–56.

Oku, H., Nakanishi, T., Shiraishi, T. and Ouchi, S. (1973) Phytoalexin induction by some agricultural fungicides and phytotoxic metabolites of pathogenic fungi. *Sci. Rep. Fac. Agr. Okayama Univ.*, **42**, 17–20.

Oku, H., Shiraishi, T. and Ouchi, S. (1977) Suppression of induction of phytoalexin, pisatin, by low-molecular weight substances from spore germination fluid of pea pathogen, *Mycosphaerella pinodes*. *Naturwissenschaften*, **64**, 643–644.

Oku, H., Shiraishi, T. and Ouchi, S. (1979) "The role of phytoalexins in host-parasite specificity", in *Recognition and Specificity in Plant Host-Parasite Interactions*, ed. J. M. Daly and I. Uritani, Jap. Sci. Soc. Press, Tokyo, 317–333.

Paradies, I., Konze, J. R., Elstner, E. F. and Paxton, J. (1980) Ethylene: indicator but not inducer of phytoalexin synthesis in soybean. *Plant Physiol.*, **66**, 1106–1109.

Paxton, J., Goodchild, D. J. and Cruickshank, I. A. M. (1974) Phaseollin production by live bean endocarp. *Physiol. Plant Pathol.*, **4**, 167–171.

Perrin, D. R. and Cruickshank, I. A. M. (1965) Studies on phytoalexins. VII. Chemical stimulation of pisatin formation in *Pisum sativum* L. *Aust. J. Biol. Sci.*, **18**, 803–816.

Preston, N. W. (1975) 2'-*O*-Methyl-phaseollidinisoflavan from infected tissue of *Vigna unguiculata*. *Phytochemistry*, **14**, 1131–1132.

Price, K. R., Howard, B. and Coxon, D. T. (1976) Stress metabolite production in potato tubers infected by *Phytophthora infestans*, *Fusarium avenaceum* and *Phoma exigua*. *Physiol. Plant Pathol.*, **9**, 189–197.

Rahe, J. E. (1973) Occurrence and levels of the phytoalexin phaseollin in relation to delimitation at sites of infection of *Phaseolus vulgaris* by *Colletotrichum lindemuthianum*. *Can. J. Bot.*, **51**, 2423–2430.

Rahe, J. E. and Arnold, R. M. (1975) Injury-related phaseollin accumulation in *Phaseolus vulgaris* and its implications with regard to specificity of host-parasite interaction. *Can. J. Bot.*, **53**, 921–928.

Rathmell, W. G. and Bendall, D. S. (1971) Phenolic compounds in relation to phytoalexin biosynthesis in hypocotyls of *Phaseolus vulgaris*. *Physiol. Plant Pathol.*, **1**, 351–362.

Reilly, J. J. and Klarman, W. L. (1980) Thymine dimer and glyceollin accumulation in U.V.-irradiated soybean suspension cultures. *Envir. Exper. Bot.*, **20**, 131–134.

Rich, J. R., Keen, N. T. and Thomason, I. J. (1977) Association of coumestans with hypersensitivity of lima bean roots of *Pratylenchus scribneri*. *Physiol. Plant Pathol.*, **12**, 329–338.

Robinson, T. J. and Wood, R. K. S. (1976) Factors affecting accumulation of pisatin by pea leaves. *Physiol. Plant Pathol.*, **9**, 285–297.

Sakai, S., Tomiyama, K. and Doke, N. (1979) Synthesis of a sesquiterpenoid phytoalexin rishitin in non-infected tissue from various parts of potato plants immediately after slicing. *Ann. Phytopath. Soc. Japan*, **45**, 705–711.

Sakai, R., Tomiyama, K., Ishizaka, N. and Sato, N. (1967) Phenol metabolism in relation to disease resistance of potato tubers. 3. Phenol metabolism in tissue neighbouring the necrogenous infection. *Ann. Phytopath. Soc. Japan*, **33**, 216–222.

Sander, C. and Hadwiger, L. A. (1979) L-Phenylalanine ammonia-lyase and pisatin induction by 5-bromodeoxyuridine in *Pisum sativum*. *Biochim. Biophys. Acta*, **563**, 278–292.

Sato, N., Kitazawa, K. and Tomiyama, K. (1971) The role of rishitin in localizing the invading hyphae of *Phytophthora infestans* in infection sites at the cut surfaces of potato tubers. *Physiol. Plant Pathol.*, **1**, 289–295.

Sato, N. and Tomiyama, K. (1969) Localized accumulation of rishitin in the potato-tuber tissue infected by an incompatible race of *Phytophthora infestans*. *Ann. Phytopath. Soc. Japan*, **35**, 202–217.

Schoppner, A. and Kindl, H. (1979) Stilbene synthase (pinosilvin synthase) and its induction by UV light. *FEBS Lett.*, **108**, 349–352.

Schwochau, M. E. and Hadwiger, L. A. (1968) Stimulation of pisatin production in *Pisum sativum* by actinomycin D and other compounds. *Arch. Biochem. Biophys.*, **126**, 731–733.

Shiraishi, T., Oku, H., Ouchi, S. and Isono, M. (1976) Pisatin production prior to the cell necrosis demonstrated in powdery mildew of pea. *Ann. Phytopath. Soc. Japan*, **42**, 609–612.

Smith, D. A., VanEtten, H. D. and Bateman, D. F. (1975) Accumulation of phytoalexins in *Phaseolus vulgaris* hypocotyls following infection by *Rhizoctonia solani*. *Physiol. Plant Pathol.*, **5**, 225–237.

Stholasuta, P., Bailey, J. A., Severin, V. and Deverall, B. J. (1971) Effect of bacterial inoculation of bean and pea leaves on the accumulation of phaseollin and pisatin. *Physiol. Plant Pathol.*, **1**, 177–183.

Stoessl, A. (1980) Phytoalexins—a biogenetic perspective. *Phytopath. Z.*, **99**, 251–272.

Stoessl, A., Unwin, C. H. and Ward, E. W. B. (1974) Post infectional inhibitors from plants. I. Capsidiol, an antifungal compound from *Capsicum frutescens*. *Phytopathology*, **74**, 141–152.

Stössel, P., Lazarovits, G. and Ward, E. W. B. (1980) Penetration and growth of compatible and incompatible races of *Phytophthora megasperma* var. *sojae* in soybean hypocotyl tissues differing in age. *Can. J. Bot.*, **58**, 2594–2601.

Sutherland, D. R. W., Russell, G. B., Biggs, D. R. and Lane, G. A. (1980) Insect feeding deterrent activity of phytoalexin isoflavonoids. *Biochem. Syst. Ecol.*, **8**, 73–75.

Swinburne, T. R. (1975) Microbial proteases as elicitors of benzoic acid accumulation in apples. *Phytopath. Z.*, **82**, 152–162.

Teasdale, J., Daniels, D., Davis, W. C., Eddy, R. Jr. and Hadwiger, L. A. (1974) Physiological and cytological similarities between disease resistance and cellular incompatibility responses. *Plant Physiol.*, **54**, 690–695.

Tomiyama, K., Doke, N., Nozue, M. and Ishiguri, Y. (1979) "The hypersensitive response of resistant plants", in *Recognition and Specificity in Plant Host-Parasite Interactions*, eds. J. M. Daly and I. Uritani, Jap. Sci. Soc. Press, Tokyo, 69–84.

Tomiyama, K. and Fukaya, M. (1975) Accumulation of rishitin in dead potato-tuber tissue following treatment with HgCl$_2$. *Ann. Phytopath. Soc. Japan*, **41**, 418–420.

Uegaki, R., Fujimori, T., Kaneko, H., Kubo, S. and Kato, K. (1980) Phytuberin and phytuberol, sesquiterpenes from *Nicotiana tabacum* treated with ethrel. *Phytochemistry*, **19**, 1543–1544.

Uritani, I., Saito, T., Honda, H. and Kim, W. K. (1975) Induction of furanoterpenoids in sweet potato roots by the larval components of the sweet potato weevils. *Agr. Biol. Chem.*, **39**, 1857–1862.

Uritani, I., Uritani, M. and Yamada, H. (1960) Similar metabolic alterations in sweet potato by poisonous chemicals and by *Ceratocystis fimbriata*. *Phytopathology*, **50**, 30–34.

Uritani, I., Oba, K., Kojima, M., Kim, W. K., Oguni, I. and Suzuki, H. (1976) "Primary and secondary defense actions of sweet potato in response to infections by *Ceratocystis fimbriata* strains", in *Biochemistry and Cytology of Plant-Parasite Interaction*, eds. Tomiyama *et al.*, Kodansha Ltd., Tokyo, 239–252.

Van den Heuvel, J. and Grootveld, D. (1980) Formation of phytoalexins within and outside lesions of *Botrytis cinerea* in French bean leaves. *Neth. J. Pl. Path.*, **86**, 27–35.

VanEtten, H. D. and Pueppke, S. G. (1976) "Isoflavonoid phytoalexins", in *Biochemical Aspects of Plant Parasite Relationships*, eds. J. Friend and D. R. Threlfall, Academic Press, 239–289.

VanEtten, H. D. and Smith, D. A. (1975) Accumulation of antifungal isoflavonoids and 1a-hydroxyphaseollone, a phaseollin metabolite in bean tissue infected with *Fusarium solani* f. sp. *phaseoli*. *Physiol. Plant Pathol.*, **5**, 225–237.

Varns, J. L., Kuć, J. and Williams, E. B. (1971) Terpenoid accumulation as a biochemical response of the potato tuber to *Phytophthora infestans*. *Phytopathology*, **61**, 174–177.

Veech, J. A. (1978) An apparent relationship between methoxy substituted terpenoid aldehydes with the resistance of cotton to *Meloidogyne incognita*. *Nematologica*, **24**, 81–87.

Wade, M. and Albersheim, P. (1979) Race-specific molecules that protect soybeans from *Phytophthora megasperma* var. *sojae*. *Proc. Nat. Acad. Sci.*, **76**, 4433–4437.

Ward, E. W. B. and Stoessl, A. (1976) On the question of "elicitors" or "inducers" in incompatible interactions between plants and fungal pathogens. *Phytopathology*, **66**, 940–941.

Ward, E. W. B., Unwin, C. H. and Stoessl, A. (1973) Postinfectional inhibitors from plants. VI. Capsidiol production in pepper fruit infected with bacteria. *Phytopathology*, **63**, 1537–1538.

Weinstein, L. I., Hahn, M. G. and Albersheim, P. (1981) Isolation and biological activity of glycinol, a pterocarpan phytoalexin synthesized by soybeans. *Plant Physiol.*, in press.

Woodward, M. D. (1981) Identification of the biosynthetic precursors of medicarpin in inoculation droplets on white clover. *Physiol. Plant Pathol.*, **18**, 33–39.

Yoshikawa, M. (1978) Diverse modes of action of biotic and abiotic phytoalexin elicitors. *Nature*, **275**, 546–547.

Yoshikawa, M., Matama, M. and Masago, H. (1981) Release of a soluble phytoalexin elicitor from mycelial walls of *Phytophthora megasperma* var. *sojae* by soybean tissues. *Plant Physiol.*, **67**, 1032–1035.

Yoshikawa, M., Yamauchi, K. and Masago, H. (1978) *De novo* messenger RNA and protein synthesis are required for phytoalexin-mediated disease resistance in soybean hypocotyls. *Plant Physiol.*, **61**, 314–317.

10 Phytoalexins:
current problems and future prospects

J. W. MANSFIELD AND J. A. BAILEY

It is appropriate to begin this concluding chapter by commenting on the valuable definition of phytoalexins given by Professor Deverall in chapter 1. This states, with slight modification, that *phytoalexins are low molecular weight antimicrobial compounds that are both synthesized by and accumulate in plants which have been exposed to microorganisms.* An important feature of this definition is that it restricts phytoalexins to compounds which are synthesized from remote precursors, probably through *de novo* synthesis of enzymes. Phytoalexins are, therefore, clearly distinguished from other antimicrobial compounds, which may also be involved in disease resistance, but which are pre-formed or formed directly from inactive precursors (Ingham, 1973; Deverall, 1977). It is equally important that this definition does not restrict the use of the term phytoalexin to compounds which have been demonstrated to have a role in the resistance of plants to disease.

Each of the chapters in this book includes many suggestions for further work on phytoalexins. In the remainder of this summary four major questions will be considered.

1. Are phytoalexins produced by all plants?

Negative results are rarely published, and in consequence there are few reports of thorough examinations concluding that phytoalexins are not produced by a particular plant species. Deverall (1977) has described his failure to detect phytoalexins in wheat or cucumber, but Cartwright and Russell (1981) and Kuć and Caruso (1977) have reported preliminary results which indicate that phytoalexins are produced by these plants. Failure to detect phytoalexins may be a consequence of using techniques inappropriate for their isolation or detection: they may not be extracted from the tissues, they may be modified during extraction or present in amounts insufficient for easy detection. Nevertheless, the difficulties encountered in detecting phytoalexins suggest that in some plants these compounds play only a minor role in disease resistance. Other processes, for example, lignification or silicification of cell walls, release of pre-formed inhibitors or accumulation of agglutinating factors may be more important (Kojima and Uritani, 1978; Heath, 1980). Further examination of the ubiquity of phytoalexins would allow an assessment of their contribution to the co-ordinated processes of resistance throughout the plant kingdom. As discussed in chapter 8, studies to date have been almost exclusively confined to angiosperms. A search for phytoalexins in gymnosperms and non-vascular plants may lead not only to the discovery of novel chemicals with antimicrobial properties, but should also provide valuable clues to the involvement of phytoalexins in the evolution of plants and their parasites.

2. Are phytoalexins active *in vivo*?

This question encompasses both the antimicrobial activity of phytoalexins under the conditions found within plant tissues and their localization around invading microorganisms. These intractable problems are crucial to the proposed role of phytoalexins as a cause of the restriction of microbial growth in plants. Attempts to determine the concentrations of phytoalexins around invading microorganisms have included rather crude dissection experiments involving, for example, analysis of epidermal strips (Mansfield, 1980) or of thin sections of tissue (Yoshikawa *et al.*, 1978). It is to be expected that more critical data will be obtained, possibly by micro-dissection coupled with very sensitive methods of quantitative analysis or by the development of immunohisto-chemical techniques.

Although the technical difficulties which need to be overcome when locating phytoalexin accumulation are undoubtedly great, they are much less formidable than those affecting attempts to reproduce the conditions which occur in infected plants during experimental assessments of antimicrobial activity. It may be more useful to adopt alternative approaches to examining the involvement of phytoalexins in resistance. A genetical approach was outlined by Deverall (1977) and has recently been successfully applied by VanEtten *et al.* (1980) to demonstrate that tolerance to pisatin is an inheritable character closely correlated with the virulence of *Nectria haematococca* to *Pisum sativum* (VanEtten *et al.*, 1980; see chapter 7). An alternative strategy would be to assess any change in virulence of mutants of a pathogen selected for their exceptional tolerance or sensitivity to the host's phytoalexin. Unfortunately such mutants have not yet been obtained.

3. What is the relationship between plant cell death and phytoalexin accumulation?

Phytoalexin accumulation, following microbial infection, is invariably associated with death of plant cells (see chapters 8 and 9). In the interactions which have been examined in detail, notably those between *Phytophthora infestans* and *Solanum tuberosum* (Tomiyama *et al.*, 1979) and *Colletotrichum lindemuthianum* and *Phaseolus vulgaris* (Bailey, 1981), plant cell death precedes accumulation of phytoalexins. This sequence of events is consistent with phytoalexin synthesis being induced in surrounding live cells by materials released from the cell(s) killed by the invading organism. A logical extension of this argument is to propose that the elicitors isolated from bacteria and fungi are merely acting as potent phytotoxins causing phytoalexins to accumulate in much the same indirect manner as proposed for mercuric chloride and Triton surfactants (Hargreaves, 1979 and 1981). Compounds released from dying plant cells are perhaps more likely to act directly by derepressing genes controlling phytoalexin biosynthesis. Clearly the mode of action of elicitors from both microorganisms and plants requires urgent consideration. In this context it is important to restate the conclusion of Grisebach and Ebel (1978), that phytoalexin biosynthesis provides great possibilities for studies on the control of gene expression in plant cells.

4. How can our knowledge of phytoalexins be used to develop new approaches to disease control?

The potential use of phytoalexins as disease control agents has been suggested on several occasions. As early as 1969, Fawcett *et al.* examined the protectant activity of wyerone and showed that it gave some protection against French bean rust and chocolate spot disease of broad bean. Ward *et al.* (1975) obtained good control of late blight of tomato with 5×10^{-4} M capsidiol, protection lasting for at least eight days. The fungitoxicities of a series of analogues of vignafuran were compared by Carter *et al.* (1978). Several compounds were fungitoxic but few gave protection against rust on broad bean or powdery mildew on wheat and none was significantly active against chocolate spot of bean. None of the vignafuran analogues showed therapeutic or systemic activity. A similar lack of therapeutic and systemic activity was also reported for ε-viniferin, although this phytoalexin protected vines from downy mildew for up to seven days (Langcake, 1981). A direct comparison of the protectant activity of seven isoflavonoid phytoalexins, and the fungicides benomyl and mancozeb, indicated that the phytoalexins have no commercial potential (Rathmell and Smith, 1980).

Results such as these suggest that while phytoalexins can provide a measure of disease control, their usefulness is limited. As discussed by Langcake (1981), the major problems appear to be firstly, that phytoalexins are less fungitoxic than most synthetic fungicides and secondly, that they are not transported around plants, a feature consistent with their highly localized role in natural resistance. It is possible that these problems could be overcome by the preparation of analogues, particularly of those groups of phytoalexins most amenable to synthesis, for example, the hydroxyflavans from narcissus (Coxon *et al.*, 1980). Possible applications of phytoalexins for pharmaceutical purposes have not been reported.

Elicitors of phytoalexin biosynthesis appear to offer scope as useful agents for disease control. It is important to distinguish between elicitors which are phytotoxic metabolites, for example, microbial elicitors such as the glycoproteins from *Cladosporium fulvum* (De Wit and Kodde, 1981) and the unidentified materials, probably of plant origin, which must act directly on the derepression of genes controlling phytoalexin biosynthesis. Use of microbial elicitors is unlikely to prove practicable, as accumulation of effective concentrations of phytoalexins would involve extensive damage to the treated plants. The derepressors may have more potential. Paradoxically, treatments with concentrations of derepressors which cause the accumulation of antimicrobial concentrations of phytoalexins within plants are perhaps unlikely to prove generally successful because many phytoalexins are themselves phytotoxic and their widespread accumulation may therefore have deleterious effects on responding tissues. However, derepressors might be used to initiate both synthesis and metabolism (turnover) of phytoalexins, without causing significant accumulation. Plants in such an activated state may, when infected, undergo an abnormally rapid phytoalexin response leading to restriction of pathogen development.

Studies on the role of phytoalexins in race-specific resistance have re-emphasized the lack of information concerning the basis of differentiation between resistant and susceptible reactions. It is now apparent that differentiation between avirulent and virulent races involves an initial recognition event which, in resistant tissues, leads to a hypersensitive reaction and the subsequent accumulation of phytoalexins. There are grounds for optimism that the recognition process provides the most likely target for the development of control measures. Support for this view comes from the finding that the rice-blast specific dichlorocyclopropane fungicide WL 28325 appears to act by modifying the process of recognition so that treated leaves respond to *Pyricularia oryzae* in a manner very similar to that of genetically resistant varieties of rice. Invaded cells undergo a hypersensitive reaction and two phytoalexins, momilactones A and B accumulate at infection sites to restrict fungal growth (Cartwright *et al.*, 1980). The dichlorocyclopropane fungicide was discovered by empirical screening, but establishment of the mode of action of this compound encourages the belief that a deeper understanding of the biochemical mechanisms involved in the process of recognition will allow the development of rational approaches to durable disease control.

REFERENCES

Bailey, J. A. (1981) "Physiological and biochemical events associated with the expression of resistance to disease", in *Active Defence Mechanisms in Plants*, ed. R. K. S. Wood, Plenum Press, New York and London, in press.

Carter, G. A., Chamberlain, K. and Wain, R. L. (1978) Investigations on fungicides. XX. The fungitoxicity of analogues of the phytoalexin 2-(2'-methoxy-4'-hydroxyphenyl)-6-methoxy benzofuran (vignafuran). *Ann. appl. Biol.*, **88**, 57–64.

Cartwright, D. W., Langcake, P. and Ride, J. P. (1980) Phytoalexin production in rice and its enhancement by a dichlorocyclopropane fungicide. *Physiol. Plant Pathol.*, **17**, 259–267.

Cartwright, D. W. and Russell, G. E. (1981) Possible involvement of phytoalexins in durable resistance of winter wheat to yellow rust. *Trans. Br. mycol. Soc.*, **76**, 323–325.

Coxon, D. T., O'Neill, T. M., Mansfield, J. W. and Porter, A. E. A. (1980) Identification of three hydroxyflavan phytoalexins from daffodil bulbs. *Phytochemistry*, **19**, 889–891.

Deverall, B. J. (1977) *Defence Mechanisms of Plants*, Cambridge University Press.

De Wit, P. J. G. M. and Kodde, E. (1981) Further characterization and cultivar specificity of glycoprotein elicitors from culture filtrates and cell walls of *Cladosporium fulvum* (syn. *Fulvia fulva*). *Physiol. Plant Pathol.*, **18**, 143–148.

Fawcett, C. H., Spencer, D. M. and Wain, R. L. (1969) The isolation and properties of a fungicidal compound present in seedlings of *Vicia faba*. *Neth. J. Pl. Path.*, **75**, 72–81.

Grisebach, H. and Ebel, J. (1978) Phytoalexins: chemical defense substances in higher plants. *Angew. Chem. Int. Ed. Engl.*, **17**, 635–647.

Hargreaves, J. A. (1979) Investigations into the mechanism of mercuric chloride stimulated phytoalexin accumulation in *Phaseolus vulgaris* and *Pisum sativum*. *Physiol. Plant Pathol.*, **15**, 279–287.

Hargreaves, J. A. (1981) Accumulation of phytoalexins in cotyledons of French bean (*Phaseolus vulgaris* L.) following treatment with Triton (t octylphenol polyethoxyethanol surfactants). *New Phytol.*, **87**, 733–741.

Heath, M. C. (1980) Reactions of nonsuscepts to fungal pathogens. *Ann. Rev. Phytopathol.*, **18**, 211–236.

Ingham, J. L. (1973) Disease resistance in higher plants. The concept of pre-infectional and post-infectional resistance. *Phytopath. Z.*, **78**, 314–335.

Kojima, M. and Uritani, I. (1978) Studies on factor(s) in sweet potato which specifically inhibits germ tube growth of incompatible isolates of *Ceratocystis fimbriata*. *Pl. Cell Physiol.*, **19**, 71–81.

Kuć, J. and Caruso, F. L. (1977) "Activated co ordinated chemical defense against disease in plants", in *Host Plant Resistance to Pests*, ed. P. A. Hedin, American Chemical Society, Washington, D.C., 78–89.

Langcake, P. (1981) Alternative chemical agents for controlling plant disease. *Proc. Royal Soc. B*, in press.

Mansfield, J. W. (1980) "Mechansims of resistance to *Botrytis*", in *The Biology of Botrytis*, eds. J. R. Coley-Smith, W. R. Jarvis and K. Verhoeff, Academic Press, London, 181–218.

Rathmell, W. G. and Smith, D. A. (1980) Lack of activity of selected isoflavonoid phytoalexins as protectant fungicides. *Pestic. Sci.*, **11**, 568–572.

Tomiyama, K., Doke, N., Nozue, M. and Ishiguri, Y. (1979) "The hypersensitive response of resistant plants", in *Recognition and Specificity in Plant Host-Parasite Interactions*, eds. J. M. Daly and I. Uritani, Japan Sci. Soc. Press, Tokyo, 317–333.

VanEtten, M. D., Matthews, P. S., Tegtmeier, K. J., Dietert, M. F. and Stein, J. I. (1980) The association of pisatin tolerance and demethylation with virulence on pea in *Nectria haematococca*. *Physiol. Plant Pathol.*, **16**, 257–268.

Ward, E. W. B., Unwin, C. H. and Stoessl, A. (1975) Experimental control of late blight of tomatoes with capsidiol, the phytoalexin from peppers. *Phytopathology*, **65**, 168–169.

Yoshikawa, M., Yamauchi, K. and Masago, H. (1978) Glyceollin: its role in restricting fungal growth in resistant soybean hypocotyls infected with *Phytophthora megasperma* var. *sojae*. *Physiol. Plant Pathol.*, **12**, 73–82.

Index

1 Subject Index

325

2 Compound Index (including named enzymes)

3 Organism Index